T0257474

# Distillation: Modeling and Applied Principles

# Distillation: Modeling and Applied Principles

Edited by **Susan Zucker**

LANRYE INTERNATIONAL

New Jersey

Published by Clanrye International,
55 Van Reypen Street,
Jersey City, NJ 07306, USA
www.clanryeinternational.com

**Distillation: Modeling and Applied Principles**
Edited by Susan Zucker

International Standard Book Number: 978-1-63240-149-6 (Hardback)

Printed in the United States of America.

# Contents

# Preface

Every book is a source of knowledge and this one is no exception. The idea that led to the conceptualization of this book was the fact that the world is advancing rapidly; which makes it crucial to document the progress in every field. I am aware that a lot of data is already available, yet, there is a lot more to learn. Hence, I accepted the responsibility of editing this book and contributing my knowledge to the community.

This book describes the process of distillation and its modeling applications. Distillation modeling and various applications generally in food processing field are elucidated in this book targeting the simulation of distillation procedure as well as the thermodynamic mathematical basics. The overviews and practical experiences presented in this book are primarily concerned with the food and beverage industry and aroma extraction. The aim of this book is to provide useful insight regarding the field of distillation to the interested researchers.

While editing this book, I had multiple visions for it. Then I finally narrowed down to make every chapter a sole standing text explaining a particular topic, so that they can be used independently. However, the umbrella subject sinews them into a common theme. This makes the book a unique platform of knowledge.

I would like to give the major credit of this book to the experts from every corner of the world, who took the time to share their expertise with us. Also, I owe the completion of this book to the never-ending support of my family, who supported me throughout the project.

**Editor**

# Part 1

## Modeling and Simulation

# Modeling and Control Simulation for a Condensate Distillation Column

Vu Trieu Minh and John Pumwa
*Papua New Guinea University of Technology (UNITECH), Lae*
*Papua New Guinea*

## 1. Introduction

Distillation is a process that separates two or more components into an overhead distillate and a bottoms product. The bottoms product is almost exclusively liquid, while the distillate may be liquid or a vapor or both.

The separation process requires three things. Firstly, a second phase must be formed so that both liquid and vapor phases are present and can contact each other on each stage within a separation column. Secondly, the components have different volatilities so that they will partition between the two phases to a different extent. Lastly, the two phases can be separated by gravity or other mechanical means.

Calculation of the distillation column in this chapter is based on a real petroleum project to build a gas processing plant to raise the utility value of condensate. The nominal capacity of the plant is 130,000 tons of raw condensate per year based on 24 operating hours per day and 350 working days per year. The quality of the output products is the purity of the distillate, $x_D$, higher than or equal to 98% and the impurity of the bottoms, $x_B$, may be less/equal than 2%. The basic feed stock data and its actual compositions are based on the other literature (PetroVietnam Gas Company,1999).

The distillation column contains one feed component, $x_F$. The product stream exiting the top has a composition of $x_D$ of the light component. The product stream leaving the bottom contains a composition of $x_B$ of the light component. The column is broken in two sections. The top section is referred to as the rectifying section. The bottom section is known as the stripping section as shown in Figure 1.1.

The top product stream passes through a total condenser. This effectively condenses all of the vapor distillate to liquid. The bottom product stream uses a partial re-boiler. This allows for the input of energy into the column. Distillation of condensate (or natural gasoline) is cutting off light components as propane and butane to ensure the saturated vapor pressure and volatility of the final product.

The goals of this chapter are twofold: first, to present a theoretical calculation procedure of a condensate column for simulation and analysis as an initial step of a project feasibility study, and second, for the controller design: a reduced-order linear model is derived such that it best reflects the dynamics of the distillation process and used as the reference model

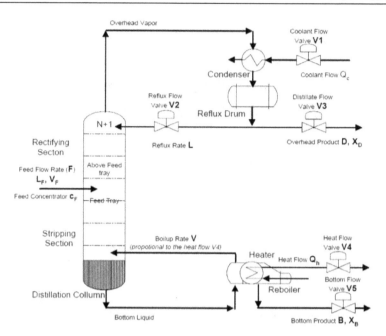

Fig. 1.1. Distillation Flow-sheet

for a model-reference adaptive control (MRAC) system to verify the ability of a conventional adaptive controller for a distillation process dealing with the disturbance and the plant-model mismatch as the influence of the feed disturbances.

In this study, the system identification is not employed since experiments requiring a real distillation column is still not implemented. So that a process model based on experimentation on a real process cannot be done. A mathematical modeling based on physical laws is performed instead. Further, the MRAC controller model is not suitable for handling the process constraints on inputs and outputs as discussed in other literature (Marie, E. *et al.*, 2008) for a coordinator model predictive control (MPC). In this chapter, the calculations and simulations are implemented by using MATLAB (version 7.0) software package.

## 2. Process data calculation

### 2.1 Methods of distillation column control

### 2.1.1 Fundamental variables for composition control

The purity of distillate or the bottoms product is affected by two fundamental variables: feed split (or cutting point) and fractionation. The feed split variable refers to the fraction of the feed that is taken overhead or out the bottom. The fractionation variable refers to the energy that is put into the column to accomplish the separation. Both of these fundamental variables affect both product compositions but in different ways and with different sensitivities.

a.  *Feed Split*: Taking more distillate tends to decrease the purity of the distillate and increase the purity of the bottoms. Taking more bottoms tends to increase distillate purity and decrease bottoms product purity.

b.  *Fractionation:* Increasing the reflux ratio (or boil-up rate) will reduce the impurities in both distillate and the bottoms product.

Feed split usually has a much stronger effect on product compositions than does fractionation. One of the important consequences of the overwhelming effect of feed split is that it is usually impossible to control any composition (or temperature) in a column if the feed split is fixed (i.e. the distillate or the bottoms product flows are held constant). Any small changes in feed rate or feed composition will drastically affect the compositions of both products, and will not be possible to change fractionation enough to counter this effect.

## 2.1.2 Degrees of freedom of the distillation process

The degrees of freedom of a process system are the independent variables that must be specified in order to define the process completely. Consequently, the desired control of a process will be achieved when and only when all degrees of freedom have been specified. The mathematical approach to determine the degrees of freedom of any process (George, S., 1986) is to sum up all the variables and subtract the number of independent equations. However, there is a much easier approach developed by Waller, V. (1992): There are five control valves as shown in Figure 1.2, one on each of the following streams: distillate, reflux, coolant, bottoms and heating medium. The feed stream is considered being set by the upstream process. So this column has five degrees of freedom. Inventories in any process must be always controlled. Inventory loops involve liquid levels and pressures. This means that the liquid level in the reflux drum, the liquid level in the column base, and the column pressure must be controlled.

If we subtract the three variables that must be controlled from five, we end up with two degrees of freedom. Thus, there are two and only two additional variables that can (and must) be controlled in the distillation column. Notice that we have made no assumptions about the number or type of chemical components being distilled. Therefore a simple, ideal, binary system has two degrees of freedom; a complex, multi-component system also has two degrees of freedom.

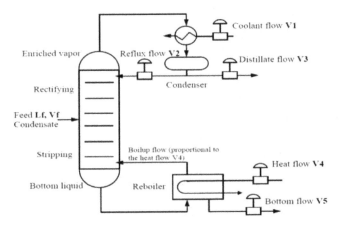

Fig. 2.1. Control Valves Location

### 2.1.3 Control structure

The manipulated variables and controlled variables of a distillation column are displayed in the Table 2.1 and in the Figure 2.1.

|   | Controlled variables | Manipulated variables | Control valve |
|---|---|---|---|
| 1 | Column pressure | Condenser duty | Coolant flow $V1$ |
| 2 | Concentration (temperature) of distillate | Reflux flow rate | Reflux flow $V2$ |
| 3 | Liquid level in the reflux drum | Distillate flow rate | Distillate flow $V3$ |
| 4 | Concentration (temperature) of bottoms | Re-boiler duty | Heat flow $V4$ |
| 5 | Liquid level in the column base | Bottoms flow rate | Bottom flow $V5$ |

Table 2.1. Manipulated variables and controlled variables of a distillation column

Selecting a control structure is a complex problem with many facets. It requires looking at the column control problem from several perspectives:

- Local perspective considering the steady state characteristics of the column.
- Local perspective considering the dynamic characteristics of the column.
- Global perspective considering the interaction of the column with other unit operations in the plant.

### 2.1.4 Energy balance control structure (L-V)

The $L$-$V$ control structure, which is called energy balance structure, can be viewed as the *standard control structure* for dual composition control of distillation. In this control structure, the reflux flow rate $L$ and the boil-up flow rate $V$ are used to control the "primary" outputs associated with the product specifications. The liquid holdups in the drum and in the column base (the "secondary" outputs) are usually controlled by distillate flow rate $D$ and the bottoms flow rate $B$.

### 2.1.5 Material balance control structure (D-V) and (L-B)

Two other frequently used control structures are the material balance structures $(D$-$V)$ and $(L$-$B)$. The $(D$-$V)$ structure seems very similar to the $(L$-$V)$ structure. The only difference between the $(L$-$V)$ and the $(D$-$V)$ structures is that the roles of $L$ and $D$ are switched.

### 2.2 Distillation process calculation

### 2.2.1 Preparation for initial data

The plant nominal capacity is 130,000 tons of raw condensate per year based on 24 operating hours per day and 350 working days per year. The plant equipment is specified with a design margin of 10% above the nominal capacity and turndown ratio of 50%. Hence, the raw condensate *feed rate* for the plant is determined as follows:

$$Feed = \frac{130,000 \; tons}{(24 \; h) \times (350 \; working \; days)} = 15.47619 \; tons \, / \, hour \tag{2.1}$$

The actual composition of the raw condensate for the gas processing plant is always fluctuates around the average composition as shown in the Table 2.2. The distillation data for given raw condensate are shown in the Table 2.3.

| Component | Mole % |
|---|---|
| Propane | 0.00 |
| Normal Butane | 19.00 |
| Iso-butane | 26.65 |
| Iso-pentane | 20.95 |
| Normal Pentane | 10.05 |
| Hexane | 7.26 |
| Heptane | 3.23 |
| Octane | 1.21 |
| Nonane | 0.00 |
| Normal Decane | 0.00 |
| n-C11H24 | 1.94 |
| n-C12H26 | 2.02 |
| Cyclopentane | 1.61 |
| Methylclopentane | 2.02 |
| Benzene | 1.61 |
| Toluen | 0.00 |
| O-xylene | 0.00 |
| E-benzen | 0.00 |
| 124-Mbenzen | 0.00 |

Table 2.2. Compositions of raw condensate

The feed is considered as a pseudo *binary mixture* of Ligas (iso-butane, n-butane and propane) and Naphthas (iso-pentane, n-pentane, and heavier components). The column is designed with $N=14$ trays. The model is simplified by lumping some components together (pseudocomponents) and modeling of the column dynamics is based on these pseudocomponents only (Kehlen, H. & Ratzsch, M., 1987). Depending on the feed composition fluctuation, the properties of pseudo components are allowed to change within the range as shown in the Table 2.4.

*Relative volatility:*

Relative volatility is a measure of the differences in volatility between two components, and hence their boiling points. It indicates how easy or difficult a particular separation will be. The relative volatility of component $i$ with respect to component $j$ is defined as:

$$\alpha_{ij} = \frac{\left[\dfrac{y_i}{x_i}\right]}{\left[\dfrac{y_j}{x_j}\right]} = \frac{K_i}{K_j} \quad \text{where} \quad \begin{cases} y_i = \text{mole fraction of component i in the vapor} \\ x_i = \text{mole fraction of component i in the liquid} \end{cases}$$

| Cut point (%) | Testing methods | |
| :---: | :---: | :---: |
| | TBP (°C) | ASTM (°C) |
| 0.00 | -1.44 | 31.22 |
| 1.00 | -0.80 | 31.63 |
| 2.00 | 1.61 | 32.94 |
| 3.50 | 6.09 | 35.33 |
| 5.00 | 10.56 | 37.72 |
| 7.50 | 18.02 | 40.29 |
| 10.00 | 24.67 | 45.29 |
| 12.50 | 28.56 | 47.32 |
| 15.00 | 29.57 | 47.84 |
| 17.50 | 30.57 | 48.35 |
| 20.00 | 31.58 | 48.86 |
| 25.00 | 33.59 | 49.89 |
| 30.00 | 35.99 | 51.09 |
| 35.00 | 39.12 | 52.92 |
| 40.00 | 43.94 | 55.83 |
| 45.00 | 50.00 | 59.64 |
| 50.00 | 58.42 | 65.19 |
| 55.00 | 66.23 | 70.38 |
| 60.00 | 69.51 | 72.55 |
| 65.00 | 70.77 | 73.34 |
| 70.00 | 75.91 | 76.68 |
| 75.00 | 86.06 | 84.11 |
| 80.00 | 98.63 | 94.20 |
| 85.00 | 100.57 | 95.91 |
| 90.00 | 115.54 | 109.54 |
| 92.50 | 125.47 | 118.90 |
| 95.00 | 131.07 | 124.24 |
| 96.50 | 138.36 | 131.05 |
| 98.00 | 148.30 | 140.20 |
| 99.00 | 159.91 | 146.78 |
| 100.00 | 168.02 | 156.75 |

Table 2.3. Distillation data

| Properties | Ligas | Naphthas |
| :--- | :--- | :--- |
| Molar weight | 54.4-55.6 | 84.1-86.3 |
| Liquid density (kg/m³) | 570-575 | 725-735 |
| Feed composition (vol %) | 38-42 | 58-62 |

Table 2.4. Properties of the pseudo components

Checking the data in the handbook (Perry, R. & Green, D., 1984) for the operating range of temperature and pressure, the relative volatility is calculated as: $\alpha = 5.68$ .

*Correlation between TBP and Equilibrium Flash Vaporization (EFV):*

The EFV curve is converted from the TBP data according to (Luyben, W., 1990). The initial data are:

$$t_{50\%(TBP)} = 58.42\ °C$$

$$t_{(30\%-10\%)(TBP)} = 35.99 - 24.67 = 11.32\ °C$$

Consulting TBP-EFV correlation chart, we obtain $t_{50\%(EFV-TBP)} = 5.2\ °C$

Therefore: $t_{50\%(EFV)} = 58.42 + 5.2 = 63.62\ °C$

Repeating the above procedure for all TBP data, the EFV (1 atm) data are determined. Then convert the EFV (1 atm) data into the EFV (4.6 atm) data by using Cox chart. The results are shown in the Table 2.5.

*Operating pressure:*

The column is designed with 14 trays, and the pressure drop across each tray is 80 kPa. Thus the pressures at feed section and top section are 4.6 atm and 4 atm respectively.

| % vol. | TBP | | EFV (1 atm) | | EFV (4.6 atm) |
|---|---|---|---|---|---|
| | t °C | Δt | Δt | t °C | t °C |
| I.B.P. | -1.44 | | | 41.62 | 93 |
| | | 1.2 | 1.5 | | |
| 5 | 10.56 | | | 43.12 | 95 |
| | | 14.11 | 4 | | |
| 10 | 24.67 | | | 47.12 | 102 |
| | | 6.91 | 3 | | |
| 20 | 31.58 | | | 50.12 | 106 |
| | | 4.41 | 2.5 | | |
| 30 | 35.99 | | | 52.62 | 110 |
| | | 7.95 | 5 | | |
| 40 | 43.93 | | | 57.62 | 116 |
| | | 14.48 | 6 | | |
| 50 | 58.42 | | | 63.62 | 125 |
| | | 11.09 | 5.5 | | |
| 60 | 69.51 | | | 69.12 | 132 |
| | | 6.4 | 6.5 | | |
| 70 | 75.91 | | | 75.62 | 141 |
| | | 22.72 | 7.5 | | |
| 80 | 98.63 | | | 83.12 | 150 |
| | | 16.91 | 7 | | |
| | 115.54 | | | 90.12 | 158 |

Table 2.5. Correlation between TBP and EFV of raw condensate

### 2.2.2 Calculation for feed section

The feed is in liquid-gas equilibrium with gas percentage of 38% volume. However it is usual to deeply cut 4% of the unexpected heavy components, which will be condensed and refluxed to the columnmore bottom. Thus there are two equilibrium phase flows: vapor $V_F$=38+4=42% and liquid $L_F$=100-42=58%.

*Operating temperature:*

Consulting the EFV curve (4.6 atm) of feed section, the required feed temperature is 118 °C corresponding to 42% volume point.

The phase equilibrium is shown in the Figure 2.2.

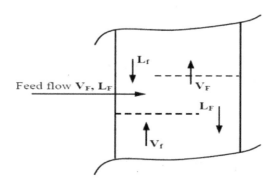

Fig. 2.2. The Equilibrium phase flows at the feed section

Where, $V_F$: Vapor phase rate in the feed flow; $L_F$: Liquid phase rate in the feed flow; $V_f$: Vapor flow arising from the stripping section; $L_f$: Internal reflux descending across the feed section.

The heavy fraction flow $L_f$ dissolved an amount of light components is descending to the column bottom. These undesirable light components shall be caught by the vapor flow $V_f$ arising to the top column. $V_f$, which can be calculated by empirical method, is equal to 28% vol. The bottoms product flow $B$ is determined by yield curve as 62% vol. Hence, the internal reflux across the feed section can be computed as: $L_f = B - L_F + V_f$ =62+28=32%vol.

Material balances for the feed section is shown in the Table 2.6. The calculation based on the raw condensate *feed rate* for the plant: 15.4762 tons/hour.

### 2.2.3 Calculation for stripping section

In the stripping section, liquid flows, which are descending from the feed section, include the equilibrium phase flow $L_F$, and the internal reflux flow $L_f$. They are contacting with the arising vapor flow $V_f$ for heat transfer and mass transfer. This process is accomplished with the aid of heat flow supplied by the re-boiler.

Main parameters to be determined are the bottoms product temperature and the re-boiler duty $Q_B$.

| Stream | Volume fraction % vol | Liquid flow rate m³/h | Liquid density ton/ m³ | Mass flow rate ton/h |
|---|---|---|---|---|
| $V_F$ | 42 | 10.9983 | 0.591 | 6.5000 |
| $V_f$ | 28 | 7.2464 | 0.598 | 4.3333 |
| $L_f$ | -32 | -8.0527 | 0.615 | -4.9524 |
| Total light fraction | 38 | 10.1923 | 0.577 | 5.8810 |
| $L_F$ | 58 | 12.3639 | 0.726 | 8.9762 |
| $V_f$ | -28 | -7.2464 | 0.598 | -4.3333 |
| $L_f$ | 32 | 8.0527 | 0.615 | 4.9524 |
| $B$ | 62 | 13.1984 | 0.727 | 9.5952 |

Table 2.6. Material Balances for the Feed Section

The column base pressure is approximately the pressure at the feed section because pressure drop across this section is negligible. Consulting the EFV curve of stripping section and the Cox chart, the equilibrium temperature at this section is 144 °C. The re-boiler duty is equal to heat input in order to generate boil-up of stripping section an increment of 144-118=26 °C.

The material and energy balances for stripping section is displayed in the Table 2.7 and with one $cal_{TC}$ = 4.184 J.

$\Rightarrow$ The re-boiler duty must be supplied: $Q_B = 6283.535 - 3983.575 = 2299.96 * 10^3$ (kJ/h).

| | ton/h | kcal/kg | kcal/h.10³ | kJ/h.10³ |
|---|---|---|---|---|
| | INLET | | | |
| $L_F$ | 8.9762 | 68 | 610.3816 | 2553.837 |
| $L_f$ | 4.9524 | 69 | 341.7156 | 1429.738 |
| Total | 13.9286 | | 952.0972 | 3983.575 |
| | OUTLET | | | |
| | ton/h | kcal/kg | kcal/h.10³ | kJ/h.10³ |
| $V_f$ | 4.3333 | 165 | 714.9945 | 2991.537 |
| $B$ | 9.5952 | 82 | 786.8064 | 3291.998 |
| Total | 13.9285 | | 1501.801 | 6283.535 |

Table 2.7. Material and Energy Balances for Stripping Section

## 2.2.4 Calculation for rectifying section

The overhead vapor flow, which includes $V_F$ from feed section and $V_f$ from stripping section, passes through the condenser (to remove heat) and then enter into the reflux drum. There exists two equilibrium phases: liquid (butane as major amount) and vapor (butane vapor, uncondensed gas – dry gas: $C_1$, $C_2$, e.g.). The liquid from the reflux drum is partly pumped back into the top tray as reflux flow $L$ and partly removed as distillate flow $D$. The top pressure is 4 atm due to pressure drop across the rectifying section. The dew point of distillate is correspondingly the point 100% of the EFV curve of rectifying section. Also consulting the Cox chart, the top section temperature is determined as 46 °C.

The equilibrium phase flows at the rectifying sections are displayed in the Table 2.8.

| INLET | | | | |
|---|---|---|---|---|
| | ton/h | kcal/kg | kcal/h.10³ | kJ/h.10³ |
| $V_F + V_f$ | 10.8333 | 115 | 1245.83 | 5212.553 |
| $L_0$ | $L_0$ | 24 | 24x$L_0$ | 100.416x$L_0$ |
| Total | 10.8333+ $L_0$ | | 1245.83+24x$L_0$ | 5212.553+100.416*$L_0$ |
| OUTLET | | | | |
| | ton/h | kcal/kg | kcal/h.10³ | kJ/h.10³ |
| Total light fraction+$L_0$ | 5.8810+$L_0$ | 97 | 570.457+97x$L_0$ | 2386.792+405.848x$L_0$ |
| $L_f$ | 4.9524 | 16 | 79.2384 | 331.533 |
| Total | 10.8334+$L_0$ | | 649.695+97x$L_0$ | 2718.326+405.848x$L_0$ |

Table 2.8. Material and Energy Balances for Rectifying Section

Calculate the internal reflux flow $L_0$: Energy balance, INLET=OUTLET:

$$5212.553 + 100.416L_0 = 2718.326 + 405.848L_0 \Rightarrow L_0 = 8.166 \text{ (ton/h)}$$
$$\Rightarrow \text{Total light fraction} + L_0 = 14.047 \text{ (ton/h)}$$

Calculate the external reflux flow L: Enthalpy data of reflux flow L looked up the experimental chart for petroleum enthalpy are corresponding to the liquid state of 40 °C (liquid inlet at the top tray) and the vapor state of 46 °C (vapor outlet at the column top).

L inlet at 40 °C: $H_{\text{liquid(inlet)}}$ = 22 kcal/kg; L outlet at 46 °C: $H_{\text{vapor(outlet)}}$ = 106 kcal/kg. Then,

$$\Delta H_{L_0} \times L_0 = \Delta H_L \times L \Rightarrow (115 - 24)(8.166) = (106 - 22)L \Rightarrow L = 8.847 \text{ (ton/h)}.$$

## 2.2.5 Latent heat and boil-up flow rate

The heat input of $Q_B$ (re-boiler duty) to the reboiler is to increase the temperature of stripping section and to generate boil-up $V_0$ as (Franks, R., 1972): $V_0 = \dfrac{Q_B - Bc_B(t_B - t_F)}{\lambda}$,

where, $Q_B$: re-boiler duty – 2299.96*10³ (kJ/h); B: flow rate of bottom product – 9595.2 (kg/h); $c_B$: specific heat capacity – 85 (kJ/kg. °C); $t_F$: inlet temperature – 118 (°C, the feed temperature); $t_B$: outlet temperature – 144 °C; $\lambda$: the latent heat or heat of vaporization.

The latent heat at any temperature is described in terms of the latent heat at the normal boiling point (Nelson, W., 1985): $\lambda = \gamma \lambda_B \dfrac{T}{T_B}$ (kJ/kg), where, $\lambda$ : latent heat at absolute temperature T (kJ/kg); $\lambda_B$: latent heat at absolute normal boiling point $T_B$ (kJ/kg); $\gamma$ : correction factor obtained from the empirical chart. The result: $\lambda$ =8500 (kJ/kg); $V_0$=4540.42 (kg/h) or 77.67 (kmole/h); $V_f$=4333.3 (kg/h) or 74.13 (kmole/h). The average vapor flow rate is rising in the stripping section $V = \dfrac{V_0 + V_f}{2} = \dfrac{77.67 + 74.13}{2} = 75.9$ (kmole/h).

## 2.2.6 Liquid holdup

Major design parameters to determine the liquid holdup on a tray, column base and reflux drum are calculated mainly based on other literature (Joshi, M., 1979; Wanrren, L. *et al.*, 2005; & Wuithier, P., 1972):

Velocity of vapor phase arising in the column: $\omega_n = C\sqrt{\dfrac{\rho_L - \rho_G}{\rho_G}}$ (m/s), where: $\rho_L$: density of liquid phase; $\rho_G$: density of vapor phase; $C$: correction factor depending flow rates of two-phase flows, obtained from the empirical chart , $C_f - P_f$ with $P_f = \dfrac{L}{G}\sqrt{\dfrac{\rho_G}{\rho_L}}$. The actual velocity $\omega$ is normally selected that $\omega = (0.80 - 0.85)\omega_n$ for paraffinic vapor. The diameter of the column is calculated with the formula: $D_k = \sqrt{\dfrac{4V_m}{3600\pi\omega}}$ (m), where, $V_m$: the mean flow of vapor in the column. Result: $D_k = 1.75$ (m).

The height of the column is calculated on distance of trays. The distance is selected on basis of the column diameter. The holdup in the column base is determined as:

$$M_B = \frac{\pi H_{NB} D_k^2}{4} \frac{d_B}{(MW)_B} \text{ (kmole)} \tag{2.2}$$

where: $H_{NB}$: normal liquid level in the column base (m); $(MW)_B$: molar weight of the bottom product (kg/kmole); $d_B$: density of the bottom product (kg/m³). Then,

$$\Rightarrow M_B = \frac{3.14(1.4)(1.75)^2}{4} \frac{726.5}{78.6} = 31.11 \text{ (kmole)}.$$

The holdup on each tray: $M = \dfrac{0.95\pi h_T D_k^2}{4} \dfrac{d_T}{(MW)_T}$ (kmole), where, $h_T$: average depth of clear liquid on a tray (m); $(MW)_T$: molar weight of the liquid holdup on a tray (kg/kmole); $d_T$: the mean density of the liquid holdup on a tray (kg/m³). Then,

$$\Rightarrow M = \frac{0.95(3.14)(0.28)(1.75)^2}{4} \frac{680}{75} = 5.80 \text{ kmole}.$$

The holdup in the reflux drum: Liquid holdup $M_D$ is equal to the quantity of distillate contained in the reflux drum, $M_D = \dfrac{5(L_f + D_f)}{60}$ (kmole), where, $M_D$: holdup in the reflux drum; $L_f$: the reflux flow rate – (4952.4 kg/h)/(60.1 mole weight) = 82.4 kmole/h; $V_f$: the distillate flow rate – (4333.3 kg/h)/(58.2 mole weight) = 74.46 kmole/h. Then,

$$\Rightarrow M_D = \frac{5(82.4 + 74.46)}{60} = 13.07 \text{ (kmole)}.$$

## 3. Mathematic model

### 3.1 Equations for flows throughout general trays

The total mole holdup in the $n$th tray $M_n$ is considered constant, but the imbalance in the input and output flows is taken into account for in the component and heat balance equations as shown in Figure 3.1.

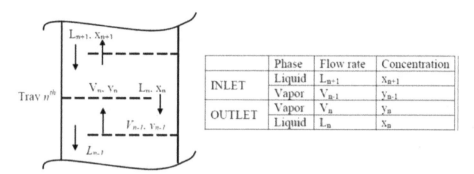

| | Phase | Flow rate | Concentration |
|---|---|---|---|
| INLET | Liquid | $L_{n+1}$ | $X_{n+1}$ |
| | Vapor | $V_{n-1}$ | $Y_{n-1}$ |
| OUTLET | Vapor | $V_n$ | $Y_n$ |
| | Liquid | $L_n$ | $X_n$ |

Fig. 3.1. A General $n$th Tray

Total mass balance:

$$\frac{d(M_n)}{dt} = L_{n+1} - L_n + V_{n-1} - V_n \tag{3.1}$$

Component balance:

$$\frac{d(M_n x_n)}{dt} = L_{n+1}x_{n+1} - L_n x_n + V_{n-1}y_{n-1} - V_n y_n \tag{3.2}$$

By differentiating (3.2) and substituting for (3.1), the following expression is obtained:

$$\frac{d(x_n)}{dt} = \frac{L_{n+1}x_{n+1} + V_{n-1}y_{n-1} - (L_{n+1} + V_{n-1})x_n - V_n(y_n - x_n)}{M_n} \tag{3.3}$$

Energy balance:

$$\frac{d(M_n h_n)}{dt} = h_{n+1}L_{n+1} - h_n L_n + H_{n-1}V_{n-1} - H_n V_n \tag{3.4}$$

or

$$M_n\frac{dh_n}{dt} + h_n\frac{dM_n}{dt} = h_{n+1}L_{n+1} + H_{n-1}V_{n-1} - h_n L_n - H_n V_n \tag{3.5}$$

Because the term $\frac{dh_n}{dt}$ is approximately zero, substituting for the change of hold up $\frac{dM_n}{dt}$ in (3.5), and rearranging the terms, the following expression is obtained:

$$V_n = \frac{h_{n+1}L_{n+1} + H_{n-1}V_{n-1} - (L_{n+1} + V_{n-1})h_n}{H_n - h_n} \tag{3.6}$$

where, n: tray $n^{th}$; $V$: vapor flow; $L$: liquid flow; $x$: liquid concentration of light component; $y$: vapor concentration of light component; $h$: enthalpy for liquid; $H$: enthalpy for vapor.

### 3.2 Equations for the feed tray: (Stage n=*f*) (See Figure 3.2)

Total mass balance:

$$\frac{d(M_f)}{dt} = F + L_{f+1} + V_{f-1} - L_f - V_f \tag{3.7}$$

Component balance:

$$\frac{d(M_f x_f)}{dt} = Fc_f + L_{f+1}x_{f+1} + V_{f-1}y_{f-1} - L_f x_f - V_f y_f$$

$$\Rightarrow \frac{dx_n}{dt} = \frac{L_{n+1}x_{n+1} + V_{n-1}y_{n-1} - (L_{n+1} + V_{n-1})x_n - V_n(y_n - x_n)}{M_n} \tag{3.8}$$

Energy balance:

$$\frac{d(M_f h_f)}{dt} = h_f F + h_{n+1}L_{n+1} + H_{n-1}V_{n-1} - h_n L_n - H_n V_n$$

$$\Rightarrow V_n = \frac{h_F F + h_{n+1}L_{n+1} + H_{n-1}V_{n-1} - (L_{n+1} + V_{n-1})h_n}{H_n - h_n} \tag{3.9}$$

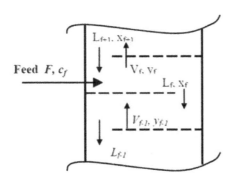

Fig. 3.2. Feed Section

## 3.3 Equations for the top section: (stage n=*N*+1) (See Figure 3.3)

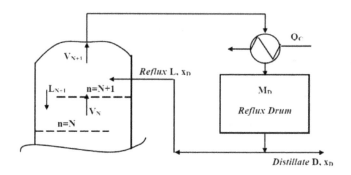

Fig. 3.3. Top Section and Reflux Drum

*Equations for the top tray (stage n=N+1)*

Total mass balance:

$$\frac{d(M_{N+1})}{dt} = L + V_N - L_{N+1} - V_{N+1} \tag{3.10}$$

Component balance:

$$\frac{d(M_{N+1}x_{N+1})}{dt} = Lx_D + V_N y_N - L_{N+1}x_{N+1} - V_{N+1}y_{N+1} \tag{3.11}$$

Energy balance:

$$\frac{d(M_{N+1}h_{N+1})}{dt} = h_D L + H_N V_N - h_{N+1}L_{N+1} - H_{N+1}V_{N+1} \tag{3.12}$$

$$\Rightarrow V_{N+1} = \frac{h_D L + H_N V_N - (L + V_N)h_{N+1}}{H_{N+1} - h_{N+1}} \tag{3.13}$$

**Reflux drum and condenser**

Total mass balance:

$$\frac{d(M_D)}{dt} = V_{N+1} - L - D \tag{3.14}$$

Component balance:

$$\frac{d(M_D x_D)}{dt} = V_{N+1}y_{N+1} - (L + D)x_D \tag{3.15}$$

Energy balance around condenser:

The condenser duty $Q_C$ is equal to the latent heat required to condense the overhead vapor to bubble point:

$$Q_C = H_{in}V_{in} - h_{out}L_{out} = V_N(H_N - h_N) \tag{3.16}$$

## 3.4 Equations for the bottom section: (Stage n=2) (See Figue 3.4)

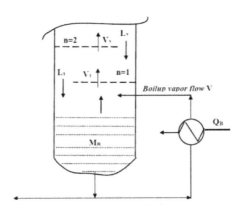

Fig. 3.4. Bottom Section and Re-boiler

*Bottom Tray (stage n=2)*

Total mass balance:

$$\frac{d(M_2)}{dt} = L_3 - L_2 + V_B - V_2 \tag{3.17}$$

Component balance:

$$\frac{d(M_2 x_2)}{dt} = L_3 x_3 - L_2 x_2 + V_B y_B - V_2 y_2 \tag{3.18}$$

Energy balance:

$$\frac{d(M_B h_B)}{dt} = h_3 L_3 + H_B V_B - h_2 L_2 - H_2 V_2 \tag{3.19}$$

$$\Rightarrow V_2 = \frac{h_3 L_3 + H_B V_B - (L_3 + V_B)h_2}{H_2 - h_2} \tag{3.20}$$

*Re-boiler and Column Bottoms (stage n=1)*

The base of the column has some particular characteristics as follows:

- There is re-boiler heat flux $Q_B$ establishing the boil-up vapor flow $V_B$.

- The holdup is variable and changes in sensible heat cannot be neglected.
- The outflow of liquid from the bottoms $B$ is determined externally to be controlled by a bottoms level controller.

Total mass balance:

$$\frac{d(M_B)}{dt} = L_2 - V_B - B \tag{3.21}$$

Component balance:

$$\frac{d(M_B x_B)}{dt} = L_2 x_2 - V_B y_B - B x_B \tag{3.22}$$

Energy balance:

$$\frac{d(M_B h_B)}{dt} = h_2 L_2 + Q_B - h_B B - H_B V_B \tag{3.23}$$

$$\Rightarrow V_B = \frac{h_2 L_2 + Q_B - h_B B - M_B \dfrac{dh_B}{dt} - h_B \dfrac{dM_B}{dt}}{H_B} \tag{3.24}$$

When all the modeling equations above are resolved, we find out how the flow rate and concentrations of the two product streams (distillate product and bottoms product) change with time, in the presence of changes in the various input variables.

## 3.5 Simplified model

To simplify the model, we make the following assumption (Papadouratis, A. *et al.* 1989):

- The relative volatility $\alpha$ is constant throughout the column. This means the vapor-liquid equilibrium relationship can be expressed by

$$y_n = \frac{\alpha x_n}{1 + (\alpha - 1)x_n} \tag{3.25}$$

where $x_n$: liquid composition on $n^{th}$ stage; $y_n$: vapor composition on $n^{th}$ stage; and $\alpha$: relative volatility.

- The overhead vapor is totally condensed in a condenser.
- The liquid holdups on each tray, condenser, and the re-boiler are constant and perfectly mixed (i.e. immediate liquid response, ($dL_2 = dL_3 = \ldots\ldots = dL_{N+2} = dL$ ).
- The holdup of vapor is negligible throughout the system (i.e. immediate vapor response, $dV_1 = dV_2 = \ldots\ldots = dV_{N+1} = dV$ ).
- The molar flow rates of the vapor and liquid through the stripping and rectifying sections are constant: $V_1 = V_2 = \ldots\ldots = V_{N+1}$ and $L_2 = L_3 = \ldots\ldots\ldots = L_{N+2}$ .
- The column is numbered from bottom ($n=1$ for the re-boiler, $n=2$ for the first tray, $n=f$ for the feed tray, $n=N+1$ for the top tray and $n=N+2$ for the condenser).

Under these assumptions, the dynamic model can be expressed by (George, S., 1986):

Condenser ($n=N+2$):

$$M_D \dot{x}_n = (V + V_F)y_{n-1} - Lx_n - Dx_n \tag{3.26}$$

Tray n ($n=f+2, ..., N+1$):

$$M \dot{x}_n = (V + V_F)(y_{n-1} - y_n) + L(x_{n+1} - x_n) \tag{3.27}$$

Tray above the feed flow ($n=f+1$):

$$M \dot{x}_n = V(y_{n-1} - y_n) + L(x_{n+1} - x_n) + V_F(y_F - y_n) \tag{3.28}$$

Tray below the feed flow ($n=f$):

$$M \dot{x}_n = V(y_{n-1} - y_n) + L(x_{n+1} - x_n) + L_F(x_F - x_n) \tag{3.29}$$

Tray n ($n=2, ..., f-1$):

$$M \dot{x}_n = V(y_{n-1} - y_n) + (L + L_F)(x_{n+1} - x_n) \tag{3.30}$$

Re-boiler ($n=1$):

$$M_B \dot{x}_1 = (L + L_F)x_2 - Vy_1 - Bx_1 \tag{3.31}$$

Flow rate are assumed as constant molar flows as

$$L_F = q_F F \tag{3.32}$$

$$V_F = F - L_F \tag{3.33}$$

Assuming condenser holdup constant

$$D = V_N - L = V + V_F - L \tag{3.34}$$

Assuming boiler holdup constant

$$B = L_2 - V_1 = L + L_F - V \tag{3.35}$$

Composition $x_F$ in the liquid and $y_F$ in the vapor phase of the feed are obtained by solving the flash equations:

$$Fc_F = L_F x_F + V_F y_F \tag{3.36}$$

And

$$y_F = \frac{\alpha x_F}{1 + (\alpha - 1)x_F} \tag{3.37}$$

where, $\alpha$ is the relative volatility.

Although the model order is reduced, the representation of the distillation system is still *nonlinear* due to the vapor-liquid equilibrium relationship in equation (3.25).

## 4. Model simulation and analysis

### 4.1 Model dynamic equations

In the process data calculation, we have calculated for the distillation column with 14 trays with the following initial data - equations (2.1), (2.2), (2.3) and (2.4): The feed mass rate of the plant: $F_{mass} = 15.47619$ (tons/hour); The holdup in the column base: $M_B = 31.11$ (kmole);

The holdup on each tray: $M = 5.80$ (kmole); The holdup in the reflux drum: $M_D = 13.07$ (kmole); The gas percentage in the feed flow: $c_F = 38\%$ ; The internal vapor flow $V_f$ selected by empirical: $V_f = 28\%$; The feed stream (m³/h) with the density $d_F$=0.670 (ton/m³);

$F = \dfrac{F_{mass}}{d_F} = \dfrac{15.47619}{0.670}$ =23.0988 (m³/h). The calculated stream data is displayed in the table 4.1.

| Stream | Formular | % | Volume (m³/h) | Density (ton/m³) | Mass (ton/h) | Molar (kg/kmol) | Molar flow (kmole/h) |
|--------|----------|---|---------------|------------------|--------------|-----------------|----------------------|
| Vapor rate in feed $V_F$ | $c_{F+4}$ | 42 | 9.7015 | 0.591 | 5.7336 | 58.2 | 98.5152 |
| Liquid rate in feed $L_F$ | $100\text{-}c_F$ | 58 | 13.3973 | 0.726 | 9.7264 | 93.3 | 104.2491 |
| Internal vapor rate $V$ | $V_f$ | 28 | 6.4677 | 0.598 | 3.8677 | 58.3 | 66.3407 |
| Internal liquid rate $L$ | $V_f\text{+}4$ | 32 | 7.3916 | 0.615 | 4.5458 | 60.1 | 75.6380 |
| Distillate flow rate $D$ | $c_F$ | 38 | 8.7775 | 0.576 | 5.0554 | 54.5 | 92.7597 |
| Bottoms flow rate $B$ | $100\text{-}c_F$ | 62 | 14.3213 | 0.727 | 10.405 | 93.8 | 110.9235 |

Table 4.1. Summary of Stream Data

Solving flash equation with the relative volatility ( $\alpha = 5.68$ ), $\Rightarrow x_F = 0.26095$; $y_F = 0.66728$ .

Reference to equations from (3.28) to (3.39) we can develop a set of nonlinear differential and algebraic equations for the simplified model can be developed as:

$$13.07\dot{x}_{16} = 164.8559y_{15} - 75.6380x_{16} - 92.7597x_{16}$$
$$\Rightarrow \dot{x}_{16} = 12.6133y_{15} - 12.8863x_{16} \tag{4.1}$$

$$5.8\dot{x}_{15} = 164.8559(y_{14} - y_{15}) + 75.6380(x_{16} - x_{15})$$
$$\Rightarrow \dot{x}_{15} = 28.4234y_{14} - 28.4234y_{15} + 13.0410x_{16} - 13.0410x_{15} \tag{4.2}$$

$$5.8\dot{x}_{14} = 164.8559(y_{13} - y_{14}) + 75.6380(x_{15} - x_{14})$$
$$\Rightarrow \dot{x}_{14} = 28.4234y_{13} - 28.4234y_{14} + 13.0410x_{15} - 13.0410x_{14} \tag{4.3}$$

$$5.8\dot{x}_{13} = 164.8559(y_{12} - y_{13}) + 75.6380(x_{14} - x_{13})$$
$$\Rightarrow \dot{x}_{13} = 28.4234y_{12} - 28.4234y_{13} + 13.0410x_{14} - 13.0410x_{13} \tag{4.4}$$

$$5.8\dot{x}_{12} = 164.8559(y_{11} - y_{12}) + 75.6380(x_{13} - x_{12})$$
$$\Rightarrow \dot{x}_{12} = 28.4234y_{11} - 28.4234y_{12} + 13.0410x_{13} - 13.0410x_{12} \tag{4.5}$$

$$5.8\dot{x}_{11} = 164.8559(y_{10} - y_{11}) + 75.6380(x_{12} - x_{11})$$
$$\Rightarrow \dot{x}_{11} = 28.4234y_{10} - 28.4234y_{11} + 13.0410x_{12} - 13.0410x_{11} \tag{4.6}$$

$$5.8\dot{x}_{10} = 164.8559(y_9 - y_{10}) + 75.6380(x_{11} - x_{10})$$
$$\Rightarrow \dot{x}_{10} = 28.4234y_9 - 28.4234y_{10} + 13.0410x_{11} - 13.0410x_{10} \tag{4.7}$$

$$5.8\dot{x}_9 = 66.3407(y_8 - y_9) + 75.6380(x_{10} - x_9) + 98.5152(0.66728 - y_9)$$
$$\Rightarrow \dot{x}_9 = 11.4381y_8 - 28.4234y_9 - 13.0410x_9 + 13.0410x_{10} + 11.3340 \tag{4.8}$$

$$5.8\dot{x}_8 = 66.3407(y_7 - y_8) + 75.6380(x_9 - x_8) + 104.2491(0.26.95 - x_8)$$
$$\Rightarrow \dot{x}_8 = 11.4381y_7 - 11.4381y_8 - 31.0150x_8 + 13.0410x_9 + 4.8440 \tag{4.9}$$

$$5.8\dot{x}_7 = 66.3407(y_6 - y_7) + 179.8871(x_8 - x_7)$$
$$\Rightarrow \dot{x}_7 = 11.4381y_6 - 11.4381y_7 + 31.0150x_8 - 31.0150x_7 \tag{4.10}$$

$$5.8\dot{x}_6 = 66.3407(y_5 - y_6) + 179.8871(x_7 - x_6)$$
$$\Rightarrow \dot{x}_6 = 11.4381y_5 - 11.4381y_6 + 31.0150x_7 - 31.0150x_6 \tag{4.11}$$

$$5.8\dot{x}_5 = 66.3407(y_4 - y_5) + 179.8871(x_6 - x_5)$$
$$\Rightarrow \dot{x}_5 = 11.4380y_4 - 11.4381y_5 + 31.0150x_6 - 31.0150x_5 \tag{4.12}$$

$$5.8\dot{x}_4 = 66.3407(y_3 - y_4) + 179.8871(x_5 - x_4)$$
$$\Rightarrow \dot{x}_4 = 11.4381y_3 - 11.4381y_4 + 31.0150x_5 - 31.0150x_4 \tag{4.13}$$

$$5.8\dot{x}_3 = 66.3407(y_2 - y_3) + 179.8871(x_4 - x_3)$$
$$\Rightarrow \dot{x}_3 = 11.4381y_2 - 11.4381y_3 + 31.0150x_4 - 31.0150x_3 \tag{4.14}$$

$$5.8\dot{x}_2 = 66.3407(y_1 - y_2) + 179.8871(x_3 - x_2)$$
$$\Rightarrow \dot{x}_2 = 11.4381y_1 - 11.4381y_2 + 31.0150x_3 - 31.0150x_2 \tag{4.15}$$

$$31.11\dot{x}_1 = 179.8871x_2 - 110.9235x_1 - 66.3407y_1$$
$$\Rightarrow \dot{x}_1 = -3.5655x_1 + 5.7823x_2 - 2.1325y_1 \tag{4.16}$$

And vapor-liquid equilibrium on each tray ($n$=1-16):

$$y_n = \frac{5.68x_n}{1 + 4.68x_n} \tag{4.17}$$

## 4.2 Model simulation with Matlab Simulink

### 4.2.1 Simulation without disturbances

The steady-state solution is determined with dynamic simulation. Figure 4.1 displays the concentration of the light component $x_n$ at each tray and Table 4.2 shows the steady state values of concentration of $x_n$ on each tray.

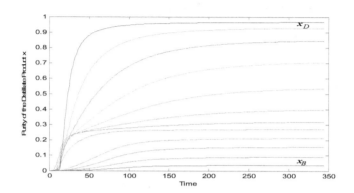

Fig. 4.1. Steady state values of concentration $x_n$ on each tray

| Tray | 1 | 2 | 3 | 4 | 5 | 6 | 7 | 8 |
|------|------|------|------|------|------|------|------|------|
| $x_n$ | **0.0375** | 0.0900 | 0.1559 | 0.2120 | 0.2461 | 0.2628 | 0.2701 | 0.2731 |
| $y_n$ | 0.1812 | 0.3597 | 0.5120 | 0.6044 | 0.6496 | 0.6694 | 0.6776 | 0.6809 |
| Tray | 9 | 10 | 11 | 12 | 13 | 14 | 15 | 16 |
| $x_n$ | 0.2811 | 0.3177 | 0.3963 | 0.5336 | 0.7041 | 0.8449 | 0.9269 | **0.9654** |
| $y_n$ | 0.6895 | 0.7256 | 0.7885 | 0.8666 | 0.9311 | 0.9687 | 0.9863 | 0.9937 |

Table 4.2. Steady state values of concentration $x_n$ on each tray

If there are no disturbance in operating condition, the system model is to achieve the steady state of product quality that the purity of the distillate product $x_D = 0.9654$ and the impurity of the bottoms product $x_B = 0.0375$.

### 4.2.2 Simulation with 10% decreasing and increasing feed flow rate

When decreasing the feed flow rate by 10%, the quality of the distillate product will get worse while the quality of the bottoms product will get better: the purity of the distillate product reduces from 96.54% to 90.23% while the impurity of the bottoms product reduces from 3.75% to 0.66%.

In contrast, when increasing the feed flow rate by 10%, the quality of the distillate product will be better while the quality of the bottoms product will be worse: the purity of the distillate product increases from 96.54% to 97.30% while the impurity of the bottoms product increases from 3.75% to 11.66%. (See Table 4.3 and Figure 4.2).

| | Purity of the Distillate Product (%) | Impurity of the Bottoms Product (%) |
|---|---|---|
| Normal Feed Rate (100%) | 96.54 | 3.75 |
| Reduced Feed Rate (90%) | 90.23 | 0.66 |
| Increased Feed Rate (110%) | 97.30 | 11.66 |

Table 4.3. Product quality depending on the change of feed rate

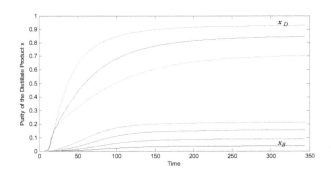

Fig. 4.2. Product qualities depending on change of feed rate

### 4.2.3 Simulation with a wave change in the feed flow rate by ±5%

When the input flow rate fluctuates in a *sine wave* by ±5% (see Figure 4.3), the purity of the distillate product and the impurity of the bottoms product will also fluctuate in a sine wave (see Figure 4.4 and Table 4.4).

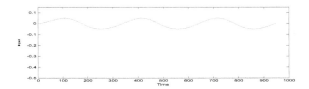

Fig. 4.3. Feed flow rate in a sine wave around ±5%

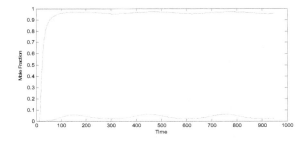

Fig. 4.4. Product quality for a sine wave feed rate

|              | Feed Flow Rate (%) | Distillate Purity (%) | Bottoms Impurity (%) |
|--------------|--------------------|-----------------------|----------------------|
| Max Value    | 105                | 96.92                 | 5.53                 |
| Min Value    | 95                 | 95.26                 | 2.06                 |

Table 4.4. Product quality depending on the input sine wave fluctuation

The product quality of this feed rate is not satisfied with $x_B \geq 96\%$ and $x_D \leq 4\%$.

## 5. Linearized control model

### 5.1 Linear approximation of nonlinear system

### 5.1.1 Vapor-Liquid equilibrium relationship in each tray

$$y_n = \frac{\alpha x_n}{1 + (\alpha - 1)x_n} = \frac{5.68x_n}{1 + 4.68x_n} \tag{5.1}$$

In order to obtain a linear mathematical model for a nonlinear system, it is assumed that the variables deviate only slightly from some operating condition (Ogata, K., 2001). If the normal operating condition corresponds to $\bar{x}_n$ and $\bar{y}_n$, then equation (5.1) can be expanded into a Taylor's series as:

$$y_n = f(\bar{x}_n) + \frac{df}{dx_n}(x_n - \bar{x}_n) + \frac{1}{2!}\frac{d^2f}{dx_n^2}(x_n - \bar{x}_n)^2 + \dots \tag{5.2}$$

where the derivatives $df/dx_n$, $d^2f/dx_n^2$,... are evaluated at $x_n = \bar{x}_n$. If the variation $x_n = \bar{x}_n$ is small, the higher-order terms in $x_n - \bar{x}_n$ may be neglected. Then equation (5.2) can be written as:

$$y_n = \bar{y}_n + K_n(x_n - \bar{x}_n) \text{ with } \bar{y}_n = f(\bar{x}_n) \text{ and } K = \frac{df}{dx_n}\bigg|_{x_n = \bar{x}_n} \tag{5.3}$$

From (5.3), equation (5.1) can be written:

$$y_n = \bar{y}_n + K_n(x_n - \bar{x}_n) \text{ with } \bar{y}_n = \frac{5.68\bar{x}_n}{1 + 4.68\bar{x}_n} \text{ and } K_n = \frac{5.68}{(1 + 4.68\bar{x}_n)^2} \tag{5.4}$$

| Tray  | 1      | 2      | 3      | 4      | 5      | 6      | 7      | 8      |
|-------|--------|--------|--------|--------|--------|--------|--------|--------|
| $x_n$ | 0.0375 | 0.0900 | 0.1559 | 0.2120 | 0.2461 | 0.2628 | 0.2701 | 0.2731 |
| $y_n$ | 0.1812 | 0.3597 | 0.5120 | 0.6044 | 0.6496 | 0.6694 | 0.6776 | 0.6809 |
| $K_n$ | 4.1106 | 2.8121 | 1.8987 | 1.4312 | 1.2268 | 1.1423 | 1.1081 | 1.0945 |
| Tray  | 9      | 10     | 11     | 12     | 13     | 14     | 15     | 16     |
| $x_n$ | 0.2811 | 0.3177 | 0.3963 | 0.5336 | 0.7041 | 0.8449 | 0.9269 | 0.9654 |
| $y_n$ | 0.6895 | 0.7256 | 0.7885 | 0.8666 | 0.9311 | 0.9687 | 0.9863 | 0.9937 |
| $K_n$ | 1.0594 | 0.9184 | 0.6970 | 0.4644 | 0.3079 | 0.2314 | 0.1993 | 0.1865 |

Table 5.1. The Concentration $x_n$, $y_n$ and the linearization coefficient $K_n$

## 5.1.2 Material balance relationship in each tray

Linearization for general trays ( $n = 2 \div 15$ ) - ACCUMULATION = INLET – OUTLET:

$$M_n \dot{x}_n = (V_{n-1} y_{n-1} + L_{n+1} x_{n+1}) - (V_n y_n + L_n x_n)$$

$$\Rightarrow \dot{x}_n = \frac{V_{n-1}}{M_n} y_{n-1} - \frac{V_n}{M_n} y_n + \frac{L_{n+1}}{M_n} x_{n+1} - \frac{L_n}{M_n} x_n \tag{5.5}$$

where,

$$L_1 = L_2 = ... = L_{F-1} = L + L_F \,; L_{F+2} = L_{F+3} = ... = L \,; V_1 = V_2 = ... = V_{F-1} = V \,;$$

$$V_{F+2} = V_{F+3} = ... = V_{N+1} = V + V_F \,; \; M_n = M = 5.8 \,; \; V_{Steady\ State} = \overline{V} = 66.3407 \,;$$

$$L_{Steady\ State} = \overline{L} = 75.6380 \,; \; L_F = 104.2491 \,; \; V_F = 98.5152 \,.$$

Substituting equation (5.4) into equation (5.5), the following expression is obtained:

$$\dot{x}_n = \frac{(\overline{y}_{n-1} - K_{n-1}\overline{x}_{n-1})V_{n-1}}{M_n} + \frac{(K_{n-1})V_{n-1}x_{n-1}}{M_n} - \frac{(\overline{y}_n - K_n\overline{x}_n)V_n}{M_n} - \frac{(K_n)V_n x_n}{M_n} + \frac{L_{n+1}x_{n+1}}{M_n} - \frac{L_n x_n}{M_n}$$

In order to obtain a linear approximation to this nonlinear system, this equation may be expanded into a Taylor series about the normal operating point from equation (5.3), and the linear approximation equations for general trays are obtained:

$$\dot{x}_n - \overline{\dot{x}}_n = \frac{(K_{n-1}\overline{V}_n)}{M} x_{n-1} - \frac{(K_n \overline{V}_n + \overline{L}_n)}{M} x_n + \frac{(\overline{L}_n)}{M} x_{n+1} + \frac{(\overline{x}_{n+1} - \overline{x}_n)}{M} L - \frac{(\overline{y}_n - \overline{y}_{n-1})}{M} V$$

$$- \frac{\overline{L}_n(\overline{x}_{n+1} - \overline{x}_n)}{M} + \frac{\overline{V}_n(K_n\overline{x}_n - K_{n-1}\overline{x}_{n-1})}{M} \tag{5.6}$$

Linearization for special trays:

Tray above the feed flow (*n=f+1*):  $\dot{x}_n = \frac{(y_{n-1} - y_n)V}{M} + \frac{(x_{n+1} - x_n)L}{M} + \frac{(y_F - y_n)V_F}{M}$  and the linear approximation equations for the tray above the feed flow:

$$\dot{x}_n - \overline{\dot{x}}_n = \frac{(K_{n-1}\overline{V})}{M} x_{n-1} - \frac{(K_n \overline{V} + L + K_n V_F)}{M} x_n + \frac{\overline{L}}{M} x_{n+1} + \frac{(\overline{x}_{n+1} - \overline{x}_n)}{M} L + \frac{(\overline{y}_n - \overline{y}_{n-1})}{M} V$$

$$- \frac{\overline{L}(\overline{x}_{n+1} - \overline{x}_n)}{M} + \frac{\overline{V}(K_n\overline{x}_n - K_{n-1}\overline{x}_{n-1}) + V_F(y_F - \overline{y}_n + K_n\overline{x}_n)}{M} \tag{5.7}$$

Tray below the feed flow (*n=f*):  $\dot{x}_n - \overline{\dot{x}}_n = \frac{(y_{n-1} - y_n)V}{M} + \frac{(x_{n+1} - x_n)L}{M} + \frac{(x_F - x_n)L_F}{M}$  and the linear approximation equations for the tray above the feed flow:

$$\dot{x}_n - \overline{\dot{x}}_n = \frac{(K_{n-1}\overline{V})}{M} x_{n-1} + \frac{(K_n \overline{V} + \overline{L} + L_F)}{M} x_n + \frac{\overline{L}}{M} x_{n+1} + \frac{(\overline{x}_{n+1} - \overline{x}_n)}{M} L - \frac{(\overline{y}_n - \overline{y}_{n-1})}{M} V$$

$$- \frac{\overline{L}(\overline{x}_{n+1} - \overline{x}_n) - L_F x_F}{M} + \frac{\overline{V}(K_n\overline{x}_n - K_{n-1}\overline{x}_{n-1})}{M} \tag{5.8}$$

Reboiler $(n=1)$: $\dot{x}_1 = \dfrac{L_2 x_2}{M_B} - \dfrac{V y_1}{M_B} - \dfrac{B x_1}{M_B}$, where, $M_B = 31.11$ a, $B = 110.9235$, and the linear approximation equations for the reboiler:

$$\dot{x}_1 - \overline{\dot{x}}_1 = -\frac{(K_1 \overline{V} + B)}{M_B} x_1 + \frac{(\overline{L} + \overline{L}_F)}{M_B} x_2 + \frac{(\overline{x}_2)}{M_B} L - \frac{(\overline{y}_1)}{M_B} V - \frac{((\overline{L} + L_F)\overline{x}_2)}{M_B} + \frac{(\overline{V} K_1 \overline{x}_1)}{M_B} \tag{5.9}$$

Condenser $(n=N)$: $\dot{x}_{16} = \dfrac{V_{15} y_{15}}{M_D} - \dfrac{L x_{16}}{M_D} - \dfrac{D x_{16}}{M_D}$, where, $M_D = 13.07$, $D = 92.7597$, and the linear approximation equations for the condenser:

$$\dot{x}_{16} - \overline{\dot{x}}_{16} = \frac{(K_{15}(\overline{V} + V_F))}{M_D} x_{15} - \frac{(\overline{L} + D)}{M_D} x_{16} - \frac{(\overline{x}_{16})}{M_D} L + \frac{(\overline{y}_{15})}{M_D} V + \frac{(\overline{L} \overline{x}_{16})}{M_D} - \frac{((\overline{V} + V_F)K_{15}\overline{x}_{15})}{M_D} \tag{5.10}$$

As a result, the model is represented in state space in terms of deviation variables:

$$\begin{aligned} \dot{z}(t) &= Az(t) + Bu(t) \\ y(t) &= Cz(t) \end{aligned}, \text{ where}$$

$$z(t) = \begin{vmatrix} x_1(t) - \overline{x}_1 \text{ Steady State} \\ x_2(t) - \overline{x}_2 \text{ Steady State} \\ \vdots \\ x_{16}(t) - \overline{x}_{16} \text{ Steady State} \end{vmatrix}, \tag{5.11}$$

$$u(t) = \begin{vmatrix} L(t) - \overline{L}_{\text{Steady State}} = dL \\ V(t) - \overline{V}_{\text{Steady State}} = dV \end{vmatrix}, \quad y(t) = \begin{vmatrix} x_1(t) - \overline{x}_1 \text{ Steady State} = dx_B \\ x_{16}(t) - \overline{x}_{16} \text{ Steady State} = dx_D \end{vmatrix}$$

The matrix $A$ elements:

For n=1, $a_{1,1} = -\dfrac{(K_1 \overline{V} + B)}{M_B}$, $a_{1,2} = \dfrac{(\overline{L} + \overline{L}_F)}{M_B}$

For n=2÷7, $a_{n,n-1} = \dfrac{(K_{n-1}\overline{V})}{M}$, $a_{n,n} = -\dfrac{(K_n \overline{V} + \overline{L} + L_F)}{M}$, $a_{n,n+1} = \dfrac{(\overline{L} + L_F)}{M}$

For n=8, $a_{8,7} = \dfrac{(K_7 \overline{V})}{M}$, $a_{8,8} = -\dfrac{(K_8 \overline{V} + \overline{L} + L_F)}{M}$, $a_{8,9} = \dfrac{(\overline{L})}{M}$

For n=9, $a_{9,8} = \dfrac{(K_8 \overline{V})}{M}$, $a_{9,9} = -\dfrac{(K_9 \overline{V} + \overline{L})}{M}$, $a_{9,10} = \dfrac{(\overline{L})}{M}$

For n=10÷15, $a_{n,n-1} = \dfrac{(K_{n-1}(\overline{V} + V_F))}{M}$, $a_{n,n} = -\dfrac{(K_n(\overline{V} + V_F) + \overline{L})}{M}$, $a_{n,n+1} = \dfrac{(\overline{L})}{M}$

For n=16, $a_{16,15} = \dfrac{(K_{15}(\overline{V} + V_F))}{M_D}$, $a_{16,16} = -\dfrac{(\overline{L} + D)}{M_D}$, then:

$$
A = \begin{vmatrix}
-12.3312 & 5.7823 & 0 & 0 & 0 & 0 & 0 & 0 & 0 & 0 & 0 & 0 & 0 & 0 & 0 & 0 \\
47.0173 & -63.1800 & 31.0150 & 0 & 0 & 0 & 0 & 0 & 0 & 0 & 0 & 0 & 0 & 0 & 0 & 0 \\
0 & 32.1649 & -52.7324 & 31.0150 & 0 & 0 & 0 & 0 & 0 & 0 & 0 & 0 & 0 & 0 & 0 & 0 \\
0 & 0 & 21.7174 & -47.3852 & 31.0150 & 0 & 0 & 0 & 0 & 0 & 0 & 0 & 0 & 0 & 0 & 0 \\
0 & 0 & 0 & 16.3701 & -45.0472 & 31.0150 & 0 & 0 & 0 & 0 & 0 & 0 & 0 & 0 & 0 & 0 \\
0 & 0 & 0 & 0 & 14.0322 & -44.0807 & 31.0150 & 0 & 0 & 0 & 0 & 0 & 0 & 0 & 0 & 0 \\
0 & 0 & 0 & 0 & 0 & 13.0657 & -43.6895 & 31.0150 & 0 & 0 & 0 & 0 & 0 & 0 & 0 & 0 \\
0 & 0 & 0 & 0 & 0 & 0 & 12.6745 & -43.5340 & 13.0410 & 0 & 0 & 0 & 0 & 0 & 0 & 0 \\
0 & 0 & 0 & 0 & 0 & 0 & 0 & 12.5189 & -43.1528 & 13.0410 & 0 & 0 & 0 & 0 & 0 & 0 \\
0 & 0 & 0 & 0 & 0 & 0 & 0 & 0 & 30.1118 & -39.1451 & 13.0410 & 0 & 0 & 0 & 0 & 0 \\
0 & 0 & 0 & 0 & 0 & 0 & 0 & 0 & 0 & 26.1041 & -32.8522 & 13.0410 & 0 & 0 & 0 & 0 \\
0 & 0 & 0 & 0 & 0 & 0 & 0 & 0 & 0 & 0 & 19.8111 & -26.2409 & 13.0410 & 0 & 0 & 0 \\
0 & 0 & 0 & 0 & 0 & 0 & 0 & 0 & 0 & 0 & 0 & 13.1998 & -21.7926 & 13.0410 & 0 & 0 \\
0 & 0 & 0 & 0 & 0 & 0 & 0 & 0 & 0 & 0 & 0 & 0 & 8.7516 & -19.6182 & 13.0410 & 0 \\
0 & 0 & 0 & 0 & 0 & 0 & 0 & 0 & 0 & 0 & 0 & 0 & 0 & 6.5772 & -18.7058 & 13.0410 \\
0 & 0 & 0 & 0 & 0 & 0 & 0 & 0 & 0 & 0 & 0 & 0 & 0 & 0 & 2.5138 & -12.8843
\end{vmatrix}
$$

The matrix B elements:

For n=1, $b_{1,1} = \dfrac{(\overline{x}_2)}{M_B}L, \quad b_{1,2} = -\dfrac{(\overline{y}_1)}{M_B}V$

For n=2÷15, $b_{n,1} = \dfrac{(\overline{x}_{n+1} - \overline{x}_n)}{M}L, \quad b_{n,2} = -\dfrac{(\overline{y}_n - \overline{y}_{n-1})}{M}V$

For n=16, $b_{16,1} = -\dfrac{(\overline{x}_{16})}{M_D}L, \quad b_{16,2}\dfrac{(\overline{y}_{15})}{M_D}V$, then:

$$
B = \begin{vmatrix}
0.0029 & -0.0058 \\
0.0114 & -0.0308 \\
0.0097 & -0.0263 \\
0.0059 & -0.0159 \\
0.0029 & -0.0078 \\
0.0013 & -0.0034 \\
0.0005 & -0.0014 \\
0.0014 & -0.0006 \\
0.0063 & -0.0015 \\
0.0136 & -0.0062 \\
0.0237 & -0.0108 \\
0.0294 & -0.0135 \\
0.0243 & -0.0111 \\
0.0141 & -0.0065 \\
0.0066 & -0.0030 \\
-0.0739 & 0.0755
\end{vmatrix}
$$

And the output matrix C is:

$$
C = \begin{vmatrix}
1 & 0 & 0 & 0 & 0 & 0 & 0 & 0 & 0 & 0 & 0 & 0 & 0 & 0 & 0 & 0 \\
0 & 0 & 0 & 0 & 0 & 0 & 0 & 0 & 0 & 0 & 0 & 0 & 0 & 0 & 0 & 1
\end{vmatrix}
$$

## 5.2 Reduced-order linear model

The full-order linear model in equation 5-11, which represents a 2 input – 2 output plant can be expressed in the $S$ domain as:

$$\begin{vmatrix} dx_D \\ dx_B \end{vmatrix} = \frac{1}{1+\tau_c s} G(0) \begin{vmatrix} dL \\ dV \end{vmatrix} \tag{5.12}$$

where $\tau_c$ is the time constant and $G(0)$ is the steady state gain

The steady state gain can be directly calculated: $G(0) = -CA^{-1}B$ or

$$G(0) = \begin{vmatrix} 0.0042 & -0.0060 \\ -0.0050 & 0.0072 \end{vmatrix} \tag{5.13}$$

The time constant $\tau_c$ can be calculated based on some specified assumptions (Skogestad, S., & Morari, M., 1987). The linearized value of $\tau_c$ is given by:

$$\tau_c = \frac{M_I}{I_s \ln S} + \frac{M_D(1-x_D)x_D}{I_s} + \frac{M_B(1-x_B)x_B}{I_s} \tag{5.14}$$

where $M_I$ is the total holdup of liquid inside the column:

$$M_I = \sum_{i=1}^{N} M_i = 5.8 * 14 = 81.2 \ (\text{kmole});$$

$I_s$ is the "impurity sum": $I_s = D(1-x_D)x_D + B(1-x_B)x_B = 7.1021$, and $S$ is the separation factor: $S = \frac{x_D(1-x_B)}{(1-x_D)x_B} = 716.1445$.

So that, the time constant $\tau_c$ in equation (5.14) can be determined: $\tau_c = 1.9588 \ (h)$.

As the result, the reduced-order model of the plant is a first order system in equation (5.12):

$$\begin{vmatrix} dx_D \\ dx_B \end{vmatrix} = \frac{1}{1+1.9588s} \begin{vmatrix} 0.0042 & -0.0060 \\ -0.0050 & 0.0072 \end{vmatrix} \begin{vmatrix} dL \\ dV \end{vmatrix} \tag{5.15}$$

or the equivalent reduced-order model in state space:

$$\dot{z}_r(t) = \begin{vmatrix} -0.5105 & 0 \\ 0 & -0.5105 \end{vmatrix} z_r(t) + \begin{vmatrix} 1 & 0 \\ 0 & 1 \end{vmatrix} u(t)$$

$$y_r(t) = \begin{vmatrix} 0.0021 & -0.0031 \\ -0.0026 & 0.0037 \end{vmatrix} z_r(t) \tag{5.16}$$

# 6. Control simulation with MRAC

The reduced-order linear model is then used as the reference model for a model-reference adaptive control (MRAC) system to verify the applicable ability of a conventional adaptive controller for a distillation column dealing with the disturbance and the model-plant mismatch as the influence of the plant feed disturbances.

Adaptive control system is the ability of a controller which can adjust its parameters in such a way as to compensate for the variations in the characteristics of the process. Adaptive control is widely applied in petroleum industries because of the two main reasons: Firstly, most processes are nonlinear and the linearized models are used to design the controllers, so that the controller must change and adapt to the model-plant mismatch; Secondly, most of the processes are non-stationary or their characteristics are changed with time, this leads again to adapt the changing control parameters.

The general form of a MRAC is based on an inner-loop Linear Model Reference Controller (LMRC) and an outer adaptive loop shown in Fig. 6.1. In order to eliminate errors between the model, the plant and the controller is asymptotically stable, MRAC will calculate online the adjustment parameters in gains $L$ and $M$ by $\theta_L(t)$ and $\theta_M(t)$ as detected state error $e(t)$ when changing $A$, $B$ in the process plant.

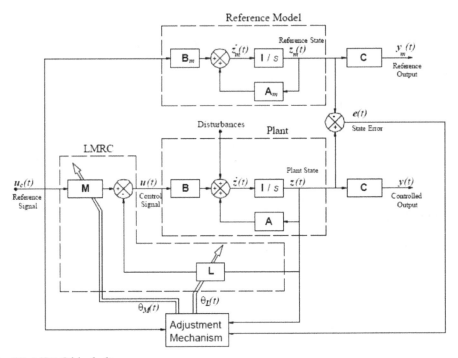

Fig. 6.1. MRAC block diagram

Simulation program is constructed using Maltab Simulink with the following data:

**Process Plant:**

$$\dot{z} = Az + Bu + noise$$
$$y = Cz$$

where $A = \begin{bmatrix} \alpha_1 & 0 \\ 0 & \alpha_2 \end{bmatrix}$, $B = \begin{bmatrix} \beta_1 & 0 \\ 0 & \beta_2 \end{bmatrix}$, $C = \begin{bmatrix} 0.004 & -0.007 \\ -0.0011 & 0.0017 \end{bmatrix}$ and $\alpha_1$, $\alpha_2$, $\beta_1$, $\beta_2$ are changing and dependent on the process dynamics.

**Reference Model:**

$$\dot{z}_m = A_m z_m + B_m u_c$$
$$y_m = C_m z_m$$

where $A_m = \begin{bmatrix} -0.2616 & 0 \\ 0 & -0.2616 \end{bmatrix}$, $B_m = \begin{bmatrix} 1 & 0 \\ 0 & 1 \end{bmatrix}$, $C_m = \begin{bmatrix} 0.004 & -0.007 \\ -0.0011 & 0.0017 \end{bmatrix}$

**State Feedback:**

$u = M u_c - L z$ where $L = \begin{bmatrix} \theta_1 & 0 \\ 0 & \theta_2 \end{bmatrix}$ and $M = \begin{bmatrix} \theta_3 & 0 \\ 0 & \theta_4 \end{bmatrix}$.

**Closed Loop:**

$$\dot{z} = (A - BL)z + BM u_c = A_c(\theta)z + B_c(\theta)u_c$$

**Error Equation:**

$e = z - z_m = \begin{bmatrix} e_1 \\ e_2 \end{bmatrix}$ is a vector of state errors,

$$\dot{e} = \dot{z} - \dot{z}_m = Az + Bu - A_m z_m - B_m u_c = A_m e + (A_c(\theta) - A_m)z + (B_c(\theta) - B_m)u_c = A_m e + \Psi(\theta - \theta^0)$$

where $\Psi = \begin{bmatrix} -\beta_1 z_1 & 0 & \beta_1 u_{c1} & 0 \\ 0 & -\beta_2 z_2 & 0 & \beta_2 u_{c2} \end{bmatrix}$

**Lyapunov Function:**

$V(e, \theta) = \frac{1}{2}\left(\gamma e^T P e + (\theta - \theta^0)^T(\theta - \theta^0)\right)$ where $\gamma$ is an adaptive gain and $P$ is a Lyapunov matrix.

**Derivative Calculation of Lyapunov Matrix:**

$\frac{dV}{dt} = -\frac{\gamma}{2} e^T Q e + (\theta - \theta^0)^T\left(\frac{d\theta}{dt} + \gamma \Psi^T P e\right)$ where $Q = -A_m^T P - P A_m$.

For the stability of the system, $\frac{dV}{dt} < 0$, we can assign the second item $(\theta - \theta^0)^T\left(\frac{d\theta}{dt} + \gamma \Psi^T P e\right) = 0$ or $\frac{d\theta}{dt} = -\gamma \Psi^T P e$. Then we always have: $\frac{dV}{dt} = -\frac{\gamma}{2} e^T Q e$. If we select a positive matrix $P > 0$, for instance, $P = \begin{bmatrix} 1 & 0 \\ 0 & 2 \end{bmatrix}$, then we have

$$Q = -A_m^T P - PA_m = \begin{bmatrix} 0.5232 & 0 \\ 0 & 1.0465 \end{bmatrix}.$$ Since matrix $Q$ is obviously positive definite, then we

always have $\dfrac{dV}{dt} = -\dfrac{\gamma}{2} e^T Qe < 0$ and the system is stable with any plant-model mismatches.

**Parameters Adjustment:**

$$\frac{d\theta}{dt} = -\gamma \begin{bmatrix} -\beta_1 z_1 & 0 \\ 0 & -\beta_2 z_2 \\ \beta c_1 u_1 & 0 \\ 0 & \beta_2 u_{2c} \end{bmatrix} [P] \begin{bmatrix} e_1 \\ e_2 \end{bmatrix} = \begin{bmatrix} d\theta_1 / dt \\ d\theta_2 / dt \\ d\theta_3 / dt \\ d\theta_4 / dt \end{bmatrix} = \begin{bmatrix} \gamma\beta_1 z_1 e_1 \\ 2\gamma\beta_2 z_2 e_2 \\ -\gamma\beta_1 u_{c1} e_1 \\ -2\gamma\beta_2 u_{c2} e_2 \end{bmatrix}$$

**Simulation results and analysis:**

It is assumed that the reduced-order linear model in equation (11) can also maintain the similar steady state outputs as the basic nonlinear model. Now this model is used as an MRAC to take the process plant from these steady state outputs ( $x_D = 0.9654$ and $x_B = 0.0375$ ) to the desired targets ( $0.98 \le x_D \le 1$ and $0 \le x_B \le 0.02$ ) amid the disturbances and the plant-model mismatches as the influence of the feed stock disturbances.

The design of a new adaptive controller is shown in Figure 6.2 where we install an MRAC and a closed-loop PID (Proportional, Integral, Derivative) controller to eliminate the errors between the reference set-points and the outputs.

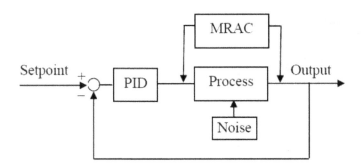

Fig. 6.2. Adaptive controller with MRAC and PID

This controller system was run with different plant-model mismatches, for instance, a plant with $A = \begin{bmatrix} -0.50 & 0 \\ 0 & -0.75 \end{bmatrix}$, $B = \begin{bmatrix} 1.5 & 0 \\ 0 & 2.5 \end{bmatrix}$ and an adaptive gain $\gamma = 25$. The operating setpoints for the real outputs are $x_{DR} = 0.99$ and $x_{BR} = 0.01$. Then, the reference set-points for the PID controller are $r_D = 0.0261$ and $r_B = -0.0275$ since the real steady state outputs are $x_D = 0.9654$ and $x_B = 0.0375$. Simulation in Figure 6.2 shows that the controlled outputs $x_D$ and $x_B$ are always stable and tracking to the model outputs and the reference set-points (the dotted lines, $r_D$ and $r_B$) amid the disturbances and the plant-model mismatches (Figure 6.3).

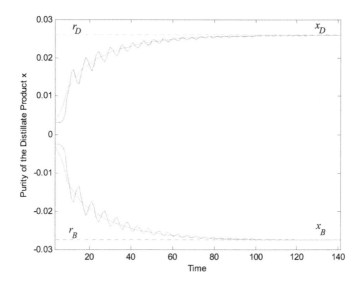

Fig. 6.3. Correlation of Plant Outputs, Model Outputs and Reference Setpoints

## 7. Conclusion

A procedure has been introduced to build up a mathematical model and simulation for a condensate distillation column based on the energy balance (L-V) structure. The mathematical modeling simulation is accomplished over three phases: the basic nonlinear model, the full order linearized model and the reduced order linear model. Results from the simulations and analysis are helpful for initial steps of a petroleum project feasibility study and design.

The reduced order linear model is used as the reference model for an MRAC controller. The controller of MRAC and PID theoretically allows the plant outputs tracking the reference set-points to achieve the desired product quality amid the disturbances and the model-plant mismatches as the influence of the feed stock disturbances.

In this chapter, the calculation of the mathematical model building and the reduced-order linear adaptive controller is only based on the physical laws from the process. The real system identifications including the experimental production factors, specific designed structures, parameters estimation and the system validation are not discussed here.

## 8. Acknowledgment

The authors would like to convey thanks for providing the opportunity to print this book chapter. Special thanks are also due to the anonymous reviewers and editor, who assisted the author in improving this book chapter significantly. Lastly, the authors would like to thank the Papua New Guinea University of Technology (UNITECH) for their support in the preparation of this book chapter.

## 9. References

Franks, R. (1972), Modeling and Simulation in Chemical Engineering, Wiley-Interscience, New York, ISBN: 978-0471275350.

George, S. (1986). Chemical Process Control: An Introduction to Theory and Practice, Prentice-Hall, *ISBN*: 9780131286290, New Jersey.

Kehlen, H. & Ratzsch, M. (1987). Complex Multicomponent Distillation Calculations by Continuous Thermodynamics. Chem. Eng. Sci., Vol. 42(2), pp. 221-232 ISSN: 0009-2509.

Luyben, W. (1990). Process Modeling, Simulation, and Control for Chemical Engineers, McGraw-Hill, ISBN: 978-0070391598, New York.

Marie, E.; Strand, S. & Skogestad S. (2008). Coordinator MPC for Maximizing Plant Throughput. Computer & Chemical Engineering, Vol. 32(2), pp. 195-204, ISSN: 0098-1354.

Nelson, W. (1985), Petroleum Refinery Engineering, McGraw-Hill, ISBN: 978-0-8247-0599-2, Singapore.

Ogata, K. (2001). Modern Control Engineering, Prentice Hall, ISBN: 978-0130609076, New York.

Papadouratis, A.; Doherty, M. & Douglas, J. (1989). Approximate Dynamic Models for Chemical Process Systems. Ind. & Eng. Ch. Re., Vol. 28(5), pp. 546-522, ISSN: 0888-5885.

Perry, R. & Green, D. (1984). Perry's Chemical Engineers Handbook, McGraw-Hill, ISBN: 0-471-58626-9, New York.

PetroVietnam Gas Company. (1999), Condensate Processing Plant Project – Process Description Document No. 82036-02BM-01, Hanoi.

Skogestad, S., & Morari, M. (1987). The Dominant Time Constant for Distillation Columns. Computers & Chemical Engineering, Vol. 11(6), pp. 607-617, ISSN: 0098-1354.

Waller, V. (1992).Practical Distillation Control, Van Nostrand Reinhold, *ISBN*: 9780442006013, New York.

Wanrren, L.; Julian, C. & Peter, H. (2005). Unit Operations of Chemical Engineering, McGraw-Hill, IBSN : 978-0072848236, New York.

Wuithier, P. (1972). Le Petrole Raffinage et Genie Chimique, Publications de l'Institut Francais du Petrole, IBSN : 978-2710801993, Paris.

# The Design and Simulation of the Synthesis of Dimethyl Carbonate and the Product Separation Process Plant

Feng Wang[1], Ning Zhao[1], Fukui Xiao[1], Wei Wei[1] and Yuhan Sun[1,2]
*[1]State Key Laboratory of Coal Conversion, Institute of Coal Chemistry,
Chinese Academy of Sciences, Taiyuan Shanxi*
*[2]Low Carbon Energy Conversion Center, Shanghai Advanced Research Institute,
Chinese Academy of Sciences, Shanghai
P.R. China*

## 1. Introduction

Dimethyl carbonate (DMC) has become a green and environmental benign chemical due to its multiple reactivity and widely potential usage in chemical industry[1]. It has been used as a substitute to replace dimethyl sulfate and methyl halides in methylation reactions and as a carbonylation agent to substitute phosgene for the production of polycarbonates and urethane polymers. It also has been evaluated to apply as non-aqueous electrolyte component in lithium rechargeable batteries. Additionally, DMC is a strong contender to help the refining industry meet the Clean Air Act specifications for oxygen in gasoline. DMC has about 3 times the oxygen content as methyl tert-butyl ether (MTBE) and its synthesis is not dependent upon FCU isobutylene yields like MTBE. DMC has a good blending octane (R + M/2 = 105), it does not phase separate in a water stream like some alcohols do, and it is both low in toxicity and quickly biodegradable[2].

Many of the properties of DMC make it a genuinely green reagent, particularly if compared to conventional alkylating agents, such as methyl halides (CH₃X) and dimethyl sulfate (DMS) or to phosgene used as a methoxycarbonylating reagent. Firstly, DMC has been proved to be a nontoxic compound. Some of the toxicological properties of DMC and phosgene and DMS are compared in Table 1. Secondly, it has been classified as a flammable liquid, smells like methanol, and does not have irritating or mutagenic effects either by contact or inhalation. Therefore, it can be handled safely without the special precautions required for the poisonous and mutagenic methyl halides and DMS and the extremely toxic phosgene. Some physicochemical properties of DMC are listed in Table 2.

The phosgene-free route for synthesis of DMC has been widely concerned by academic and industrial researchers, such as the oxidative carbonylation of methanol, the transesterification of propylene or ethylene carbonate (PC or EC), the methanolysis of urea and direct synthesis of carbon dioxide with methanol. Recently, the newly derived route of the synthesis of DMC by urea methanolysis method was considered as a novel routine for the DMC synthesis because of the advantages of easily obtained materials, moderate

| property | DMC | phosgene | DMS |
|----------|-----|----------|-----|
| oral acute toxicity (rats) | LD50 13.8 g/kg | | LD50 440 mg/kg |
| acute toxicity per contact (cavy) | LD50 > 2.5 g/kg | | |
| acute toxicity per inhalation (rats) | LC50 140 mg/L; (4 h) | LC50 16 mg/m3; (75 min) | LC50 1.5 mg/L (4 h) |
| mutagenic properties | none | | mutagenic |
| irritating properties (rabbits, eyes, skin) | none | corrosive | |
| biodegradability (OECD 301 C) | > 90%(28 days) | rapid hydrolysis | rapid hydrolysis |
| acute toxicity (fish) (OECD 203) | NOEC[a] 1000 mg/L | | LC50 10-100 mg/L (96 h) |
| acute toxicity on aerobial bacteria of wastewaters (OECD 209) | EC50 > 1000 mg/L | | |

[a] NOEC=Concentration which does not produce any effect.

Table 1. Comparison between the Toxicological and Ecotoxicological Properties of DMC, Phosgene, and DMS[1]

| | |
|---|---|
| mp (°C) | 4.6 |
| bp (°C) | 90.3 |
| density ($D^{20}_4$) | 1.07 |
| viscosity ($\mu^{20}$, cps) | 0.625 |
| flashing point (°C) | 21.7 |
| dielectric constant ($\varepsilon^{25}$) | 3.087 |
| dipole moment ($\mu$, D) | 0.91 |
| $\Delta H^{vap}$ (kcal/kg) | 88.2 |
| solubility H2O (g/100 g) | 13.9 |
| azeotropical mixtures | with water, alcohols, hydrocarbons |

Table 2. Some Physical and Thermodynamic Properties of DMC[1]

reaction conditions and low investment for equipment. As a result, the separation of the reacted mixture which contain an azeotropic mixture of DMC and methanol became very important for the production of DMC.

## 2. The properties of DMC

The vapor pressure data of DMC

The vapor pressure of DMC has been measured by Rodriguez[3]. The data of the experiment of DMC has been showed in table 3, which could be regressed by the extended Antoine equation.

| T (K) | P⁰(kPa) | T (K) | P⁰(kPa) | T (K) | P⁰(kPa) |
|-------|---------|-------|---------|-------|---------|
| 326.06 | 26.66 | 372.06 | 133.29 | 393.98 | 247.92 |
| 328.41 | 29.32 | 372.67 | 135.96 | 394.39 | 250.58 |
| 330.58 | 31.99 | 373.27 | 138.62 | 394.80 | 253.25 |
| 332.61 | 34.66 | 373.91 | 141.29 | 395.18 | 255.92 |
| 334.53 | 37.32 | 374.54 | 143.95 | 395.58 | 258.58 |
| 336.34 | 39.99 | 375.15 | 146.62 | 395.99 | 261.25 |
| 338.03 | 42.65 | 375.73 | 149.28 | 396.38 | 263.91 |
| 339.64 | 45.32 | 376.33 | 151.95 | 396.77 | 266.58 |
| 341.20 | 47.98 | 376.89 | 154.62 | 397.16 | 269.24 |
| 342.68 | 50.65 | 377.37 | 157.28 | 397.54 | 271.91 |
| 344.09 | 53.32 | 377.94 | 159.95 | 397.95 | 274.58 |
| 345.45 | 55.98 | 378.47 | 162.61 | 398.29 | 277.24 |
| 346.75 | 58.65 | 379.02 | 165.28 | 398.67 | 279.91 |
| 348.04 | 61.31 | 379.63 | 167.94 | 398.93 | 282.57 |
| 349.26 | 63.98 | 380.05 | 170.61 | 399.30 | 285.24 |
| 350.43 | 66.64 | 380.62 | 173.28 | 399.66 | 287.91 |
| 351.57 | 69.31 | 381.16 | 175.94 | 400.02 | 290.57 |
| 352.69 | 71.98 | 381.71 | 178.61 | 400.38 | 293.24 |
| 353.77 | 74.64 | 382.23 | 181.27 | 400.66 | 295.90 |
| 354.81 | 77.31 | 382.71 | 183.94 | 400.99 | 298.57 |
| 355.86 | 79.97 | 383.22 | 186.61 | 401.35 | 301.23 |
| 356.85 | 82.64 | 383.73 | 189.27 | 401.66 | 303.90 |
| 357.81 | 85.31 | 384.40 | 191.94 | 402.03 | 306.57 |
| 358.72 | 87.97 | 384.98 | 194.60 | 402.39 | 309.23 |
| 359.63 | 90.64 | 385.48 | 197.27 | 402.76 | 311.90 |
| 360.54 | 93.30 | 385.95 | 199.93 | 403.11 | 314.56 |
| 361.73 | 95.97 | 386.45 | 202.60 | 403.46 | 317.23 |
| 362.14 | 97.30 | 386.92 | 205.27 | 403.80 | 319.89 |
| 362.55 | 98.63 | 387.40 | 207.93 | 404.50 | 325.23 |
| 362.98 | 99.97 | 387.91 | 210.60 | 405.15 | 330.56 |
| 363.46 | 101.30 | 388.33 | 213.26 | 405.45 | 333.22 |
| 363.84 | 102.63 | 388.77 | 215.93 | 406.12 | 338.56 |
| 364.24 | 103.97 | 389.25 | 218.59 | 406.75 | 343.89 |
| 364.65 | 105.30 | 389.70 | 221.26 | 407.03 | 346.55 |
| 365.04 | 106.63 | 390.14 | 223.93 | 407.69 | 351.88 |
| 365.63 | 109.30 | 390.59 | 226.59 | 408.31 | 357.22 |
| 366.37 | 111.96 | 391.04 | 229.26 | 408.56 | 359.88 |
| 367.10 | 114.63 | 391.45 | 231.92 | 409.22 | 365.21 |
| 367.89 | 117.29 | 391.89 | 234.59 | 409.82 | 370.54 |
| 368.65 | 119.96 | 391.89 | 234.59 | 410.13 | 373.21 |
| 369.38 | 122.63 | 392.31 | 237.26 | 410.71 | 378.54 |
| 370.10 | 125.29 | 392.75 | 239.92 | 411.09 | 381.21 |
| 370.80 | 127.96 | 393.15 | 242.59 | 411.29 | 383.87 |
| 371.47 | 130.62 | 393.55 | 245.25 | | |

Table 3. Experimental vapor pressure and boiling point for DMC

## 3. The vapor-liquid equilibrium

Vapor–liquid equilibrium (VLE) data are always required for engineering, such as designing in distillation tower, which is the most common operation performed in the chemical industry for the separation of liquid mixture. Dimethyl carbonate and methanol constitute azeotropic mixture in a composition ratio of 30:70 (weight ratio), and thus it is difficult to separate the mixture by distillation under normal pressure. ENIChem has a German patent showing that the percentage of methanol in the binary methanol–DMC azeotrope increases with pressure: going from 70% methanol at 101.33 kPa, up to 95% methanol at 1013.3 kPa.

### 3.1 The VLE for methanol and DMC

The thermodynamic properties of the binary methanol (1)–DMC (2) under atmosphere pressure have been reported, as well as the relationship of temperature and the binary azeotrope. Zhang and Luo reported the only calculated the binary vapor liquid equilibrium (VLE) data under normal pressure based on group contribution UNIFAC method. Li et al. measured the related binary VLE data with an Ellis Cell at 101.325 kPa. Rodriguez et al. also measured the vapor–liquid equilibria of dimethyl carbonate with linear alcohols by a dynamic re-circulating method under normal pressure, and estimated the new interaction parameters for UNIFAC and ASOG method. Theoretically, the predictive group contribution methods may be applicable until 0.5MPa. Based on the above methods, both of the vapor and liquid phases were directly sampled and analyzed.

Vapour–liquid equilibrium data for methanol (1) +DMC (2) system at normal pressure has been presented in table coming from A. Rodriguez [4]. The results reported in these tables indicate that the binary systems of methanol – DMC exhibited a positive deviation from ideal behaviour and a minimum boiling azeotrope.

| T (K) | x | y | $\gamma 1$ | $\gamma 2$ |
|---|---|---|---|---|
| 361.99 | 0.0103 | 0.0523 | 2.219 | 0.993 |
| 359.93 | 0.0252 | 0.1258 | 2.326 | 0.992 |
| 357.45 | 0.0457 | 0.2065 | 2.280 | 0.996 |
| 355.71 | 0.0620 | 0.2669 | 2.298 | 0.992 |
| 354.69 | 0.0709 | 0.2950 | 2.297 | 0.996 |
| 352.38 | 0.0958 | 0.3613 | 2.249 | 1.002 |
| 349.83 | 0.1291 | 0.4379 | 2.204 | 0.999 |
| 347.97 | 0.1582 | 0.4818 | 2.110 | 1.017 |
| 346.56 | 0.1834 | 0.5202 | 2.064 | 1.021 |
| 344.85 | 0.2210 | 0.5687 | 1.989 | 1.023 |
| 343.99 | 0.2472 | 0.5915 | 1.906 | 1.035 |
| 342.57 | 0.2913 | 0.6238 | 1.795 | 1.067 |
| 341.74 | 0.3251 | 0.6488 | 1.724 | 1.079 |
| 340.99 | 0.3619 | 0.6703 | 1.644 | 1.102 |

| T (K) | x | y | γ1 | γ2 |
|-------|------|------|-------|-------|
| 340.11 | 0.4247 | 0.6960 | 1.502 | 1.165 |
| 339.18 | 0.4916 | 0.7206 | 1.390 | 1.256 |
| 338.69 | 0.5386 | 0.7394 | 1.325 | 1.316 |
| 338.21 | 0.5800 | 0.7547 | 1.279 | 1.386 |
| 337.55 | 0.6622 | 0.7806 | 1.187 | 1.582 |
| 337.18 | 0.7181 | 0.7955 | 1.131 | 1.794 |
| 336.97 | 0.7684 | 0.8123 | 1.088 | 2.022 |
| 336.95 | 0.8160 | 0.8332 | 1.051 | 2.265 |
| 336.88 | 0.8617 | 0.8560 | 1.025 | 2.612 |
| 336.89 | 0.8824 | 0.8736 | 1.021 | 2.698 |
| 336.98 | 0.9104 | 0.8931 | 1.008 | 2.988 |
| 337.11 | 0.9341 | 0.9166 | 1.003 | 3.158 |
| 337.25 | 0.9549 | 0.9406 | 1.002 | 3.273 |
| 337.39 | 0.9726 | 0.9614 | 1.000 | 3.487 |
| 337.60 | 0.9889 | 0.9833 | 0.998 | 3.699 |

Table 4. Vapour–liquid equilibrium data for methanol (1) +DMC (2) system at 101.3 kPa[4]

The azeotrope data for methanol-DMC on the high pressure has been show on the following table, which was a comparison of the data from different literature. The data has exhibited the composition of DMC in an azeotrope of DMC-methanol decreased with the increases of pressure. These thermodynamic data showed that the separation of the mixture of methanol and DMC would be difficult with the normal distillation.

| T (K) | p (kPa) | x1 | w1 | p(kPa) | x1 | w1 |
|-------|---------|--------|--------|--------|--------|--------|
| 337.35 | 102.73 | 0.8500 | 0.6684 | 101.33 | 0.8677 | 0.7000 |
| 377.15 | 405.70 | 0.9100 | 0.7824 | 405.2 | 0.9150 | 0.7929 |
| 391.15 | 613.00 | 0.9150 | 0.7929 | 607.8 | 0.9298 | 0.8249 |
| 411.15 | 1077.00 | 0.9200 | 0.8036 | 1013.0 | 0.9521 | 0.8761 |
| 428.15 | 1576.00 | 0.9625 | 0.9013 | 1519.5 | 0.9739 | 0.9300 |

Table 5. Comparisons of azeotrope data for methanol (1)–dimethyl carbonate (2) binary system at different temperatures from different literature.

## 3.2 The VLE for DMC with other compound

| T (K) | x | y | γ1 | γ2 |
|-------|------|------|-------|-------|
| 362.37 | 0.0180 | 0.0494 | 1.853 | 0.992 |
| 361.70 | 0.0293 | 0.0790 | 1.864 | 0.993 |
| 361.00 | 0.0394 | 0.1044 | 1.877 | 0.997 |

| $T$ (K) | $x$ | $y$ | $\gamma 1$ | $\gamma 2$ |
|---------|-----|-----|------------|------------|
| 359.81 | 0.0591 | 0.1520 | 1.900 | 1.001 |
| 358.27 | 0.0882 | 0.2157 | 1.909 | 1.004 |
| 357.07 | 0.1128 | 0.2628 | 1.899 | 1.008 |
| 355.97 | 0.1386 | 0.3071 | 1.879 | 1.011 |
| 355.07 | 0.1621 | 0.3413 | 1.846 | 1.018 |
| 354.07 | 0.1906 | 0.3776 | 1.802 | 1.030 |
| 353.19 | 0.2193 | 0.4151 | 1.779 | 1.033 |
| 352.20 | 0.2564 | 0.4507 | 1.714 | 1.054 |
| 351.56 | 0.2871 | 0.4786 | 1.665 | 1.066 |
| 350.70 | 0.3352 | 0.5160 | 1.588 | 1.093 |
| 350.45 | 0.3539 | 0.5315 | 1.564 | 1.098 |
| 350.00 | 0.3902 | 0.5549 | 1.507 | 1.123 |
| 349.68 | 0.4141 | 0.5696 | 1.475 | 1.143 |
| 349.40 | 0.4504 | 0.5842 | 1.406 | 1.189 |
| 349.17 | 0.4832 | 0.6031 | 1.365 | 1.216 |
| 349.00 | 0.5097 | 0.6156 | 1.329 | 1.249 |
| 348.86 | 0.5383 | 0.6266 | 1.288 | 1.295 |
| 348.75 | 0.5671 | 0.6396 | 1.253 | 1.339 |
| 348.66 | 0.5945 | 0.6538 | 1.226 | 1.377 |
| 348.61 | 0.6125 | 0.6625 | 1.208 | 1.408 |
| 348.57 | 0.6330 | 0.6728 | 1.189 | 1.443 |
| 348.45 | 0.6721 | 0.6925 | 1.158 | 1.525 |
| 348.46 | 0.7173 | 0.7101 | 1.112 | 1.668 |
| 348.57 | 0.7481 | 0.7286 | 1.089 | 1.746 |
| 348.70 | 0.7824 | 0.7491 | 1.065 | 1.861 |
| 348.93 | 0.8297 | 0.7818 | 1.039 | 2.053 |
| 349.06 | 0.8472 | 0.7938 | 1.028 | 2.153 |
| 349.34 | 0.8740 | 0.8184 | 1.016 | 2.278 |
| 349.60 | 0.8976 | 0.8429 | 1.008 | 2.405 |
| 349.83 | 0.9166 | 0.8662 | 1.006 | 2.496 |
| 350.23 | 0.9417 | 0.8984 | 1.000 | 2.677 |
| 350.71 | 0.9667 | 0.9335 | 0.993 | 3.020 |
| 350.95 | 0.9775 | 0.9543 | 0.995 | 3.048 |
| 351.13 | 0.9875 | 0.9730 | 0.997 | 3.223 |

Table 6. Vapor–liquid equilibrium data for ethanol (1) + DMC (2) system at 101.3 kPa[4]

| $T$ (K) | $x$ | $y$ | $\gamma 1$ | $\gamma 2$ |
|---------|-----|-----|-----------|-----------|
| 369.72 | 0.0124 | 0.0356 | 2.368 | 0.998 |
| 369.19 | 0.0229 | 0.0622 | 2.275 | 1.000 |
| 368.66 | 0.0342 | 0.0894 | 2.223 | 1.002 |
| 368.20 | 0.0458 | 0.1165 | 2.193 | 1.001 |
| 367.51 | 0.0628 | 0.1543 | 2.161 | 1.001 |
| 366.86 | 0.0818 | 0.1872 | 2.052 | 1.006 |
| 366.21 | 0.1021 | 0.2217 | 1.985 | 1.009 |
| 365.43 | 0.1303 | 0.2596 | 1.865 | 1.021 |
| 365.05 | 0.1440 | 0.2801 | 1.841 | 1.024 |
| 364.49 | 0.1669 | 0.3089 | 1.782 | 1.031 |
| 364.05 | 0.1859 | 0.3308 | 1.736 | 1.040 |
| 363.37 | 0.2262 | 0.3695 | 1.627 | 1.058 |
| 362.96 | 0.2504 | 0.3922 | 1.580 | 1.070 |
| 362.59 | 0.2767 | 0.4120 | 1.519 | 1.088 |
| 362.31 | 0.2975 | 0.4317 | 1.493 | 1.095 |
| 361.89 | 0.3407 | 0.4596 | 1.406 | 1.127 |
| 361.49 | 0.3796 | 0.4855 | 1.349 | 1.159 |
| 361.31 | 0.4089 | 0.5057 | 1.312 | 1.177 |
| 361.12 | 0.4365 | 0.5232 | 1.279 | 1.200 |
| 360.94 | 0.4721 | 0.5424 | 1.233 | 1.238 |
| 360.70 | 0.5064 | 0.5620 | 1.200 | 1.280 |
| 360.59 | 0.5363 | 0.5790 | 1.171 | 1.315 |
| 360.49 | 0.5616 | 0.5931 | 1.149 | 1.350 |
| 360.35 | 0.5916 | 0.6101 | 1.127 | 1.396 |
| 360.30 | 0.6213 | 0.6279 | 1.106 | 1.440 |
| 360.27 | 0.6558 | 0.6473 | 1.081 | 1.504 |
| 360.20 | 0.6846 | 0.6682 | 1.071 | 1.549 |
| 360.25 | 0.7185 | 0.6900 | 1.052 | 1.618 |
| 360.31 | 0.7524 | 0.7142 | 1.038 | 1.693 |
| 360.55 | 0.7885 | 0.7458 | 1.026 | 1.746 |
| 360.82 | 0.8269 | 0.7793 | 1.014 | 1.834 |
| 361.07 | 0.8643 | 0.8163 | 1.008 | 1.929 |
| 361.31 | 0.8840 | 0.8380 | 1.004 | 1.972 |
| 361.78 | 0.9163 | 0.8738 | 0.995 | 2.091 |
| 362.34 | 0.9494 | 0.9168 | 0.990 | 2.232 |
| 362.99 | 0.9774 | 0.9613 | 0.989 | 2.268 |

Table 7. Vapor–liquid equilibrium data for DMC (1) + 1-propanol (2) system at 101.3 kPa[4]

| $T$ (K) | $x$ | $y$ | $\gamma 1$ | $\gamma 2$ |
|---|---|---|---|---|
| 389.73 | 0.0120 | 0.0582 | 2.314 | 0.992 |
| 387.85 | 0.0315 | 0.1389 | 2.207 | 0.987 |
| 386.14 | 0.0490 | 0.1985 | 2.118 | 0.994 |
| 384.59 | 0.0678 | 0.2563 | 2.058 | 0.994 |
| 383.07 | 0.0862 | 0.3072 | 2.019 | 0.998 |
| 381.06 | 0.1158 | 0.3785 | 1.954 | 0.996 |
| 379.14 | 0.1454 | 0.4314 | 1.868 | 1.013 |
| 377.02 | 0.1845 | 0.4946 | 1.789 | 1.022 |
| 374.93 | 0.2287 | 0.5499 | 1.701 | 1.043 |
| 373.03 | 0.2797 | 0.5976 | 1.596 | 1.075 |
| 371.60 | 0.3265 | 0.6355 | 1.514 | 1.103 |
| 370.59 | 0.3654 | 0.6617 | 1.451 | 1.131 |
| 370.02 | 0.3897 | 0.6768 | 1.415 | 1.150 |
| 369.00 | 0.4397 | 0.7044 | 1.344 | 1.194 |
| 368.46 | 0.4694 | 0.7194 | 1.307 | 1.223 |
| 368.05 | 0.4941 | 0.7312 | 1.277 | 1.250 |
| 367.53 | 0.5268 | 0.7464 | 1.242 | 1.288 |
| 366.97 | 0.5638 | 0.7638 | 1.207 | 1.332 |
| 366.34 | 0.6087 | 0.7838 | 1.169 | 1.396 |
| 365.81 | 0.6478 | 0.8021 | 1.142 | 1.451 |
| 365.12 | 0.7029 | 0.8218 | 1.101 | 1.595 |
| 364.41 | 0.7646 | 0.8492 | 1.069 | 1.756 |
| 363.89 | 0.8175 | 0.8677 | 1.038 | 2.032 |
| 363.63 | 0.8497 | 0.8829 | 1.024 | 2.209 |
| 363.49 | 0.8911 | 0.9008 | 1.001 | 2.600 |
| 363.31 | 0.9335 | 0.9328 | 0.995 | 2.910 |
| 363.28 | 0.9678 | 0.9662 | 0.995 | 3.030 |

Table 8. Vapor–liquid equilibrium data for DMC(1) + 1-butanol (2) system at 101.3 kPa[4]

| $T$ (K) | $x$ | $y$ | $\gamma 1$ | $\gamma 2$ |
|---|---|---|---|---|
| 409.75 | 0.0082 | 0.0469 | 1.687 | 1.006 |
| 408.28 | 0.0180 | 0.0970 | 1.642 | 1.009 |
| 405.75 | 0.0373 | 0.1863 | 1.611 | 1.008 |
| 403.40 | 0.0557 | 0.2587 | 1.581 | 1.013 |
| 400.41 | 0.0824 | 0.3516 | 1.558 | 1.010 |
| 397.81 | 0.1081 | 0.4229 | 1.520 | 1.013 |
| 393.86 | 0.1493 | 0.5209 | 1.492 | 1.016 |

| $T$ (K) | $x$ | $y$ | $\gamma 1$ | $\gamma 2$ |
|---|---|---|---|---|
| 390.21 | 0.1955 | 0.6031 | 1.446 | 1.018 |
| 386.97 | 0.2443 | 0.6677 | 1.392 | 1.025 |
| 384.83 | 0.2811 | 0.7065 | 1.354 | 1.034 |
| 383.23 | 0.3127 | 0.7340 | 1.319 | 1.043 |
| 381.83 | 0.3431 | 0.7573 | 1.288 | 1.053 |
| 380.09 | 0.3845 | 0.7843 | 1.247 | 1.071 |
| 378.47 | 0.4274 | 0.8095 | 1.211 | 1.086 |
| 376.75 | 0.4790 | 0.8336 | 1.167 | 1.119 |
| 375.97 | 0.5027 | 0.8437 | 1.150 | 1.138 |
| 374.58 | 0.5509 | 0.8615 | 1.115 | 1.183 |
| 373.25 | 0.5998 | 0.8786 | 1.085 | 1.232 |
| 371.89 | 0.6466 | 0.8945 | 1.065 | 1.285 |
| 370.46 | 0.6979 | 0.9115 | 1.049 | 1.341 |
| 369.29 | 0.7370 | 0.9244 | 1.042 | 1.386 |
| 368.36 | 0.7775 | 0.9345 | 1.027 | 1.479 |
| 366.72 | 0.8413 | 0.9530 | 1.016 | 1.601 |
| 365.10 | 0.9063 | 0.9724 | 1.0 1 | 1.713 |
| 364.64 | 0.9371 | 0.9814 | 1.001 | 1.757 |
| 363.96 | 0.9693 | 0.9904 | 0.997 | 1.917 |

Table 9. Vapor–liquid equilibrium data for DMC(1) + 1-pentanol (2) system at 101.3 kPa[4]

The azeotropic mixture of DMC with some common compounds has been listed in table 10[4].

| Component | T(K) | Composition (mol%) |
|---|---|---|
| Methanol | 336.90 | 0.8503 |
| Ethanol | 348.46 | 0.7055 |
| 1-propanol | 360.29 | 0.6364 |
| 1-butanol | 363.32 | 0.9306 |

Table 10. the azeotropic mixtures of DMC with some compounds at 101.3 kPa

## 4. The calculation of VLE

Rigorous thermodynamic model is the base of the process simulation and optimization. The vapor–liquid equilibrium relations for a binary system are:

$$\hat{f}_i^G(T,p,y_i,k_{i,i}) = \hat{f}_i^L(T,p,x_i,k_{i,j})$$

These correlations could be resolved by the Equation of State (EOS) functions. Although, Shi has correlated the vapor liquid equilibrium of methanol and DMC from the experiment data by a modified Peng-Robinson equation of stage both for the liquid and vapor phase, there had none of the EOS now available can simultaneously describe both of the liquid and vapor phase thermo-dynamical properties accurately, especially for liquid or liquid mixtures. Although EOS well expresses the p–V–T relationship of vapor or gas phase, the calculation for liquid density now is an unsubstantial domain for EOS. That is said that we cannot directly use EOS to predict the molar volume and fugacity of a liquid phase accurately.

Nowadays, the commonly used for the calculation of vapor liquid equilibrium was the combination of EOS + γ method, which the EOS computed for the vapor phase and γ for the liquid phase. And also the Henry's method was used to describe the gas liquid equilibrium.

Here listed one of EOS for the vapor or gas phase. The Peng-Robinson equation of state can be used to evaluate the compressibility factor and species fugacity coefficient.

$$P = \frac{RT}{v-b} - \frac{a}{V(V+b)+b(V-b)}$$

$$a = 0.45724\alpha(T_r)R^2Tc^2 \, / \, Pc$$

$$b = 0.077880RTc \, / \, Pc$$

Shi et al used the follow correlation for the calculation of parameters of methanol and DMC:

$$\alpha(T_r) = 1 + (1 - T_r)\left(m + n/T_r^2\right)$$

The parameter of m and n for methanol and DMC was show below.

Methanol: m 1.1930; n 0.09370

DMC:   m 1.0236; n 0.06463

The parameter also can be estimated by the following correlation:

$$\alpha(T_r,\omega) = \left[1 + \left(0.37464 + 1.54226\omega - 0.26992\omega^2\right)\left(1 - T_r^{1/2}\right)\right]^2 ; \; 0 < \omega < 0.5$$

$$\alpha(T_r,\omega) = \left[1 + \left(0.3796 + 1.4850\omega - 0.1644\omega^2 + 0.0166\omega^3\right)\left(1 - T_r^{1/2}\right)\right]^2 ; \; 0.2 < \omega < 2.0$$

For the $0.2 < \omega < 0.5$, the function get the similar estimated value.

The mixing rule for the function used is as follow:

$$b = \sum x_i b_i$$

$$a = \sum\sum x_i x_j a_{i,j}$$

$$a_{i,j} = a_i^{1/2} a_j^{1/2}(1 - k_{ij})$$

The Peng-Robinson equation of state may be written in compressibility factor form:

$$Z^3 - (1-B)Z^2 + (A - 2B - 3B^2)Z - (AB - B^2 - B^3) = 0$$

$$A = ap/(RT)^2, \quad B = bp/(RT)$$

$\varphi_i^G$ is expressed as follow:

$$\ln\varphi_i^G = Z_i - 1 - \ln(Z_i - \beta_i) - q_i I_i$$

where $Z_i$ is the compressibility factor and obtained from Eq. (4);

$$\beta = \Omega P_r / T_r ; \quad q = \Psi\alpha(T_r)P_r / (\Omega T_r); \quad I = \frac{1}{\sigma - \varepsilon}\ln\left(\frac{Z + \sigma\beta}{Z + \varepsilon\beta}\right)$$

| EoS | $\alpha(T_r)$ | $\sigma$ | $\varepsilon$ |
|---|---|---|---|
| PR | $\alpha(T_r,\omega)^a$ | $1+\sqrt{2}$ | $1-\sqrt{2}$ |

$^a$ $\alpha(T_r,\omega) = \left[1 + \left(0.37464 + 1.54226\omega - 0.26992\omega^2\right)\left(1 - T_r^{1/2}\right)\right]^2$

Table 11. Parameter assignments for PR EoS[5].

| Substance | Tc/K | pc/MPa | $\omega$ |
|---|---|---|---|
| CO | 132.85 | 3.494 | 0.045 |
| O2 | 154.58 | 5.043 | 0.022 |
| CO2 | 304.12 | 7.374 | 0.225 |

Table 12. Critical constants and acentric factors[5]

1.  the method for the liquid activity coefficient

    a.  the Wilson method

$$\ln\gamma_i = 1 - \ln\sum_j x_j\Lambda_{ij} - \sum_j (x_j\Lambda_{ji} / \sum_l x_l\Lambda_{jl})$$

*where:* $\Lambda_{ii} = 1$

$$\Lambda_{ij} = (\frac{V_j^L}{V_i^L})\exp(-\frac{\lambda_{ij} - \lambda_{ii}}{RT})$$

Where V represents the liquid molar volume of pure component; $\lambda_{ij} - \lambda_{ij}$ represents the energy parameter.

The Wilson model is not supported for the prediction of the liquid-liquid equilibria.

b. the NRTL method

The non-random two-liquid model (NRTL equation) is an activity coefficient model that correlates the activity coefficients $\gamma_i$ of a compound i with its mole fractions xi in the liquid phase concerned. The concept of NRTL is based on the hypothesis of Wilson that the local concentration around a molecule is different from the bulk concentration. This difference is due to a difference between the interaction energy of the central molecule with the molecules of its own kind $U_{ii}$ and that with the molecules of the other kind $U_{ij}$. The energy difference also introduces a non-randomness at the local molecular level.

The general equation is:

$$\ln(\gamma_i) = \frac{\sum_{j=1}^{n} x_j \tau_{ji} G_{ji}}{\sum_{k=1}^{n} x_k G_{ki}} + \sum_{j=1}^{n} \frac{x_j G_{ij}}{\sum_{k=1}^{n} x_k G_{kj}} \left( \tau_{ij} - \frac{\sum_{m=1}^{n} x_m \tau_{mi} G_{mi}}{\sum_{k=1}^{n} x_k G_{kj}} \right)$$

with

$$G_{ij} = \exp(-\alpha_{ij} \tau_{ij})$$

$$\alpha_{ij} = \alpha_{ij0} + \alpha_{ij1} T$$

$$\tau_{ij} = A_{ij} + \frac{B_{ij}}{T} + \frac{C_{ij}}{T^2} + D_{ij} \ln(T) + E_{ij} T^{F_{ij}}$$

c. the UNIFAC method

The UNIversal Functional Activity Coefficient (UNIFAC) method is a semi-empirical system for the prediction of non-electrolyte activity estimation in non-ideal mixtures, which was first published in 1975 by Fredenslund, Jones and Prausnitz. UNIFAC uses the functional groups present on the molecules that make up the liquid mixture to calculate activity coefficients. By utilizing interactions for each of the functional groups present on the molecules, as well as some binary interaction coefficients, the activity of each of the solutions can be calculated.

In the UNIFAC model the activity coefficients of component i of a mixture are described by a combinatorial and a residual contribution.

$$\ln \gamma_i = \ln \gamma_i^C + \ln \gamma_i^R$$

Combinatorial part

$$\ln \gamma_i^C = \ln \frac{J_i}{x_i} + 1 - \frac{J_i}{x_i} - \frac{1}{2} Z q_i \left( \ln \frac{\varphi_i}{\theta_i} + 1 - \frac{\varphi_i}{\theta_i} \right)$$

where $J_i = x_i r_i^{2/3} / \sum x_j r_j^{2/3}$ ; $\varphi_i = x_i r_i / \sum x_j r_j$ ; $\theta_i = x_i q_i / \sum x_j q_j$ and $Z = 10$.

The coordination number, z, i.e. the number of close interacting molecules around a central molecule, is frequently set to 10. It can be regarded as an average value that lies between cubic (z=6) and hexagonal packing (z=12) of molecules that are simplified by spheres.

Where $r_i$ is the volume parameters of component i, computed by $r_i = \sum v_j^{(i)} R_j$ . Where the $v_j^{(i)}$ is the number of groups of type k in component i, and $R_j$ is the volume parameter for group k; $q_i$ is the area parameter for component i, calculated as $q_i = \sum v_j^{(i)} Q_j$ .

Residual part

$$\ln \gamma_i^R = \sum_k v_k^{(i)} \left( \ln \Gamma_k - \ln \Gamma_k^{(i)} \right)$$

k is the group activity coefficient of group k in the mixture and (i) k is the group activity coefficient of group k in the pure substance.

$$\ln \Gamma_k = Q_k \left[ 1 - \ln \left( \sum_m \theta_m \tau_{mk} \right) - \sum_m \frac{\theta_m \tau_{mk}}{\sum_n \theta_n \tau_{nk}} \right]$$

$$\theta_m = \frac{Q_m X_m}{\sum Q_n X_n} ; \; X_m = \frac{\sum_j v_m^{(j)} x_j}{\sum_j \sum_n v_n^{(j)} x_j}$$

$X_m$ represents the fraction of group m in the mixture.

$$\tau_{nm} = \exp \left( \frac{-\left( u_{nm} - u_{mm} \right)}{T} \right)$$

Where the energy parameter of $u_{ij}$ characterize the interaction between group i and j, and estimated from experiment data.

Alternatively, in some process simulation software $\tau_{ij}$ can be expressed as follows:

$$\ln \tau_{ij} = A_{ij} + B_{ij} / T + C_{ij} \ln T + D_{ij} T^2 + E_{ij} / T^2$$

The "C", "D", and "E" coefficients are primarily used in fitting liquid–liquid equilibria (with "D" and "E" rarely used at that). The "C" coefficient is useful in vapor-liquid equilibria as well.

For the system mixture of DMC-Phenol and Phenol-Methanol

Wang[6] measured the DMC-Phenol and Phenol-Methanol mixture system and predicted the VLE by the Wilson, NRTL, UNIQUAC equations with considering the ideal vapor behavior. The Wilson, NRTL, UNIQUAC equations energy parameters can be obtained using the following expressions where $a_{ij}$ , $b_{ij}$ are the binary parameters regressed.

Wilson: $\Lambda_{ij} = \exp(a_{ij} + b_{ij} / T)$

NRTL: $\tau_{ij} = a_{ij} + b_{ij} / T$

UNIQUAC: $\tau_{ij} = \exp(a_{ij} + b_{ij} / T)$

The regressed parameter from the experiment obtained by Wang listed in table 13.

| equation | parameters | Phenol(1)-DMC(2) | Phenol(1)-methanol(2) |
|---|---|---|---|
| Wilson | $a_{12}$; $b_{12}$ | -3.215767; 1465.520 | -4632.789; 9.657878 |
| | $a_{21}$; $b_{21}$ | 1.213508; -423.7649 | 1777.066; -3.209978 |
| NRTL | $a_{12}$; $b_{12}$ | -0.9630158; 386.8243 | 1712.322; -5.054467 |
| | $a_{21}$; $b_{21}$ | 3.061205; -1467.521 | -3715.243; 9.690805 |
| | $\alpha$ | 0.300 | 0.300 |
| UNIQUAC | $a_{12}$; $b_{12}$ | 0.7670656; -273.3901 | -7802.509; -85.39368 |
| | $a_{21}$; $b_{21}$ | -1.505834; 691.9294 | 289.7973; 0.2346874 |

Table 13. Binary parameters of Wilson, NRTL and UNIQUAC equations

2.  Calculating the fugacity of gas in liquid by Henry's method

For estimating the fugacity of component i in the liquid (L) phase, Eq. (6) was proposed by Sander et al.

$$\hat{f}_1^L = x_i H_{i,r} \gamma_i / \gamma_{i,r}^{\infty}$$

where the subscript i and r represent a gas and a reference solvent, respectively. $H_{i,r}$ is Henry's constant for gas i in a reference solvent. $\gamma_i$ is called the activity coefficient in the unsymmetric convention. $\gamma_{i,r}^{\infty}$ is the activity coefficient at infinite dilution in the symmetric convention.

The reference Henry's constant $H_{i,r}$ is calculated as a function of temperature from the following expression,

$$\ln\left(H_{i,r} / Pa\right) = \left(A + \frac{B}{T} + C\ln T\right) \times 101325$$

Wang et. al.[5] has studied the gas liquid equilibrium of CO, O2 and CO2 with DMC by the Henry method with the UNIFAC model for liquid system. Table 14 and 15 presents the chosen reference solvents for studied gases (CO, $O_2$ and $CO_2$) in the Wang's work and the estimated parameters A, B and C for calculating the reference Henry's constant. The activity coefficients i is obtained from the modified UNIFAC model. The UNIFAC energy parameter that was obtained by Wang has listed in Table 16. And also the parameter data can be obtained in Wang's article.

| Gas | A | B | C | Reference solvent |
|-----|---|---|---|-------------------|
| CO | 7.53116 | -6.36893 | 0.0 | Propanol |
| O2 | 26.1577 | -924.307 | -2.73771 | Ethanol |
| CO2 | 27.5146 | -1846.89 | -2.99332 | Hexadecane |

Table 14. Constants for calculation of reference Henry's constant according to Henry constant equation.

| Group | Rk | Qk |
|-------|----|----|
| CO | 2.094 | 2.120 |
| O2 | 1.764 | 1.910 |
| CO2 | 2.592 | 2.522 |
| CH3 | 1.8022 | 1.696 |
| CH2 | 1.3488 | 1.080 |
| OH | 1.060 | 1.168 |
| –OCOO– | 3.1642a | 2.7874b |

[a] R – OCOO – = 2[Van der Waal's volume from Bondi [20]]/15.17cm3 mol-1.
[b] Q – OCOO – = 2[Van der Waal's area from Bondi [20]]/(2.5×109) cm2 mol-1.

Table 15. The group parameters Rk and Qk values for GLE calculation.

| Energy parameters, unm(K) | CO | O2 | CO2 | –CH3 |
|---------------------------|------|------|-----|-------|
| –OCOO– | –364.4 | –328.5 | –4.4 | –786.5 |

Table 16. The new UNIFAC interaction-energy parameters obtained by Wang

## 5. Model for catalytic distillation for the synthesis of DMC by urea and methanol

The current routes for the DMC synthesis are the oxy-carbonylation of methanol (EniChem process and UBE process) and the trans-esterification method (Texaco process). Recently, an attractive route for the synthesis of DMC by a urea methanolysis method over solid bases catalyst has been carried out in a catalytic distillation

The catalytic distillation (CD) [7], which is also known as reactive distillation (RD) that combines the heterogeneous catalyzed chemical reaction and the distillation in a single unit, has attracted more interest in academia and become more important in the chemical processing industry as it has been successfully used in several important industrial processes. The CD provides some advantages such as high conversion in excess of the chemical equilibrium, energy saving, overcoming of the azeotropic limitations and prolonging the catalyst lifetime.[8] The number of the contributions both for the simulative and experimental investigations about catalytic distillation are greatly increasing in recent years, especially for the modeling and simulation studies. And the applications of the catalytic distillation in its field are expanding. The modeling analysis approach for the design, synthesis, and feasibility analysis of the reactive distillation process have been parallely developed since the equilibrium stage model was used for process analysis through computer in late 1950s.

On current knowledge, the real distillation process always operates away from equilibrium and for multi-component mass transfer in the distillation, and the stage efficiency is often different for each component.[9] In recent years, the non-equilibrium model, also called rate-based model, has been developed for reactive distillation column to describe the mass transfer between vapor and liquid phase using the Maxwell-Stefan equations.[10] Always, the two phase non-equilibrium model is used for the prediction of the catalytic distillation, which treats the solid catalyst as a pseudo-liquid phase for the reaction in the catalyst. Also a more complex three-phase model [11] have been developed in some contributions in recent years to rigorously describe the reaction kinetics and mass transfer rate between the liquid and the solid catalytic phase in the catalytic distillation. However, a pseudo-homogeneous non-equilibrium model might adequately simulate the temperature profile, yield and selectivity for a CD process for a kinetically controlled reaction system. Additionally, the difficulties are related to the determination of additional model parameters required when using such models, and good estimation methods for the calculation of the diffusion coefficients and the non-ideal thermodynamic behavior inside a catalyst are also absent.

In our former work[12], modeling and simulation of such a catalytic distillation process for the DMC synthesis from urea and methanol was carried out based on the non-equilibrium model. The heterogeneously catalyzed reactions in the liquid bulk phase are considered as pseudo-homogeneous for the synthesis of DMC. Furthermore, the effect of distillation total pressure and the reaction temperature was studied. The interaction between the chemical reaction and the product separation were illustrated with the non-equilibrium model. And the mass transfer rate between the liquid and vapor phase have been taken into account by using the Maxwell-Stefan equations.

### 5.1 The configuration of the simulated catalytic distillation

The simulated column, a two meter tall stainless steel reactive distillation with an inner diameter of 22 mm, was configured with two feeding inlets and a side outlet. The materials were fed into the distillation column through preheater with volumes of 500ml for each feed stream. It would take about 2-5 hours for the feed material to pass though the preheater to the distillation column, which was enough for the complete conversion of urea to MC in the preheater, as the first reaction for DMC synthesis by urea methanolysis method could take place with high yield even in the absence of catalyst. The distillation column was divided into three sections, the rectifying section, the reaction section and the stripping section. 100 ml catalyst pellets weighted 103g with an average diameter of 3 mm were randomly packed in the reaction zone and the grid metal rings with a diameter of 3.2 mm were packed into the non-reaction zones. The distillation configured with a partial condenser to release the non-condensing gas of ammonia and a partial reboiler to discharge the heavy component of MC. The temperature in the reaction zone was set to 454.2 K for the synthesis reaction and the process was carried out under the pressure of 9-13 atm.

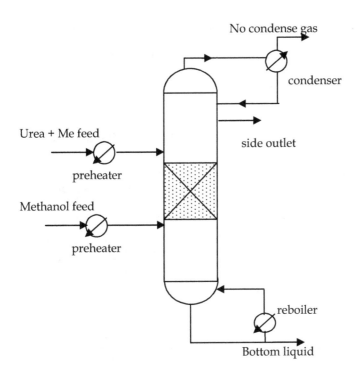

Fig. 1. The scheme of the catalytic distillation for synthesis of DMC

## 5.2 Chemical reactions

The synthesis of DMC from urea and methanol is catalyzed by the solid base catalysts shown in the scheme.

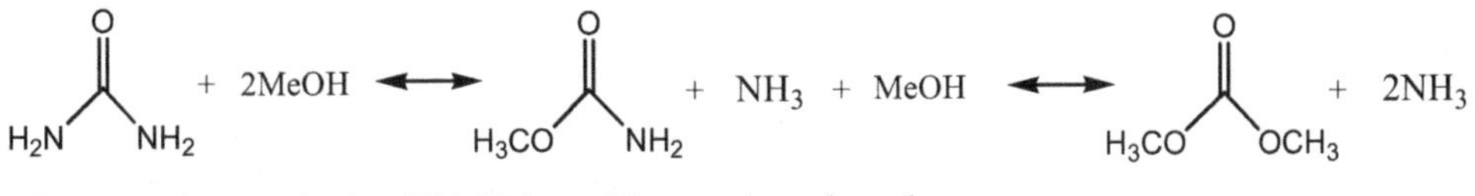

Scheme 1. the synthesis of DMC from Urea and methanol

The synthesis of DMC is a two-step reaction. The intermediate methyl carbamate (MC) is produced with high yield in the first step and further converted to DMC by reacting with methanol on catalyst in the second step. Our co-workers have developed the ZnO catalyst to catalyze the DMC synthesis reaction in CD process, which exhibited high activity toward the reactions. It was found by our workers that the reaction of the first step took place with high yield even in the absence of catalyst, and the catalyst was mainly effective for the second step. In CD process for the synthesis of DMC, the material mixture of urea and methanol was fed in the CD column through a preheater which has been heated to 423K and the materials stayed in the preheater for sufficient time to convert the urea to MC. As a result, only second step of DMC synthesis reaction, where MC converting to DMC, took place in the catalytic distillation column (shown as follow).

$$MC + MeOH \quad \longleftrightarrow \quad DMC + NH_3 \tag{1}$$

The macro-kinetic model for the forward and reverse reactions by Arrhenius equations are represented as follows:

$$R = \omega \cdot k_1 \exp(-\frac{Ea_1}{R_gT})C_{MC}C_{Me} - \omega \cdot k_2 \exp(-\frac{Ea_2}{R_gT})C_{DMC}C_{NH_3} \tag{2}$$

Where $\omega$ represents the amount of catalyst presented in the column section. $k_1$ and $k_2$ represent the Arrhenius frequency factors, and $Ea_1$ and $Ea_2$ are activation energy for the forward and reverse reactions, respectively. The values of Arrhenius parameters for the synthesis of DMC by urea and methanol over the solid base catalyst are listed in Table 17.

| $k_1$ (g$^{-1}$mol$^{-1}$Ls$^{-1}$) | $k_2$ (g$^{-1}$mol$^{-1}$Ls$^{-1}$) | $Ea_1$ (J/mol) | $Ea_2$ (J/mol) |
|---|---|---|---|
| 1.104E3 | 1.464E-3 | 1.01E5 | 4.90E4 |

Table 17. Arrhenius parameters for DMC synthesis catalyzed by solid base catalyst

The system of DMC synthesis process in a CD column mainly involved four components: methanol, DMC, MC and ammonia, as the first step reaction was omitted in the distillation column. The boiling points of the pure components at atmospheric pressure was ranged as follows: methanol (Me) 337.66 K; DMC 363.45 K; MC 450.2 K; ammonia (NH$_3$) 239.72 K, respectively. It could be seen that MC should almost exist in the liquid phase in CD process under high pressure and the reactions would take place in the liquid phase in a CD reaction zone. The system included a binary azeotrope of Me-DMC and the predicted data have been shown in Table 2, with respective boiling points at different pressures. Since the system included a no condenser component of ammonia and a binary azeotropic pair of methanol-DMC, it shows the strong non-ideal properties and the vapor liquid equilibrium was calculated by the EOS + activity method.

## 5.3 The non-equilibrium model

The non-equilibrium model is schematically shown in Fig.2. This NEQ stage represents a section of packing in a packed column. The heterogeneously catalyzed synthesis of DMC in CD process is treated as pseudo homogenous. Mass transfers at the vapor-liquid interface are usually described via the well-known two-film model. A rigorous model for catalytic distillation processes have been presented by Hegler, Taylor and Krishna. In the present contribution the two-phase non equilibrium model have been developed to investigate the steady state of the DMC synthesis process in catalytic distillation.

The follow assumptions have been made for the non-equilibrium model: (1) the process reached steady state; (2) the first reaction has been omitted as it took place with high yield in the preheater; (3) the reactions occurred entirely in the liquid bulk; (4) the reactions have been considered as pseudo-homogeneous; (5) the pressure in the CD column has been treated as constant.

The model equations composed of material balance, energy balance, mass transfer, energy transfer, phase equilibria, pressure drop equations and summation equations, which had been showed under the follows:

The material balances both for vapor and liquid phase are defined as:

$$V_{j+1}y_{i,j+1} - (1+S_j^V)V_jy_{i,j} + F_j^V z_{i,j}^V - N_{i,j}^V = 0 \tag{3}$$

$$L_{j-1}x_{i,j-1} - (1+S_j^L)L_jx_{i,j} + F_j^L z_{i,j}^L + N_{i,j}^L + R_{i,j}^L = 0 \tag{4}$$

The multi-component mass transfer rates are described by the generalized Maxwell-Stefan equations. The mass transfer equations for liquid phase are described as follow:

$$-\frac{x_i}{R_gT}\nabla_T\mu_i = \sum_{\substack{j=1\\j\neq i}}^{c} \frac{x_jN_i^L - x_iN_j^L}{C_t^L k_{ij}^L a} \tag{5}$$

where $\mu_i$ represent the chemical potential, $k_{ij}^L$ is liquid mass transfer coefficient. Only $c-1$ of these equations are independent. The vapor phase mass transfer has a similar relation to the liquid phase.

The energy balances for both vapor and liquid phase are defined as:

$$V_{j+1}H_{j+1}^V - (1+S_j^V)V_jH_j^V + F_j^V H_j^{VF} - e_j^V + Q_j^V = 0 \tag{6}$$

$$L_{j-1}H_{j-1}^L - (1+S_j^L)L_jH_j^L + F_j^L H_j^{LF} + e_j^L + H_j^{LR} + Q_j^L = 0 \tag{7}$$

where the vapor and liquid energy transfer rate is considered as equal. The vapor heat transfer rate is defined as:

$$e^V = -h^V a\frac{\partial T^V}{\partial \eta} + \sum_{i=1}^{C} N_i^V H^V \tag{8}$$

The Vapor-liquid equilibrium occurs at the vapor-liquid interface:

$$y_{i,j}^I - K_{i,j}^I x_{i,j}^I = 0 \tag{9}$$

where the superscript $I$ denotes the equilibrium compositions at the vapor-liquid interface and $K_{i,j}^I$ represents the vapor liquid equilibrium ratio for component i on stage j. And the equilibrium constant is computed by:

$$K_i^I = \frac{P_i^0 \gamma_i f_i^0}{P f_i} \tag{10}$$

The Wilson equations for the liquid phase have been selected to calculate the liquid activity coefficient.

In addition to the above equations, there also have the summation equations for the mole fractions:

$$\sum_{i=1}^{C} x_{i,j} - y_{i,j} = 0 \tag{11}$$

Thermo-physical constants such as density, enthalpy, heat conductivity, viscosity, and surface tension have been calculated based on the correlations suggested by Reid et al. (1987) and by Danbert and Danner (1989). Furthermore, the mass transfer coefficients are computed by the empirical Onda relations.

$$k_{ik}^L = 0.0051(\frac{w^L}{a_w g})^{2/3} \left(\frac{\mu_m^L}{\rho_m^L D_{ik}^L}\right)^{-0.5} \left(\frac{\mu_m^L g}{\rho_m^L}\right)^{1/3} \left(a_t d_p\right)^{0.4} \tag{12}$$

$$k_{ik}^V = \alpha(\frac{w^V}{a_t \mu_m^V})^{0.7} \left(Sc_{ik}^V\right)^{1/3} \left(a_t d_p\right)^{-2} \frac{a_t D_{ik}^V p}{p_{Bm} R_g T} \tag{13}$$

where $\alpha$ is 2.0 for the non-reaction packing of 3.2 mm metal grid ring. The wet area of the packing is estimated using the equations developed by onda et al shown as follow:

$$\frac{a_w}{a_t} = 1 - \exp[-1.45(\frac{w^L}{a_t \mu_m^L})^{0.1} \left(\frac{w^{L2} a_t}{\rho_m^{L2} g}\right)^{-0.05} \left(\frac{w^{L2}}{\rho_m^L \sigma a_t}\right)^{0.2} (\sigma_C / \sigma)^{0.75}] \tag{14}$$

The effective interfacial area is estimated using the empirical relation developed by Billet:

$$\frac{a}{a_t} = \frac{1.5}{\left(a_t 4\varepsilon/a_t\right)^{0.5}} \left(\frac{u^L 4\varepsilon/a_t}{\mu_m^L / \rho_m^L}\right)^{-0.2} \left(\frac{u^{L2} \rho_m^L 4\varepsilon/a_t}{\sigma_m^L}\right)^{0.75} \left(\frac{u^{L2}}{g 4\varepsilon/a_t}\right)^{-0.45} \tag{15}$$

The mass transfer coefficients for the reaction zone are estimated using the equations developed by Billet, as shown as follow:

$$k_{ik}^L a = 1.13 \frac{a_t^{2/3}}{\left(4\varepsilon/a_t\right)^{0.5}} D_{ik}^{L\,0.5} \left(\frac{g}{\mu^L / \rho^L}\right)^{1/6} \mu_m^{L\,1/3} \frac{a}{a_t} \tag{16}$$

$$k_{ik}^L a = 0.275 \frac{a_t^{3/2}}{\left(4\varepsilon/a_t\right)^{0.5}} D_{ik}^V \frac{1}{\left(\varepsilon - h_L\right)^{0.5}} \left(\frac{\mu_m^V / \rho_m^V}{D_{ik}^V}\right)^{1/3} \left(\frac{u_m^V}{a\mu_m^V / \rho_m^V}\right)^{3/4} \frac{a}{a_t} \tag{17}$$

Heat transfer coefficients are predicted using Chilton-Colburn analogy as follow:

$$h^V = k_{av}^V C_{pm}^V \left(Le^V\right)^{2/3} \text{ for vapor phase}$$

$$h^L = k_{av}^L C_{pm}^L \left(Le^L\right)^{1/2} \text{ for liquid phase.} \tag{18}$$

## 5.4 The treatment of the reaction for the synthesis of DMC

Commonly, the reaction rates are determined by the concentration of the component, not the volume of the component. And this factor could cause a negative composition of a component during the iteration for the solving of a catalytic distillation model. Consequently, the reaction of a system can be considered as the combination of the positive reactions and the negative reactions.

$$R_{i,j}^L = R_{i,j}^+ + R_{i,j}^- = \sum_{m=1}^{nr}\left(\varepsilon_j \cdot v_{i,m}^+ r_{m,j}^+ - \varepsilon_j \cdot v_{i,m}^- r_{m,j}^-\right) \tag{19}$$

Defined the consumptive coefficient as:

$$E_{i,j}^R = -R_{i,j}^- / x_{i,j} \tag{20}$$

For the condition of $x_i$ equal zero, the consumptive coefficient is set to zero.

## 5.5 The method of Maxwell-Stefan equations

For the multi-component mass transfer in the catalytic distillation can be considered as the one dimensional mass transfer behavior[13]. And the vapor liquid equilibria is achieved at the vapor liquid interface. It can be noticed that there are no accumulation on the vapor liquid interface and the mass transfer of vapor and liquid are equal to each other.

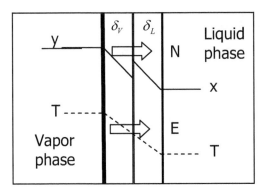

Fig. 2. The multi-components mass transfer

According to the two-film theory, the Maxwell-Stefan equations for vapor and liquid phase are shown as:

$$-\frac{x_i}{RT}\nabla_T \mu_i = \sum_{\substack{j=1 \\ j\neq i}}^{c} \frac{x_j N_i - x_i N_j}{C_t k_{ij}} = \sum_{\substack{j=1 \\ j\neq i}}^{n} \frac{x_j J_i - x_i J_j}{C_t k_{ij}} \tag{21}$$

$$-\frac{y_i}{RT}\nabla_T \mu_i = \sum_{\substack{j=1 \\ j\neq i}}^{c} \frac{y_j N_i - y_i N_j}{C_t k_{ij}} = \sum_{\substack{j=1 \\ j\neq i}}^{n} \frac{y_j J_i - y_i J_j}{C_t k_{ij}} \tag{22}$$

For the liquid mass transfer equations, it could be rearranged by n-1 matrix as:

$$\left(N^L\right) = -C_t^L\left[k^L\right]\frac{\partial x}{\partial \eta} + (x)N_t^L \tag{23}$$

Where:

$$\left[k^L\right] = [R]^{-1}[\Gamma]$$

$$R_{ii} = \frac{x_i}{k_{in}} + \sum_{\substack{k=1 \\ k\neq i}}^{n} \frac{x_k}{k_{ik}} \quad , \quad R_{ij} = -x_i\left(\frac{1}{k_{ij}} - \frac{1}{k_{in}}\right) \tag{24}$$

$\Gamma$ represent the matrix of Thermal factor for multi-component. The elements of the thermal factor matrix are the partial molar difference of activity coefficient of the component mixture, which could be computed as follow:

$$\Gamma_{ij} = \delta_{ij} + x_i\left.\frac{\partial \ln \gamma_i}{\partial x_j}\right|_{\sum x_j = 1} \tag{25}$$

$$i, j = 1, 2, 3, \cdots, n-1$$

Note: This solution of partial molar difference of activity coefficient was restricted by the summation equation. This should be especially noted.

For the liquid mass transfer equation, it could be written as:

$$\left(J^L\right) = C_t^L\left[k^L\right]\frac{\partial x}{\partial \eta} \tag{26}$$

The mass transfer rates were consistent along the film distance, and the differential equation could be written as:

$$\left(\frac{dx}{d\eta}\right) = [\Phi](x) + (\zeta) \tag{27}$$

$$\Phi_{ii} = \frac{N_i}{C_t^L k_{in}} + \sum_{\substack{k=1 \\ k\neq i}}^{n} \frac{N_k}{C_t^L k_{ik}} \quad , \quad \Phi_{ij} = -N_i\left(\frac{1}{C_t^L k_{ij}} - \frac{1}{C_t^L k_{in}}\right)$$

$$i, j = 1, 2, 3, \cdots, n-1$$

There were two boundary conditions as follow.

$$\eta = 0, (bulk), (y) = (y_b)$$

$$\eta = 1, (film), (y) = (y_I) \tag{28}$$

The differential equations could be solved as:

$$(x_\eta - x_b) = \{\exp([\Phi]\eta) - [I]\} \cdot \{\exp[\Phi] - [I]\}^{-1} \cdot (x_I - x_b) \tag{29}$$

Additionally, the mass transfer fluxes were expressed as:

$$\left. \frac{d(x)}{\eta} \right|_{\eta=0} = [\Phi] \cdot \{\exp[\Phi] - [I]\}^{-1} \cdot (x_I - x_b) \tag{30}$$

$$(J^L) = C_t^L [k^L][\Phi] \cdot \{\exp[\Phi] - [I]\}^{-1} \cdot (x_I - x_b) \tag{31}$$

The energy balance should occur on the vapor liquid interface, and these could be written as:

$$\sum_{i=1}^{n} N_i \cdot (\overline{H_i^V} - \overline{H_i^L}) = 0 \tag{32}$$

$$\sum_{i=1}^{n} (J_i + x_i N_t) \cdot (\overline{H_i^V} - \overline{H_i^L}) = 0 \tag{33}$$

$$\beta_{ik} \equiv \delta_{ik} - x_i \Lambda_k$$

$$\Lambda_k = (\nu_k - \nu_n) / \sum_{j=1}^{n} \nu_j x_j \tag{34}$$

As a result, the mass transfer rate could be described as:

$$N_t = -\sum_{i=1}^{n} J_i \cdot (\overline{H_i^V} - \overline{H_i^L}) \Big/ \sum_{i=1}^{n} y_i \cdot (\overline{H_i^V} - \overline{H_i^L}) \tag{35}$$

$$N^L = c_t^L [\beta^V] \cdot [R^L]^{-1} \cdot [\Gamma^L] \cdot [\Xi^V] \cdot (x^I - x^b) \tag{36}$$

Where the high flux correction factor were defined as:

$$[\Xi] = [\Gamma]^{-1} [\Phi] \cdot \left[ \exp[\Gamma]^{-1} [\Phi] - [I] \right]^{-1} \tag{37}$$

Similarly, the mass transfer for vapor phase could be described as:

$$N^V = c_t^V [\beta^V] \cdot [R^V]^{-1} \cdot [\Gamma^V] \cdot [\Xi^V] \cdot (y^b - y^I) \tag{38}$$

The total mass transfer rates described by vapor phase were showed as:

$$N = c_t^V [\beta^V] \cdot [K_{ov}] \cdot [\Xi^V] \cdot (y^b - y^*) \tag{39}$$

The solving for these mass transfer equations involved an iterative method and the more mathematic knowledge were needed for the calculation of high flux correction factor, which exponent calculations for a matrix were involved. The computational codes have been developed by the author. Otherwise, these equations were more complicated for the mass transfer in a catalytic distillation system.

Power et al. (1988) found that the high flux correction factor, which is for the calculation of multi-component mass transfer, is not important in distillation and it has been ignored in the calculation for distillation system. In the non-equilibrium stage model of Krishnamurthy and Taylor (1993), the total mass transfer rates are obtained by combining the liquid and vapor mass transfer equations, which the high flux correction factor has been ignored, and multiplying by the interfacial area available for mass transfer. As a result, the total mass transfer rates for the vapor phase described as matrix form are: [14]

$$N_j^L = N_j^V = C_{t,j}^V \left[ \beta_j^V \right] \left[ K_j^{OV} \right] a_j \left( y_j - y_j^* \right) \tag{40}$$

$$\left[ K_j^{OV} \right]^{-1} = \left[ K_j^V \right]^{-1} + \frac{C_{t,j}^V}{C_{t,j}^L} \left[ K_j^E \right] \left[ K_j^L \right]^{-1} \left[ \beta_j^L \right]^{-1} \left[ \beta_j^V \right] \tag{41}$$

$$\left[ K_j^V \right] = \left[ \Omega_j^V \right]^{-1}$$

$$\Omega_{ii,j}^V = \frac{y_{i,j}}{k_{iC,j}^V} + \sum_{\substack{l=1 \\ l \neq i}}^{C} \frac{y_{l,j}}{k_{il,j}^V} \quad , \quad \Omega_{ik,j}^V = -y_{i,j} \left( \frac{1}{k_{ik,j}^V} - \frac{1}{k_{iC,j}^V} \right) \tag{42}$$

Where i ,k=1, 2, $\cdots$, c-1.

$$\beta_{ik,j}^V = \delta_{ik} - y_{i,j} \left( \Delta II_{k,j}^{Vap} - \Delta H_{n,j}^{Vap} \right) \Big/ \sum_{l=1}^{c} y_{l,j} \Delta H_{l,j}^{Vap} \tag{43}$$

The similar correlation can be described for liquid phase. Where $\delta$ is the unit matrix.

## 5.6 Solving of the non-equilibrium model

By combining the above mentioned equations, the vapor component can be determined by the liquid bulk composition, which described by the matrix form as: [15]

$$Y_j = \left( c_{t,j}^V \left[ \beta_j^V \right] \left[ K_j^{OV} \right] a_j + (1 + S_j^V) V_j [\delta] \right)^{-1} \left( V_{j+1} Y_{j+1} + F_j^V z_j^V + c_{t,j}^V \left[ \beta_j^V \right] \left[ K_j^{OV} \right] a_j Y_j^* \right) \tag{44}$$

Defined the single stage vaporizing coefficient:

$$E_{i,j}^V = y_{i,j} / y_{i,j}^* = y_{i,j} / K_{i,j}^E x_{i,j} \tag{45}$$

For $x_{i,j} = 0$, $E_{i,j}^V = 1$.

And then we can get a modified tri-diagonal matrix method for solving the non-equilibrium stage model of the catalytic distillation.

$$
\begin{vmatrix}
B_{i1} & C_{i1} & & & & \\
A_{i2} & B_{i2} & C_{i2} & & & \\
 & & \cdots & & & \\
 & & A_{i,j} & B_{i,j} & C_{i,j} & \\
 & & & \cdots & & \\
 & & & & A_{i,n} & B_{i,n}
\end{vmatrix}
\cdot
\begin{vmatrix}
x_{i1} \\ x_{i2} \\ \vdots \\ x_{ij} \\ \vdots \\ x_{in}
\end{vmatrix}
=
\begin{vmatrix}
b_{i1} \\ b_{i2} \\ \vdots \\ b_{ij} \\ \vdots \\ b_{in}
\end{vmatrix}
\tag{46}
$$

$$A_{i,j} = L_{j-1}$$
$$B_{i,j} = -(1+S_j^V)V_j K_{i,j}^E E_{i,j}^V - (1+S_j^L)L_j x_{i,j} - E_{i,j}^R$$
$$C_{i,j} = V_{j+1} K_{i,j+1}^E E_{i,j+1}^V$$
$$b_{i,j} = -F_j^V z_{i,j}^V - F_j^L z_{i,j}^L - R_{i,j}^+$$

## 5.7 The model results

The typical modeling data for the catalytic distillation of DMC synthesis process from urea and methanol over solid base catalyst, which was operated under the pressure of 9 atm. [16-17]

| Parameter | Measured | Estimated |
|---|---|---|
| Temperature of top (°C) | 91.0 | 94.8 |
| Temperature of reboiler (°C ) | 165.3 | 169.1 |
| T in reaction zone (°C) | 180 | 181.1 |
| Material feed (mL/h) | 20 | 20 |
| Methanol feed (mL/h) | 60 | 60 |
| Yield of DMC (%) | 45 | 45 |
| Reflux ratio | 4 | 4 |
| Condenser, mass fraction | | |
| Me | 0.768 | 0.782 |
| DMC | 0.052 | 0.053 |
| MC | 0 | 0 |
| Ammonia | 0.18 | 0.165 |
| Rebloiler, mass fraction | | |
| Me | 0.260 | 0.271 |
| DMC | 0 | 0 |
| MC | 0.740 | 0.729 |
| Ammonia | 0 | 0 |

| Parameter | Measured | Estimated |
|---|---|---|
| Product, mass fraction | | |
| Me | 0.927 | 0.927 |
| DMC | 0.070 | 0.070 |
| MC | 0 | 0 |
| Ammonia | 0.003 | 0.003 |

Table 18. Typical results from experiment and predictions for the synthesis of DMC

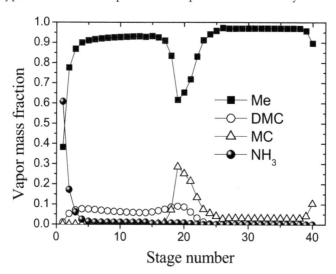

Fig. 3. Vapor concentration distribution in distillation column.

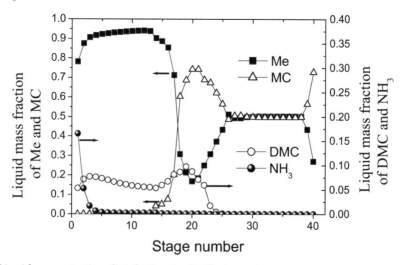

Fig. 4. Liquid concentration distribution in distillation column

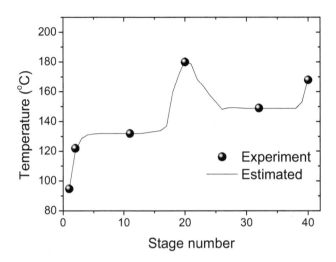

Fig. 5. Temperature distribution in distillation column

## 6. Separation of the mixture of methanol and DMC

Separation of azeotropic mixtures is a challenge commonly encountered in commodity and fine chemical processes. Many techniques suitable for separation of azeotropic mixtures have been developed recently, such as pressure-swing distillation (PSD), extractive and azeotropic distillation, liquid-liquid extraction, adsorption, prevaporation using membrane, crystallization and some new coupling separation techniques. Despite of the newly developed membrane separation process or adsorption process, it was very important to properly design of the traditional separation of DMC from the reaction mixture using the distillation tower with the existence of the azeotrope of methanol-DMC for large scale of DMC production.

Zhang[18] has developed a process model for atmospheric-pressurized rectification to simulate the separation of DMC and methanol with low concentration DMC, which came from the DMC synthesis through urea methanolysis method. The simulation was carried out based on the Aspen Plus platform with the Wilson liquid activity coefficient model.

Li[4] has given a pressurized-atmospheric separation process for separating the product of DMC from the mixture of methanol and DMC with 30 wt.% of DMC based on the simulation model. They have optimized the operating conditions based on the developed process and the optimized condition: 40 of ideal stages, 29th of feed stage, 7~10 of reflux ratio and 1.3 MPa of pressure for pressurized distillation. However, this process has a drawback of high investment of the equipment and lower stability of operation. Our

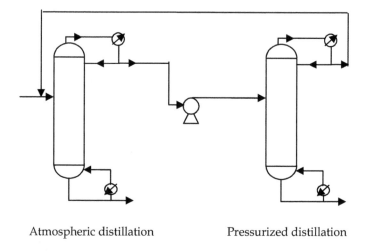

Atmospheric distillation                    Pressurized distillation

Fig. 6. The schematic diagram of the pressure swing process.

workers, Zhang et al., also developed a novel separation process of atmospheric-pressurized separation process which had the ability to separate the low concentration of DMC for the separation of the mixture with 12 wt.% of DMC base on the 500t/a pilot plant by the simulation model. The sensitive study and the optimization to this process had shown that the reflux ratio for the atmospheric and pressurized distillation had been 3.4 and 1.0 respectively, and 0.65, 0.93 of the distillate to feed ratio for the atmospheric and pressurized distillation. In this process, 99.5wt.% or higher concentration of methanol could be recovered, while the pressurized-atmospheric separation process could only obtained a solution containing 13.3 wt.% of DMC in the recovered methanol stream.

The model simulation and process design for the separation of DMC and methanol has been lettered in many literatures. Much number of the literatures had been presented on the simulation for the DMC synthesis with trans-esterification method, in addition with the detailed research on the catalytic distillation for the DMC synthesis by urea and methanol. However, the simulation work for the other DMC synthesis process had been little reported. Furthermore, the pressure-swing distillation process, the extractive distillation process and the azeotropic distillation process had been developed in the open reported simulative literatures for the product separation of DMC and methanol mixture. Among the derived separation process, the pressure swing distillation process and the extractive distillation process had been considered suitable for the product separation of DMC and methanol mixture.

# 7. References

[1] Tundo, P.; Selva, M. The Chemistry of Dimethyl Carbonate. Acc. Chem. Res. 2002, 35, 706.

[2] Ono, Y. Dimethyl Carbonate for Environmentally Benign Reactions. Catal. Today 1997, 35, 15.

[3] Rodriguez, J. Canosa, A. Dominguez, J. Tojo. Isobaric vapour–liquid equilibria of dimethyl carbonate with alkanes and cyclohexane at 101.3 kPa. Fluid Phase Equilibria 198 (2002) 95–109.

[4] Rodriguez, J. Canosa, A. Dominguez, J. Tojo. Vapour–liquid equilibria of dimethyl carbonate with linear alcohols and estimation of interaction parameters for the UNIFAC and ASOG method. Fluid Phase Equilibria 201 (2002) 187–201

[5] Ding Wang, Aiguo Xuan, Yuanxin Wua, Zhiguo Yana, Qimei Miao. Study on gas – liquid equilibria with the UNIFAC model for the systems of synthesizing dimethyl carbonate. Fluid Phase Equilibria, 302 (2011) 269 – 273

[6] Wang-Ming Hu, Lv-Ming Shen, Lu-Jun Zhao. Measurement of vapor–liquid equilibrium for binary mixtures of phenol–dimethyl carbonate and phenol–methanol at 101.3 kPa. Fluid Phase Equilibria, 219 (2004) 265–268

[7] Agreda, V. H.; Partin, P. H.; Heise, W. H. High Purity Methyl Acetate via Reactive Distillation. Chem. Eng. Process. 1990, 86, 40.

[8] Pilavachi, P. A.; Schenk, M.; Perez-Cisneros, E.; Gani, R. Modeling and Simulation of Reactive Distillation Operations. Ind. Eng. Chem. Res. 1997, 36, 3188.

[9] Malone, M. F.; Doherty, M. F. Reactive Distillation. Ind. Eng. Chem. Res. 2000, 39, 3953.

[10] Tuchlenski, A.; Beckmann, A.; Reusch, D.; DuKssel, R.; Weidlich, U.; Janowsky, R. Reactive Distillation - Industrial Applications, Process Design & Scale-up. Chem. Eng. Sci. 2001, 56, 387.

[11] Xu, Y.; Zheng, Y.; Ng, F. T. T.; Rempel, G. L. A Three-phase Nonequilibrium Dynamic Model for Catalytic Distillation. Chem. Eng. Sci. 2005, 60, 5637.

[12] Wang Feng, Zhao Ning, Li Junping, et al. Modeling of the Catalytic Distillation Process for the Synthesis of Dimethyl Carbonate by Urea Met hanolysis Method[J]. Ind Eng Chem Res,2007,46(26):8972–8979.

[13] Taylor, R.; Krishna, R. Multicomponent Mass Transfer; John Wiley and Sons: New York, 1993.

[14] Wang Feng, Zhao Ning, LI Jun-ping, Xiao Fu-kui, Wei Wei, Sun Yu-han. Modeling analysis for process operation of synthesis of Dimethyl carbonate. CHEMICAL ENGINEERING(CHINA), 2009, 37(2): 71-74

[15] Wang Feng, Zhao Ning, Li Junping, Xiao Fukui, Wei Wei, Sun Yuhan. Simulation of Catalyst distillation Using Non-Equilibrium model. PETROCHEMICAL TECHNOLOGY, 2007, 36(11): 1128-1133

[16] Wang Feng, Zhao Ning, Li Junping, et al. Non-Equilibrium Model for Catalytic Distillation Process[J].Rront Chem Eng Chin,2008, 2(4): 379-384.

[17] Wang Feng, Zhao Ning, Li Junping, Wu Dudu, Wei Wei, Sun Yuhan. Non-Equilibrium Stage Model for Dimethyl Carbonate Synthesis by Urea Methanolysis in Catalytic Distillation Tower. PETROCHEMICAL TECHNOLOGY, 2008, 37(4): 359-363

[18] Zhang Junliang, Wang Feng, Peng Weicai, Xiao Fukui, Wei Wei, Sun Yuhan. Process Simulation for Separation of Dimethyl Carbonate and Methanol Through Atmospheric-Pressurized Rectification. PETROCHEMICAL TECHNOLOGY, 2010, 39(6): 646-650

# 3

# Energy Conservation in Ethanol-Water Distillation Column with Vapour Recompression Heat Pump

Christopher Enweremadu
*University of South Africa, Florida Campus*
*South Africa*

## 1. Introduction

Ethanol or ethyl alcohol $CH_3CH_2OH$, a colorless liquid with characteristic odor and taste; commonly called grain alcohol has been described as one of the most exotic synthetic oxygen-containing organic chemicals because of its unique combination of properties as a solvent, a germicide, a beverage, an antifreeze, a fuel, a depressant, and especially because of its versatility as a chemical intermediate for other organic chemicals. Ethanol could be derived from any material containing simple or complex sugars. The sugar-containing material is fermented after which the liquid mixture of ethanol and water is separated into their components using distillation.

Distillation is the most widely used separation operation in chemical and petrochemical industries accounting for around 25-40% of the energy usage. One disadvantage of distillation process is the large energy requirement. Distillation consumes a great deal of energy for providing heat to change liquid to vapour and condense the vapour back to liquid at the condenser. Distillation is carried out in distillation columns which are used for about 95% of liquid separations and the energy use from this process accounts for an estimated 3% of the world energy consumption (Hewitt et al, 1999). It has been estimated that the energy use in distillation is in excess. With rising energy awareness and growing environmental concerns there is a need to reduce the energy use in industry. The potential for energy savings therefore exists and design and operation of energy efficient distillation systems will have a substantial effect on the overall plant energy consumption and operating costs.

The economic competitiveness of ethanol has been heightened by concerns over prices and availability of crude oil as well as greenhouse gas emissions which have stimulated interest in alternatives to crude oil to provide for automotive power and also by the use of bioethanol in the production of hydrogen for fuel cells. Therefore, there is the need to explore ways of producing ethanol at competitive costs by the use of energy efficient processes. To cope with the high energy demand and improve the benefits from the process, the concept of polygeneration and hydrothermal treatment especially when dealing with small scale ethanol plants is fast gaining interest. However, the analysis of the bioethanol process shows that distillation is still the most widely used.

Over the years, there have been many searches for lower energy alternatives or improved efficiencies in distillation columns. One such search led to the use of heat pumps, the idea which was introduced in the 1950s. Also, Jorapur and Rajvanshi (1991) have used solar energy for alcohol distillation and concluded that it was not economically viable. Heat pumping, however, has been known as an economical energy integration technology for reduction in consumption of primary energy and to minimize negative impact of large cooling and heating demands to the environment. One of the heat pump cycles which have been widely studied is the recompression of the vapours where the reboiler is heated by adding a compressor to the column to recover some of the heat lost in the distillate.

Most studies have concluded that heat pumping is an effective means of saving energy and reducing column size without estimating the actual energy consumption and the parameters that are likely to have significant effect on energy consumption. Estimating the actual energy consumption is an important aspect towards the determination of the viability of the system in ethanol–water separation.

The purpose of this chapter was to study how previously neglected and/or assumed values of different parameters (the pressure increase across the compressor was ignored, column heat loss was assumed to be 10% of the reboiler heat transfer rate, and the overall heat transfer coefficient was determined without considering it as an explicit function of dimensionless numbers, and its dependence on fluid viscosity and thermal conductivity neglected) affect the process efficiency, energy consumption and the column size of a vapour recompression heat pump.

## 2. Energy requirements in ethanol distillation

Ethanol distillation, like any other distillation process requires a high amount of thermal energy. Studies carried out by several authors reveal that the distillation process in ethanol distilleries consumes more than half of the total energy used at the distillery (Pfeffer et al. 2007). It has been estimated that distillation takes up about 70-85% of total energy consumed in ethanol production. Pfeffer et al (2007) estimated that distillation consumes half of the total production energy 5.6 MJ/Liter out of 11.1 – 12.5 MJ/Liter.

The energy requirements for ethanol production have improved markedly during the past decade due to a variety of technology and plant design improvements. The energy needed to produce a liter of ethanol has decreased nearly 50% over the past decade and that trend is likely to continue as process technology improves ( Braisher *et al*, 2006).

## 3. Energy conservation schemes in distillation column

Distillation columns are usually among the major energy-consuming units in the food, chemical, petrochemical and refining industries. According to Danziger (1979), the most effective method of economizing energy in a distillation column is energy recovery of which direct vapour recompression has been regarded as the best solution.

### 3.1 Heat pumping distillation systems

Basically, the heat pump can be regarded simply as reverse heat engine. The heat pump requires either work input or external driving thermal energy to remove the heat from a low temperature source and transform it to a higher level.

The conventional heat pumps are electrically driven vapour recompression types, which work on the principle that a liquid boils at a higher temperature if its pressure is increased. A low-pressure liquid passes into the evaporator, where it takes in heat causing the liquid to boil at low temperature. The low-pressure vapour is passed to the compressor where it is compressed by the application of work to a higher pressure. The resulting high pressure vapour flows to the condenser where it condenses, giving up its latent heat at a high temperature, before expanding back to a low pressure liquid.

The heat pump cycle may be connected to a distillation column in three ways (Fonyo and Benko, 1998) . The simplest alteration is to replace steam and cooling water with refrigerant (closed system). The other two types of heat pump system apply column fluids as refrigerant . When the distillate is a good refrigerant the vapour recompression can be used. If the bottom product is a good refrigerant the bottom flashing can be applied.

In this work, the direct vapour recompression system is studied due to its good economic figures ( Emtir et al, 2003). Also the vapour recompression is the most suitable as the boiling points of both key components (ethanol and water) are close to each other (Danziger, 1979) and the appropriate heat transfer medium (ethanol vapour) is available.

### 3.2 Use of vapour recompression in distillation columns

Vapour recompression system has been extensively studied since 1973, the year of drastic rise in energy (Null, 1976). The vapour recompression system is accomplished by using compressor to raise the energy level of vapour that is condensed in reboiler–condenser by exchange of heat with the bottoms. The condensate distillate is passed into reflux drum while the bottom product is vaporised into the column.

Vapour recompression consists of taking the overhead vapour of a column, condensing the vapour to liquid, and using the heat liberated by the condensation to reboil the bottoms liquid from the same column. The temperature driving force needed to force heat to flow from the cooler overhead vapours to the hotter bottoms product liquid is set up by either compressing the overhead vapour so that it condenses at a higher temperature, or lowering the pressure on the reboiler liquid so it boils at a lower temperature, then compressing the bottoms vapour back to the column pressure. While conventional column has a separate condenser and reboiler, each with its own heat transfer fluid such as cooling water and steam, the vapour recompression column has a combined condenser–reboiler, with external heat transfer fluids.

The advantage of vapour recompression lies in its ability to move large quantities of heat between the condenser and reboiler of the column with a small work input. This results from cases where there is only a small difference between the overhead and bottoms temperature. Also, the temperature, and therefore the pressure, at any point may be set where desired to achieve maximum separation. This effect is of particular importance where changing the pressure affects the relative volatility. By operating at more favourable conditions, the reflux requirement can be reduced and therefore the heat duties. These advantages can reduce a large amount of energy.

## 4. Ethanol-water vapour recompression distillation column

Figure 1 shows a schematic illustration of the distillation column with direct vapour recompression heat pump. An ethanol-water solution in a feed storage tank (FST) at

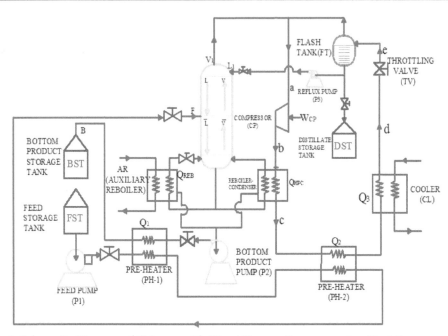

Fig. 1. Schematic Diagram of Column with Direct Vapour Recompression Heat Pump

ambient conditions, is preheated with bottom product and condensate in heat exchangers, preheaters PH1 and PH2, and fed to the column. An auxiliary reboiler (AR) is used to start the unit. This reboiler supplies the auxiliary heat duty, which is the heat of vaporization because the main reboiler can work only if there is some compressed vapour already available. The overhead vapours from the top are compressed in the compressor (CP) up to the necessary pressure in such a way that its condensing temperature is greater than the boiling temperature of the column bottom product. The vapour is then condensed by exchanging heat within the tubes of the reboiler-condenser (RC). In a condenser, the inlet temperature is equal to the outlet temperature. Ethanol vapour will only lose its latent heat of condensation. At the same time, the cold fluid (ethanol-water mixture) in the reboiler will absorb this latent heat and its temperature will increase to boil up the mixture to temperature $T_{CEV}$. The liberated latent heat of condensation provides the boil-up rate to the column while the excess heat extracted from the condensate is exchanged with the feed in preheater PH2. The condensate, which is cooled in the cooler (CL) up to its bubble point at the column operating pressure, expands through the throttling valve (TV) at the same pressure and reaches the flash tank (FT). After expansion, the output phases are a vapour phase in equilibrium with a liquid phase. One part of the product in the liquid phase is removed as distillate and stored in the tank (DST), while the remainder is recycled into the column as reflux L1. The excess of vapour is recycled to the compressor.

## 4.1 Methodology

Like this work, nearly all publications in this field are based on modelling and simulation (Brousse et al., 1985; Ferre et al., 1985; Collura and Luyben, 1988; Muhrer *et al*, 1990; Oliveira et al. 2001). The mathematical modeling of the distillation system is derived by applying energy, composition and overall material balances together with vapour-liquid equilibrium under some assumptions (see Muhrer *et al*, 1990 and Enweremadu, 2007). These and other assumptions are aimed at simplifying the otherwise cumbersome heat-and mass-transfer, and the fluid flow equations Mori et al (2002).

## 4.2 Calculation of the distillation column

In this system, there is a direct coupling between the distillation column and the rest of the system, as the heat pump working fluid is the column's own fluid which, is a binary mixture of ethanol and water at composition $X_D$. Therefore, the set of equations are not solved separately as in distillation column assisted by an external heat pump.

The detailed calculation of the overall material and component material balance such as the bottom flow rate, B and distillate flow rate, D; reflux ratio, $R_r$; the molar vapour flow rate which leaves the column top and feeds the condenser, $V_1$; feed vapour flow rate, $V_F$; feed vapour fraction, q; vapour molar flow rate remaining at the bottom of the column, $L_2$ are given (see Enweremadu, 2007).

The overall (global) energy balance equation applied to a control volume comprising the distillation column and the feed pre-heaters provides the total energy demand in the reboiler:

$$Q_{reb} = Dh_D + Bh_B + L_1h_{LV,e} + Q_{losses} - Fh_F - Q_1 - Q_2 \qquad (1)$$

where $Q_{reb}$ is the total heat load added to the reboiler, $Q_{losses}$ represents the heat losses in the column, which are to be determined; $Q_1$ and $Q_2$ are the heat loads of the pre-heaters; $h_{LV,e}$ is latent heat of vaporisation downstream of throttling valve; $h_D$, is the enthalpy of the distillate; $h_B$ is the enthalpy of the bottom product; $h_F$, is the enthalpy of the feed. The details of the mass balance variables are determined in Enweremadu (2007).

The first step in the design of a distillation column is the determination of the number of theoretical plates required for the given separation. The theoretical trays are numbered from the top down, and subscripts generally indicate the tray from which a stream originates with n and m standing for rectifying and stripping sections respectively. The design procedure for a tray distillation column consists of determining the liquid and vapour composition or fraction from top to bottom, along the column. In calculating the composition profile of the column two equations relating liquid mole fraction to temperature and vapour mole fraction to the liquid fraction are used. The compositions at the top ($X_D$) and bottom ($X_B$) of the column are previously pre-established data. In this work, the minimum number of theoretical stages ($N_{min}$) is calculated using Fenske's equation:

$$N_{min} = \frac{\log\left(\dfrac{X_D}{1-X_D} \cdot \dfrac{1-X_B}{X_B}\right)}{\log \alpha} \qquad (2)$$

where $\alpha$ is the relative volatility in the column. The actual number of plates is given by:

$$N = \frac{N_{min}}{\eta_T} \qquad (3)$$

where $\eta_T$ is the tray efficiency.

## 4.2.1 Heat losses from distillation column

The heat loss from the distillation column is the main factor that affects heat added and removed at the reboiler and condenser respectively. Most distillation columns operate above ambient temperature, and heat losses along the column are inevitable since insulating materials have a finite thermal conductivity. Heat loss along the distillation column increase condensation and reduces evaporation. Thus, the amount of vapour diminishes in the upward part of the column, where the flow of liquid is also less than at the bottom.

To prevent loss of heat, the distillation column should be well insulated. Insulation of columns using vapour recompression varies with the situation. Where the column is hot and extra reboiler duty is used, the column should be insulated (Sloley, 2001). The imperfect insulation of the column causes some heat output.

In determining the heat loss from the distillation column, it is assumed that the temperature is uniform in the space between two plates. The heat transfer between the column wall and the surrounding is then determined from the well-known relationship for overall heat transfer coefficient:

$$Q_{Losses} = U_p A_o \Delta T_p \tag{4}$$

where $U_p$, the overall heat transfer coefficient is given by Gani, Ruiz and Cameron (1986), as

$$U_p = f\left(h_o, h_i, K_p, A_o, A_1, A_m, t_{ins}\right) \tag{5}$$

where the temperature difference, $\Delta T_p$, is given as $\Delta T_p = T_p - T_{amb}$

$h_o$, the heat transfer coefficient between the surroundings and the column external surface, is given as

$$h_o = f(Nu, K_{ins}, d_o, t_{ins}) \tag{6}$$

$h_i$ is the heat transfer coefficient inside the column; $K_p$ is the thermal conductivity of the tray material; $A_o$ is the external area of heat exchange; $A_i$ is the internal area of heat exchange; $A_m$ is the logarithmic mean area; $t_{ins}$ is the thickness of insulation.

The heat output is calculated with the general expression for convection around cylindrical objects.

$$Q_{loss} = \frac{T_p - T_{amb}}{\dfrac{1}{h_i A_i} + \dfrac{\ln\left(r_{owall}/r_{iwall}\right)}{K_{wall} \cdot A_{wall}} + \dfrac{\ln\left(r_{ins}/r_{owall}\right)}{K_{ins} \cdot A_m} + \dfrac{1}{h_o A_o}} \tag{7}$$

The column inner surface heat transfer resistance is neglected as the heat transfer coefficient for condensing vapor is large and therefore will have little effect on the overall heat transfer.

Based on the assumptions in Enweremadu (2007), the heat transfer due to free convection between the surroundings and the external column wall and due to conduction through the insulation materials is predicted.

Also, from geometry of the insulated cylinder (Fig.2), the external diameter of insulation is given as

$$d_{ins} = d_o + 2t_{ins} \tag{8}$$

Details of how the logarithmic mean diameter of the insulating layer ($d_{ins,m}$), external area of heat exchange ($A_o$) and the logarithmic mean area ($A_m$) can be found in Enweremadu (2007).

From dimensional analysis,

$$h_o = \frac{K_{ins} \cdot Nu}{d_o + 2t_{ins}} \tag{9}$$

where, $t_{ins}$ is the thickness of insulation; $K_{ins}$ – thermal conductivity of the insulation materials; Nu – Nusselt number; $d_o$ – external diameter of column; $T_{amb}$ – temperature of the surrounding; $T_p$ – plate temperature.

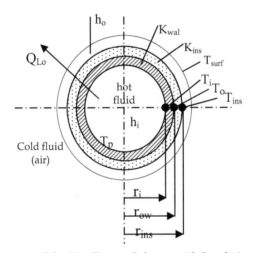

Fig. 2. Hypotethical Section of the Distillation Column with Insulation

For vertical cylinders, the commonly used correlations for free convection are adapted from Rajput (2002) as:

For laminar flow,

$$Nu = 0.59(Gr.Pr)^{1/4} \text{ for } (10^4 < Gr.Pr < 10^9) \tag{10}$$

For turbulent flow,

$$Nu = 0.10(Gr.Pr)^{1/3} \text{ for } (10^9 < Gr.Pr < 10^{12}) \tag{11}$$

where Gr is the Grashof number and Pr is Prandtl.

Based on the assumptions of neglecting $h_i$, $A_i$ and the effect of thermal resistance, equation (5) reduces to:

$$U_P = f(h_o, K_p, A_o, A_m, t_{ins}) \tag{12}$$

while equation (7) is given as

$$Q_{losses} = \frac{(T_P - T_{amb})\pi P_s N}{\dfrac{\ln(r_{ins} / r_{owall})}{K_p \cdot A_m} + \dfrac{1}{h_o A_o}} \tag{13}$$

where $U_P = \dfrac{1}{\dfrac{\ln(r_{ins}/r_{owall})}{K_p \cdot A_m} + \dfrac{1}{h_o A_o}}$

The heat loss from the column trays is given by

$$Q_{\text{loss from trays}} = \frac{(T_P - T_{amb})2\pi P_s N}{\dfrac{\ln(r_{ins}/r_o)}{K_p \cdot \dfrac{2\pi P_s . t_{ins}}{\ln\left(1 + \dfrac{2t_{ins}}{do}\right)}} + \dfrac{1}{\dfrac{K_{ins} \cdot Nu}{d_o + 2t_{ins}} \cdot (\pi d_{ins}, m.P_s)}} \tag{14}$$

The total heat loss from the column is expressed as

$$Q_{\text{loss}} = Q_{\text{loss from trays}} + \text{Heat loss from the two cylinder heads} \tag{15}$$

Based on the assumptions made, heat loss through the cylinder heads is given by

$$Q_{\text{loss at cylinder heads}} = \frac{2(T_P - T_{amb})\pi r_o^2}{\dfrac{t_{ins}}{K_p} + \dfrac{1}{h_o}} \tag{16}$$

Therefore,

$$Q_{\text{loss}} = \frac{(T_P - T_{amb})2\pi P_s N}{\dfrac{\ln(r_{ins}/r_o)}{K_p \cdot \dfrac{2\pi P_s . t_{ins}}{\ln\left(1 + \dfrac{2t_{ins}}{do}\right)}} + \dfrac{1}{\dfrac{K_{ins} \cdot Nu}{d_o + 2t_{ins}} \cdot (\pi d_{ins}, m.P_s)}} + \frac{2(T_P - T_{amb})\pi r_o^2}{\dfrac{t_{ins}}{K_p} + \dfrac{1}{h_o}} \tag{17}$$

## 4.3 Calculation of heat pump and compressor parameters

The heat pump is thermodynamically linked to the column through the heat load from the pump to the column $Q_{HPC}$ and from the column to the pump $Q_{CHP}$, and reboiler–condenser temperature. These parameters provide the basis for the heat pump calculation.

The calculation of the heat pump parameters begins with the estimation of the working fluid condensation temperature obtained from the reboiler temperature and temperature drop across the heat exchangers.

$$T_{CHP} = T_{CEV} + \Delta T_{CHP} \tag{18}$$

where $T_{CEV}$ is the column vapourization (reboiler) temperature and $\Delta T_{CHP}$, a pre-established mean temperature difference across the heat exchangers (temperature drop in reboiler-condenser). Next is the estimation of the relevant thermodynamic properties of the working fluid. These are obtained from thermodynamic correlations.

The thermodynamic properties are determined as functions of temperature. The relationships used for calculating the working fluid density, viscosity, thermal conductivity and heat capacity for input at various locations are presented in Enweremadu (2007). The condensation pressure, $P_{CHP}$ is expressed as a function of condensation temperature as

$$P_{CHP} = f(T_{CHP}) \qquad (19)$$

while the condensation pressure is determined from ideal gas equation.

The latent heat of condensation from column to heat pump is numerically exactly equal to the latent heat of vaporisation, but has the opposite sign: latent heat of vaporisation is always positive (heat is absorbed by the substance), whereas latent heat of condensation is always negative (heat is released by the substance). Latent heat of condensation is expressed as a function of condensation temperature and is determined from the relationship (Ackland, 1990):

$$h_{LV,CHP} = \Delta H_{vap} \left[ \frac{1 - \frac{T_{CHP}}{T_C}}{1 - \frac{T_{bp}}{T_C}} \right]^{\left[ A + B\left( 1 - \frac{T_{CHP}}{T_C} \right) \right]} \qquad (20)$$

where $\Delta H_{vap}$ is the heat of vapourisation at the boiling point of ethanol; $T_C$ is the critical temperature; $A$ and $B$ are constants.

The vapour specific volume at location "a" (entrance to the compressor) is expressed as a function of column condensation temperature, $T_{CC}$ and pressure at the top of the column, $P_{TOP}$:

$$v_a = \frac{T_{cc} \times R}{P_{TOP}} \qquad (21)$$

Since compression is polytropic, at location "b" (compressor discharge), the vapour specific volume is determined by:

$$V_b = v_a \left( \frac{P_{TOP}}{P_{CHP} + \Delta P} \right)^{\frac{1}{n}} \qquad (22)$$

where n is the polytropic index and $\Delta P$ is the pressure increase across the compressor.

The vapour specific enthalpy at "a" is a function of top pressure, $P_{TOP}$ and top temperature, $T_{TOP}$.

$$h_a = f(T_{TOP}, P_{TOP}) \qquad (23)$$

The vapour specific enthalpy at "b" may be determined as a function of compressor discharge temperature $T_b$ and condensation pressure $P_{CHP}$ but in this study, it is determined by the development of numerical computation with calculations utilizing the Redlich – Kwong equation of state. The Redlich – Kwong equation of state is given as

$$P = \frac{RT}{V-b} - \frac{a}{T^{\frac{1}{2}}V(V+b))}$$

(24)

Where

$$a = 0.42747 \left( \frac{R^2 T_c^{\frac{5}{2}}}{P_c} \right)$$

$$b = 0.08664 \left( \frac{RT_c}{P_c} \right)$$

and P = pressure (atm); V = molar volume (liters/g-mol); T = temperature (K); R = gas constant (atm. Liter/g-mol.K); $P_c$ = critical pressure (atm).

Taking the reference state for the enthalpy of liquid ethanol $h_L^o$, temperature, $T_o$ and the enthalpy of vaporisation $\Delta H_{vap}^o$, then the enthalpy of ethanol vapour as an ideal gas at temperature T can be calculated from

$$h_b^o = h_L^o + \Delta H_{vap}^o + \int_{T_o}^{T} C_p^o dT$$

(25)

Using the isothermal enthalpy departure and the Redlich-Kwong equation of state, the enthalpy of ethanol vapour at T and P can be calculated from

$$h_b = h_L^o + \Delta H_{vap}^o + \int_{T_o}^{T} C_p^o dT + RT \left[ Z - 1 - \frac{1.5a}{bRT^{1.5}} \ln \left( 1 + \frac{b}{V} \right) \right]$$

(26)

where Z is the compressibility factor, $C_p^o$ is the molar specific heat capacities of gases at zero pressure given as a polynomial in temperature.

Equations 24–26 are then solved with POLYMATH(R) Simultaneous Algebraic Equation Solver (See Enweremadu, 2007).

The vapour specific heat at location "e" is calculated thus (Oliveira et al, 2002):

$$h_e = h_{L,c} - C_{pl}\Delta T_{SC}$$

(27)

The specific liquid enthalpies have been assumed to be simple functions of temperature.

The liquid specific enthalpy at location "c" is determined from EZChemDB Thermodynamic Properties Table for Ethanol (AM Cola LLC, 2005) using the expression

$$h_{L,c} = f(T_{CHP})$$

(28)

while the liquid specific enthalpy at location e (at the exit of the throttling valve) is determined from

$$h_{L,e} = f(T_{CC}) \tag{29}$$

The temperature at compressor discharge is determined from the knowledge of the compressor efficiency. The ideal discharge temperature (the temperature that gives an overall change in entropy equal to zero) is calculated before correcting with the compressor efficiency.

$$T_b = T_{TOP} + \frac{T_{TOP}\left[ -1 + \left( \dfrac{P_{CHP} + \Delta P}{P_{TOP}} \right)^{0.263} \right]}{\eta_{pol}} \tag{30}$$

The dryness fraction after the isenthalpic expansion is given by

$$\beta_e = \frac{h_e - h_{L,e}}{h_{LV,e}} \tag{31}$$

where the molar latent heat of vaporization at location "e" is adapted from Ackland (1990):

$$h_{LV,e} = \Delta H_{vap_{bp}} * \left( \left(1 - T_d/T_C\right) \middle/ \left(1 - T_{bp}/T_C\right) \right)^{\wedge} A + B\left( \left(1 - T_d/T_C\right) \right) \tag{32}$$

Since this is a throttling process, $T_d = T_e$ and $h_d = h_e$

The molar vapour flow rate which is recycled in the flash tank and conveyed to the compressor is calculated by

$$V_R = \frac{V_1 \beta_e}{1 - \beta_e} \tag{33}$$

Therefore, the molar flow rate across the compressor is expressed as

$$\dot{M} = V_1 + V_R \tag{34}$$

While the dryness fraction at condenser exit is determined by

$$\beta_c = \frac{Q_{23}/\dot{M} - C_{P_L}\left(T_{CHP} - T_d\right)}{h_{LV,CHP}} \tag{35}$$

where $Q_{23}$ is the distribution of excess heat between the pre-heater $Q_2$ and the cooler $Q_3$ and $Cp_L$ is the molar specific heat of the working fluid in the liquid phase.

The energy balance, applied to the heat pump working fluid, yields the available energy for exchange at the condenser, as follows:

$$Q_{cd} = \dot{M}\left[ Cp_v\left(T_b - T_{CHP}\right) + \left(1 - \beta_c\right)h_{LV,CHP} \right] \tag{36}$$

A comparison is made between this energy available at the condenser, $Q_{cd}$, with the energy required by the column reboiler, $Q_{reb}$. This brings about the following heat load control.

i.   If the rate of energy available at the heat pump condenser, $Q_{cd}$, is greater than the rate of energy required by the reboiler $Q_{reb}$, then the condenser gives up $Q_{reb}$ to the reboiler and the remaining energy is conveyed to the preheaters ($Q_2$) and cooler ($Q_3$)

$$\text{if } Q_{cd} > Q_{reb} \text{ then } Q_{HPC} = Q_{reb} \tag{37}$$

ii.  But if $Q_{cd}$ is smaller than or equal to $Q_{reb}$, then all energy available is transferred to the reboiler and the auxiliary reboiler will provide the "extra" $Q_{reb}$ i.e.

$$\text{if } Q_{cd} \leq Q_{reb} \text{ then } Q_{HPC} = Q_{cd} \tag{38}$$

where $Q_{HPC}$ is the energy yield by the heat pump to the distillation column. The factor by which the heat pump contributes to the heat load of the reboiler is given as

$$f = \frac{Q_{HPC}}{Q_{reb}} \tag{39}$$

For a distillation column with vapour recompression, driving the compressor uses the most energy. Thus, the power consumption must be known so as to assess the feasibility of such a system. For a perfect gas, that is, a gas having a constant specific heat, Cp = Cp°, then the specific enthalpy rise between the compressor inlet and outlet is

$$\Delta h = h_b - h_a = C_p^o \left( T_b - T_a \right) \tag{40}$$

And if the change of state is isentropic,

$$\Delta h = \int_a^b v dp = \frac{\gamma}{\gamma - 1} \cdot \frac{Ru \cdot T}{M} \left[ \left( \frac{P_b}{P_a} \right)^{\frac{\gamma - 1}{\gamma}} - 1 \right] \tag{41}$$

In reality, ideal gases do not exist and therefore improvements are made on equation (41).

Therefore, compression is polytropic and the isentropic index $\gamma$, is replaced by the polytropic index, n (see Enweremadu, 2007). The compressor polytropic efficiency $\eta_{pol}$ = 0.7 - 0.8 is used.

Also, because a saturated vapour, especially at higher pressures, shows deviations from the ideal gas behaviour, the compressibility factor, Z is used. Hence equation (41) becomes

$$\Delta h_{eff} = \frac{n}{n-1} \cdot \frac{Z \cdot R_u \cdot T_a}{\eta_{pol} \cdot M} \left[ \left( \frac{P_b}{P_a} \right)^{\frac{n-1}{n}} - 1 \right] \tag{42}$$

Therefore, the power input for driving the compressor is the energy that increase the enthalpy of the gas

$$\dot{W}_{cp} = \frac{\dot{M}}{\eta_{pol}} \frac{n}{n-1} P_a \upsilon_a \left[ \left( \frac{P_b}{P_a} \right)^{\frac{n}{n-1}} - 1 \right] \tag{43}$$

Equation (43) shows that the pressure ratio $\dfrac{P_b}{P_a}$ is crucial to the power requirement. This ratio or the pressure increase to be provided by the compressor of a column with vapour recompression is influenced by the following (Meili, 1990; Han et al, 2003):

- Pressure drop in vapour ducts (pipes) and over valves and fittings, $\Delta P_p$.
- Pressure drop across the column, $\Delta P_{cl}$.
- The difference in boiling points between the top and bottom products, $\Delta P_b$.
- Temperature difference in the reboiler, $\Delta P_{CHP}$.

### 4.3.1 Determination of the pressure increase over the compressor

Pressure drops in the vapour ducts may be caused by frictional loss, $\Delta P_f$; static pressure difference, due to the density and elevation of the fluid, $\Delta P_s$; and changes in the kinetic energy, $\Delta P_k$. Since, there are elbows, valves and other fittings along the pipes then the pressure drop is calculated with resistance coefficients specifically for the elements. Therefore, the pressure drop along a circular pipe with valves and fittings is given by

$$\Delta P_P = \Delta P_s + \Delta P_f + \Delta P_K = \frac{\rho u^2}{2} \left( 1 + \frac{\lambda l_p}{d_p} + \sum \xi \right) \tag{44}$$

and u is the fluid velocity; $d_p$ is the pipe diameter and $\rho$ is the fluid density; $\lambda$ is the Fanning friction factor which is a function of Reynolds number; $l_p$ is the pipe length; $\mu$ is the dynamic viscosity of the fluid and $\xi$ is the resistance coefficient.

The pressure drop over the entire distillation column, $\Delta P_{cl}$ is caused by losses due to vapour flowing through the connecting pipes and through pressure drop over the stages in rectifying and stripping section. This depends mainly on the column internals, number of stages, gas load and operating conditions. $\Delta P_{cl} = 0$, if zero vapour boil up is assumed. But constant pressure drop is assumed in this work. The pressure drop over a stage consists of dry and wet pressure drop. The dry pressure is caused by vapour passing through the perforation of the sieve tray. The aerated liquid (static head) on the tray causes the wet pressure drop. Constant pressure drop per tray have been estimated from several authors to be equal to 5.3mmHg per tray (Muhrer, Collura and Luyben 1990). The total column pressure drop has been found by summing plate pressure drops $\Delta P_{cl}$

$$\Delta P_{cl} = 0.13332 x 5.3 x N = 0.707 N \tag{45}$$

The top and bottom products have different compositions and boiling points. For a fixed bottom temperature of the column, there is a vapour – pressure difference, $\Delta P_b$ due to the difference in boiling points.

$$\Delta P_b = P_{TOP} - P_{BOTTOM} \tag{46}$$

where,

$$P_{TOP} = 10^{\left[A_{TOP} - \frac{B_{TOP}}{T_{TOP} + C_{TOP}}\right]}$$

(47)

$$P_{BOTTOM} = 10^{\left[A_{BOTTOM} - \frac{B_{BOTTOM}}{T_{BOTTOM} + C_{BOTTOM}}\right]}$$

(48)

The temperature difference in the reboiler- condenser is expressed by means of the vapour – pressure equation as a pressure difference, $\Delta P_{CHP}$. Temperature differences of 8 – 17°C are quite common for ethanol-water distillation (Gopichand et al, 1988; Canales and Marquez, 1992). Using the Clausius – Clapeyron equation for a two- point fit,

$$\Delta P_{CHP} = e^{-\frac{\Delta Hvap}{R}\left(\frac{-\Delta T_{CHP}}{T_{CHP}.T_{CEV}}\right)}$$

(49)

Therefore the total pressure increase over the compressor becomes

$$\Delta P = \Delta P_b + \Delta P_{cl} + \Delta P_{CHP} + \Delta P_p$$

(50)

For this distillation system, the compression (pressure) ratio is

$$\frac{P_b}{P_a} = \frac{P_{CHP} + \Delta P}{P_{TOP}}$$

(51)

where $P_{TOP}$ is inlet pressure (vapour pressure at top temperature).

Other compressor parameters are calculated by the following equations:

i.    Compressor power input is determined from equation (43) and (51)

$$\dot{W}_{cp} = \frac{\dot{M}}{\eta_{pol}} \frac{n}{n-1} P_{TOP} \cdot \upsilon_a \left[\left(\frac{P_{CHP} + \Delta P}{P_{TOP}}\right)^{\frac{n-1}{n}} - 1\right]$$

(52)

ii.   Compressor heat load rate (energy balance)

$$Q_{cp} = \eta_{pol}\dot{W}_{cp} - \dot{M}\left(h_b - h_a\right)$$

(53)

iii.  Compressor volumetric efficiency

$$\eta_v = C_v\left\{1 - r\left[\left(\frac{P_{CHP} + \Delta P}{P_{TOP}}\right)^{\frac{1}{m}} - 1\right]\right\}$$

(54)

where Cv is empirical volumetric coefficient and r is the compressor clearance ratio.

iv.   Compressor nominal capacity or compressor displacement rate

$$V_{.}\omega = \frac{\dot{M}\upsilon_a}{\eta_v}$$

(55)

where $V_c$ is compressor displacement volume ($m^3$) and $\omega$ is angular velocity (rad $s^{-1}$).

### 4.3.2 Determination of the reboiler-condenser parameters

The overall heat transfer coefficient between condenser and reboiler is given by

$$(UA)_{HPC} = \frac{Q_{HPC}}{\Delta T_{CHP}} \tag{56}$$

However, a careful analysis reveals that the overall heat transfer coefficient U is an explicit function of Prandtl, Reynolds and Nusselt numbers, and depends on other properties such as viscosity and thermal conductivity. The overall heat transfer coefficient referenced to inner surface is given by

$$\frac{1}{U} = \frac{1}{hi} + (ri / ro)\frac{1}{ho} + \frac{ri\ln(ro / ri)}{2K_{wall}} \tag{57}$$

As thermal resistance of the wall is negligible, ($K_{wall}$ is large and $\ln(r_o/r_i)) \approx 0$, it is then compared with the inner tube diameter ($r_i/r_o \approx 1$)

Then

$$\frac{1}{U} = \frac{1}{hi_{ex}} + \frac{1}{ho_{ex}} \tag{58}$$

$$ho_{ex} = \frac{0.023 K_m}{do_{ex}}(\mathrm{Re}\,m)^{0.8}(\mathrm{Pr}\,m)^{0.4}(\frac{\mu_{wall}}{\mu_m})^{0.14} \tag{59}$$

where $\mu_m$ is the mean bulk fluid viscosity and $\mu_{w\,all}$ is the viscosity of the liquid at the wall.

The expression for condensation at low velocities inside tubes is adapted from (Holman, 2005).

$$hi_{ex} = 0.555\left[\frac{\rho_l(\rho_l - \rho_v)K_L^3 gh'_{fg}}{\mu_L di_{ex}(T_{CHP} - T_{wall})}\right]^{0.25} \tag{60}$$

where $h'_{fg} = h_{fg} + 0.375 C_{P_L}(T_{CHP} - T_{wall})$

where $K_L$ is thermal conductivity of the liquid, $di_{ex}$ is the inside diameter of the reboiler-condenser tubes and $\mu_L$ is the density of the condensate (liquid).

Therefore, the overall heat transfer coefficient may be determined from

$$\frac{1}{UA_{HPC}} = \frac{1}{\dfrac{0.023 K_m}{do_{ex}}\left(\dfrac{\rho_m u do_{ex}}{\mu_m}\right)^{0.8}\left(\dfrac{C_{pm}}{K_m}\right)^{0.4}\left(\dfrac{\mu_{wall}}{\mu_m}\right)^{0.14}} + \frac{1}{0.555\left[\dfrac{\rho_L(\rho_L - \rho_v)K_L^3 gh'_{fg}}{\mu_L di_{ex}(T_{CHP} - T_{wall})}\right]} \tag{61}$$

Assuming adiabatic expansion at the throttling valve, then

$$h_e = h_d = h_{L,c} - Cp_L.\Delta T_{SC} \tag{62}$$

From the condenser prescribed degree of sub-cooling, the temperature of the working fluid after cooling and before throttling is given by

$$T_d = T_{CHP} - \Delta T_{SC} \tag{63}$$

where $\Delta T_{SC}$ is the degree of sub-cooling (K)

The corresponding latent heat (enthalpy) is given as

$$h_d = h_{L,c} - C_{P_L}\Delta T_{SC} \tag{64}$$

## 4.4 Analysis of distribution of excess heat rate

The distillation system uses the column's working fluid as refrigerant and does not execute a closed cycle. Therefore the excess heat which may occur is not assessed by an overall energy balance but by the method of Oliveira *et al* (2001). When the energy available at the condenser $Q_{cd}$, is greater than the energy required by the reboiler $Q_{reb}$, the column receives the amount $Q_{cd}$ and the energy left over corresponds to the excess. But if $Q_{cd}$ is smaller than or equal to $Q_{reb}$, then all the energy available is transferred to the reboiler, i.e. there will be no excess. Thus,

$$\text{if } Q_{cd} > Q_{reb} \text{ then } Q_{23} = Q_{cd} - Q_{reb} \tag{65}$$

$$\text{if } Q_{cd} \leq Q_{reb} \quad \text{then} \quad Q_{23} = 0$$

where $Q_{23}$ is the excess heat due to energy interactions between the heat pump and the reboiler.

The distribution of the excess heat rate, $Q_{23}$, between the pre-heater ($Q_2$) and cooler ($Q_3$) is accomplished by controlling the feed condition pre-heated by $Q_2$. In other words, the value of $Q_2$ should be such that the feed reaches a prescribed condition. The pre-heating of the feed is carried out by $Q_1$ (heat exchanged between the bottom product and the feed) and $Q_2$ (heat exchange between the heat pump working fluid and the feed), in the heat exchangers. The heat provided by the bottom product is determined as follows:

$$Q_1 = B.C_{P_B}\left(T_{CEV} - T_{BE}\right) \tag{66}$$

where, $T_{CEV}$ is the column evaporation temperature. $Cp_B$ is the specific heat of the bottom product. $T_{BE}$ is the temperature at the bottom product flow exit.

The best feed condition is that of saturated liquid (Halvorsen, 2001). The energy required to pre-heat the feed to reach saturated condition is expressed as

$$Q_{FSL} = F.C_{PF}(T_{sat,F} - T_F) \tag{67}$$

where, $T_{sat,F}$ and $T_F$ are the saturation temperature of the feed source and the temperature of the feed source respectively. $Cp_F$ is the specific heat of the feed source.

To make the feed a saturated vapour, the energy required is given as

$$Q_{FSV} = Q_{FSL} + Fh_{LV,F} \tag{68}$$

It is important to verify whether $Q_1$ alone is capable of pre-heating the feed to reach the desired condition, otherwise the amount of heat that should be withdrawn from the second pre-heater $Q_2$, will be determined as

$$Q_{withdrawn} = F.C_{P_F}(T_{sat,F} - T_F) + 0.5Fh_{LV,F} \tag{69}$$

The value of heat at the second pre-heater $Q_2$ should be, at the most equal to $Q_{withdrawn}$ to prevent the feed reaching 50% dry. Therefore, a convenient heat load control could be made as follows:

$$\text{If} \quad Q_{23} > Q_{withdrawn}, \quad \text{then} \quad Q_2 = Q_{withdrawn} \quad Q_3 = Q_{23}\text{-}Q_2 \tag{70}$$

$$\text{If} \quad Q_{23} < Q_{withdrawn}, \quad \text{then} \quad Q_2 = Q_{23}, \quad Q_3 = 0$$

## 4.5 Thermodynamic analysis

Since vapour recompression uses a refrigeration cycle rather than a Carnot cycle, the performance of the heat pump is defined according to the following relation;

$$COP_h = \frac{Q_{HPC} + Q_{23}}{\dot{W}_{cp}} \tag{71}$$

The thermodynamic efficiency of a separation process is the ratio of the minimum amount of thermodynamic work required for separation to the minimum energy required for the separation (Olujic *et al*, 2003). For a vapour recompression distillation column, the energy required for separation process is composed of the reboiler heat load, $Q_{reb}$, and the compressor power input, $\dot{W}_{cp}$

$$Q_T = Q_{reb} + \dot{W}_{cp} \tag{72}$$

For the separation of a binary mixture by distillation the minimum thermodynamic energy required to achieve complete separation is given by (Liu and Quian, 2000):

$$W_{min} = -RT_{TOP}\left(X_F \ln(X_F) + (1 - X_F)\ln(1 - X_F)\right) \tag{73}$$

Then the thermodynamic efficiency is expressed as:

$$\eta_{VRC} = \frac{W_{min}}{Q_T} \tag{74}$$

## 4.6 Solution method and error analysis

The equations that model the system components were grouped together in one single system. The analyses of the status of the variables were carried out to identify those that were the input data and those which were the unknowns. The equations were then grouped together, resulting in a set of non-linear algebraic equations, which were solved iteratively based on the step by step use of the successive substitution method. Solution was obtained when convergence was attained. The convergence was checked by using the criterion:

$$\varepsilon_a = \frac{X_{i+1} - X_i}{X_{i+1}} \times 100\% \tag{75}$$

The model is coded in MATLAB environment and used to evaluate the unknowns. A control programme for column VRC was written to compare the actual column (column in which the parameters studied were considered, $VRC_{\Delta P}$).

## 5. Discussion of results

### 5.1 Effects of pressure increase over the compressor

Figure 3 shows how the compressor power input, $W_{cp}$ varies with the pressure increase over the compressor, $\Delta P$. It is obvious from the plots that an increase in pressure over the compressor increases the compression (pressure) ratio leading to increase in compressor power input. The curve in Figure 3 shows the effect of pressure increase over the compressor on coefficient of performance. As the pressure increase over the compressor, $\Delta P$ increases, the compression ratio increases and the coefficient of performance, COP decreases due to increase in compressor power input.

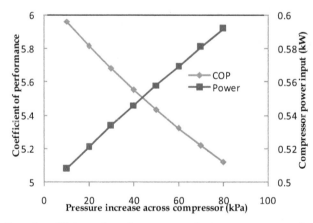

Fig. 3. The variation of coefficient of performance and compressor power input with pressure increase across compressor

In Figure 4, a negative non-linear relationship exists between the compressor volumetric efficiency and pressure increase over the compressor. For a given compressor nominal capacity, when the pressure increase across the compressor, $\Delta P$ increases, the pressure ratio also increases resulting in the reduction in volumetric efficiency.

The influence of pressure increase over the compressor, $\Delta P$ on the required compressor displacement rate ($Vc\omega$) is shown in Figure 4. The linear relationship that exists between the

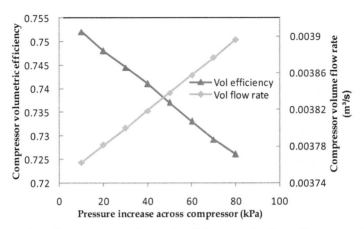

Fig. 4. The variation of compressor volumetric efficiency and volume flow rate with pressure increase across compressor

variables indicates that the compressor displacement rate is directly proportional to $\Delta P$. A reduction in compressor volumetric efficiency caused by increase in $\Delta P$, increases the compressor displacement rate. This implies an increase in the displacement volume required although this cannot be observed in the figures. The compressor displacement required for a given speed is related to compressor size. For the large specific volume obtained, $9.74 m^3$ $kg^{-1}$ ethanol as the heat pump working fluid will require a compressor of greater capacity i.e large displacement rate.

It can be observed from Figure 5 that an increase in pressure increase across the compressor, $\Delta P$ increases the total energy consumption.

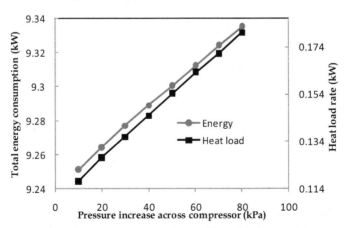

Fig. 5. The variation of total energy consumption and heat load rate with pressure increase across compressor

For a given reboiler heat transfer rate, $Q_{reb}$, it is obvious that as $\Delta P$ increases, the total energy consumption increases. The total energy consumption and hence the energy savings from the work of Oliveira *et al* (2001) in heat pump distillation gave lower values. This could be attributed to non-consideration of the effect of pressure increase over the compressor and the subsequent increase in compression ratio and in compression power input respectively.

## 5.2 Effects of column heat loss

The direct effect of column heat loss could be seen from equation (17). From the equation, it follows that, for a given reboiler heat load or heat expenditure, fewer trays are required for a given separation if heat losses are reduced. Where heat loss occurs, more vapour has to be produced in the reboiler, since the reboiler must provide not only the heat removed in the condenser but also the heat loss. The effect of this is a decrease in process and energy efficiency. Indirectly, heat loss affects the column size in terms of number of plates. In the control system, the reflux ratio was as high as 7.5 compared with 5.033 obtained for the actual system. Therefore, if heat losses are properly accounted for, there may not be any need for downward review of the number of plates in order to reduce the reflux ratio (Enweremadu and Rutto, 2010). Therefore, pressure drop across the column, $\Delta P_{cl}$ and the difference in boiling points between the top and bottom products, $\Delta P_b$ which have the most profound effects on the pressure increase across the compressor, $\Delta P$ will be properly predicted. The overall implication of this is that the column size would be determined properly.

## 5.2.1 Overall heat transfer coefficient

Analysis of the overall heat transfer coefficient, U of the heat pump reboiler-condenser revealed an increase in the value of U. This was expected as the value of U in boiling and condensation processes are high. Also, the value of the heat transfer coefficient of the condensing ethanol is dominated the relationship used in determining U. A low value of overall heat transfer coefficient U will result in an increase in the heat exchanger surface area which may be a disadvantage to ethanol-water system. But the results from this work showed an increase in the value of U with the implication of a reduction in the reboiler-condenser heat transfer area.

The variation of the reboiler-condenser thermal conductance (UA) with the heat pump distribution factor, f, is shown in Figure 6. The plots show that the greater the heat load taken by the heat pump i.e. larger f's, the larger the thermal conductance, UA, and the larger the heat exchanger area. However, for better performance of any heat transfer system, the thermal resistance ($R_{th}$) which is the inverse of thermal conductance ($R_{th} = 1/UA$), should be as low as possible. Therefore the value of the heat transfer area for the $VRC_{\Delta P}$ system will be smaller compared to the VRC system. Hence, the reboiler-condenser studied has a better performance.

The relationship between the thermal conductance, UA and the reboiler-condenser temperature difference, $\Delta T_{CHP}$ shows that the higher the reboiler-condenser temperature difference, the lower the thermal conductance. The implication of this is that a higher $\Delta T_{CHP}$ causes a reduction of the necessary heat transfer area. However, beyond a certain limit of the thermal driving force, the heat transfer area and the performance of the heat pump reboiler-condenser reduces. This is expected as higher $\Delta T_{CHP}$ leads to higher compression ratio, higher compressor power input and higher energy consumption.

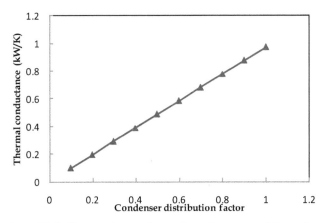

Fig. 6. Variation of Reboiler-condenser thermal conductance with condenser distribution factor

### 5.3 Comparison of the vapour recompression distillation systems

Table 1 summarises the comparison of some parameters of the two vapour recompression columns studied, the control (VRC) and the actual (VRC$_{\Delta P}$). Both systems operate at 101.2 kPa and the effect of pressure drop effect is considered to account properly for variations in heat duty. A pressure drop of 0.707kPa per tray is assumed here as a reasonable estimate for the purposes of this study. The pressure ratio as used throughout this work is the ratio of the condensation pressure, $P_{CHP}$ to the top pressure, $P_{TOP}$, for the VRC system and  sum of the condensation pressure, $P_{CHP}$ and pressure increase over the compressor, $\Delta P$, to the top pressure, $P_{TOP}$ for the VRC$_{\Delta P}$ system. Since the energy consumption changes linearly with the feed flow rate, and as the present work makes a comparison of the relative performances, the base case flow rate was taken to be $1.098 \times 10^{-4}$kmol s$^{-1}$ · Also the column heat loss and the effect of such parameters as pressure increase across the compressor, the overall heat transfer coefficient of reboiler–condenser as an explicit function of Prandtl, Reynolds and Nusselt numbers which in turn depend on fluid properties are considered to account properly for variations in heat duty (Enweremadu, 2007; Enweremadu et al, 2009).

Table 1 shows that the VRC enables some energy savings when compared with VRC$_{\Delta P}$. The VRC has a slightly lower compression ratio and consumes less energy than VRC$_{\Delta P}$ system. Although there was a marginal increase in $\Delta P$, which increased the compressor power input slightly, the performances of the two systems differ greatly by 27.5%. However, in addition to the pressure increase, the energy consumption appears to depend more on the rate of heat transfer in the reboiler. Hence the total energy consumption is indirectly related to column heat loss and pressure increase across the compressor through reboiler heat transfer and compressor power input respectively.

Also from Table 1, the heat transfer duties of the reboiler-condenser in terms of the overall heat transfer coefficient for the two systems show that the VRC$_{\Delta P}$ has a higher value when compared with the VRC system. This implies that the VRC$_{\Delta P}$ will require a smaller heat

| Parameter | Actual column | Control column |
|---|---|---|
| Compressor power input (kW) | 0.518 | 0.495 |
| Total energy consumption (KW) | 9.26 | 7.26 |
| Rate of column heat loss (kW): $$Q_{loss} = \frac{(T_p - T_{amb})2\pi P_s N}{\dfrac{\ln(r_{ins}/r_o)}{K_P \cdot \dfrac{2\pi P_s \cdot t_{ins}}{\ln\left(1 + \dfrac{2t_{ins}}{d_o}\right)}} + \dfrac{1}{\dfrac{K_{ins} \cdot Nu}{d_o + 2t_{ins}} \cdot (\pi d_{ins}, m.P_s)}} + \frac{2(T_P - T_{amb})\pi r_o^2}{\dfrac{t_{ins}}{K_P} + \dfrac{1}{h_o}}$$ | 2.6 | 0.7 |
| Overall heat transfer coefficient (kW/m²K): $$\frac{1}{UA_{HPC}} = \frac{1}{\dfrac{0.023 K_m}{d_{o_{ex}}}\left(\dfrac{\rho_m u d_{o_{ex}}}{\mu_m}\right)^{0.8}\left(\dfrac{C_{pm}}{K_m}\right)^{0.4}\left(\dfrac{\mu_{wall}}{\mu_m}\right)^{0.14}} + \frac{1}{0.555\left[\dfrac{\rho_L(\rho_L - \rho_v)K_L^3 gh'}{\mu_L di_{ex}(T_{CHP} - T_{wall})}\right]}$$ | 1.4 | 0.3 (U=Q/A ΔT) |
| Reboiler heat transfer rate (kW): $Q_{reb} = Dh_D + Bh_B + I_{.1}h_{LV,c} + Q_{losses} - Fh_\Gamma - Q_1 - Q_2$ | 8.7 | 6.8 (10% of reboiler heat rate) |
| Coefficient of performance: $$COP = \frac{Q_{HPC} + Q_{23}}{\dot{W}_{cp}}$$ | 5.85 | 6.15 |
| Condenser heat distribution factor: $$f = \frac{Q_{HPC}}{Q_{reb}}$$ | 0.3 | 0.5 |
| Compression ratio | 1.22 | 1.12 |
| Compressor displacement rate (m³/s) | 3.88x10⁻³ | 3.74x10⁻³ |
| Compressor heat load rate (kW) | 0.13 | 0.12 |
| Vapour specific volume after compression (m³/kg) | 9.49 | 9.93 |
| Temperature at compressor discharge (K) | 395 | 390 |
| Compressor volumetric efficiency | 0.749 | 0.756 |
| Thermodynamic efficiency | 15.8 | 20.2 |
| Reflux ratio | 5.033 | 7.5 |

Table 1. Model results of the actual column and control column (Enweremadu, 2007; Enweremadu et al, 2008 & 2009)

transfer area which is economical in terms of material conservation. Also, with higher U, the $VRC_{\Delta P}$ will have a better performance.

The column heat losses for the two vapour recompression distillation columns are shown in Table 1. The heat losses in distillation columns with heat pumps have been assumed to correspond to around 3% of the energy supplied to the reboiler (Danziger, 1979). Oliveira, Marques and Parise (2002) assumed it to be as high as 10%. However, in this study, the heat exchanged by the distillation column with the surroundings is considered and its effect included in the balance equation (1). The results obtained for the $VRC_{\Delta P}$ showed a marked difference between the two systems.

A comparison of the main reboiler heat transfer rate for the two systems is presented in Table 1. It is evident that neglecting and /or assuming a value for column heat loss instead of determining it had a significant effect on the values of $Q_{reb}$ in both systems. The heat pump distribution factor, f, for the $VRC_{\Delta P}$ system is 0.346 which is slightly less than that for the VRC system (0.451). Since low value of heat load taken by the heat pump, f implies lower thermal conductance, then the value of the heat transfer area for the $VRC_{\Delta P}$ system is smaller when compared to the VRC system. However, lower value of thermal conductance, UA, for the $VRC_{\Delta P}$ system indicates that the VRC system will have a better performance.

The simulation results also show that the coefficient of performance and the thermodynamic efficiency of the $VRC_{\Delta P}$ system is lower when compared to the VRC system. Fonyo and Benko (1998) have shown that the electrically-driven compression heat pump should work with a COP not lower than 3-5. The value of 5.85 obtained from this study has shown that although there is a decrease in the effectiveness of the $VRC_{\Delta P}$ system when compared with the VRC system with COP of 6.15, it is within the acceptable range. This may be due to the fact that the compressor power input and the total energy consumption in the $VRC_{\Delta P}$ system is higher than in the VRC system.

## 6. Conclusions

From the outcome of the study, the following conclusions may be drawn:

1. Pressure increase across the compressor, $\Delta P$ increases the compression ratio, the compressor power input, temperature in the heat pump reboiler-condenser while the compressor volumetric efficiency decreases. The effect of these is the reduction in the heat pump coefficient of performance and the use of a compressor of greater capacity.
2. Neglecting the effects of pressure increase across the compressor, $\Delta P$ reduces the compression ratio and hence maximizes the energy efficiency. However, this leads to a substantial decrease in temperature in the heat pump reboiler-condenser. The overall effect of this is a decrease in the overall heat transfer coefficient, U resulting in an increase in heat transfer area.
3. From the comparison between the $VRC_{\Delta P}$ and VRC systems, there was a profound difference in the overall heat transfer coefficient while the column heat loss was substantial.
   The increase in the total energy consumption, reboiler heat transfer rate and the thermodynamic efficiency were appreciable, while there was only a marginal increase

in the compressor power input resulting in a difference of 5.12% in the coefficient of performance between the $VRC_{AP}$ and VRC systems.

4.  Calculation of the column heat loss in contrast to the assumed value of certain percentage of the reboiler heat transfer rate gives a higher value resulting in higher energy consumption and lower thermodynamic efficiency.

## 7. References

Ackland, T. (1990). *Physical Properties of Ethanol-Water Binary Mix.* 20.07.2006, Available from http://www.homedistillers.org

AM Cola LLC (2005). EZChemDB Thermodynamic Properties Table for Ethanol

Braisher, M.; Gill, S.; Treharne, W.; Wallace, M. & Winterburn, J. (2006). Design Proposal: Bio-ethanol Production Plant. 15.02.2007, Available from http://www.ethanol.org

Brousse, E.; Claudel, B. & Jallut, C. (1985). Modeling and Optimisation of the Steady Operation of a Vapour Recompression Distillation Column. *Chemical Engineering Science*, Vol. 40, No. 11, pp, 2073–2078, ISSN 0009-2509

Canales, E. & Marquez, F. (1992). Operation and Experimental Results on a Vapor Recompression Pilot Plant Distillation Column. *Industrial & Engineering Chemistry Research*, Vol. 31, pp. 2547 -2555, ISSN 0888-5885

Collura, M. & Luyben, W. (1988). Energy – Saving Distillation Designs in Ethanol Production. *Industrial & Engineering Chemistry Research,* Vol. 27, pp. 1686–1696, ISSN 0888-5885

Danziger, R. (1979). Distillation Column with Vapor Recompression. *Chemical Engineering Progress* Vol. 75, No. 9, pp. 58-64, ISSN 0360-7275

Emtir, M.; Miszey, P.; Rev, E. & Fonyo, Z. (2003). Economic and Controllability Investigation and Comparison of Energy-Integrated Distillation Schemes. *Chemical and Biochemical Engineering Quarterly.* Vol. 17, No.1, pp. 31-42, ISSN 0352-9568

Enweremadu, C.(2007). Simulation of an ethanol–water distillation column with direct vapour recompression heat pump. PhD Thesis. Ladoke Akintola University of Technology, Ogbomoso, Nigeria

Enweremadu, C.; Waheed, M. & Ojediran, J. (2008). Parametric study of pressure increase across a compressor in ethanol-water vapour recompression distillation, *Scientific Research and Essay*, Vol. 3, No.9, pp. 231-241, ISSN 1992-2224

Enweremadu, C.; Waheed, M. & Ojediran, J. (2009). Parametric study of an ethanol-water distillation column with direct vapour recompression heat pump, *Energy for Sustainable Development.* Vol. 13, No. 2, pp. 96-105, ISSN 0973-0826

Enweremadu, C. & Rutto, H. (2010). Investigation of heat loss in ethanol-water distillation column with direct vapour recompression heat pump, *Proceedings of World Academy of Science, Engineering & Technology 2010, International Conference on Thermal Engineering,* pp. 69-76, ISSN 1307-6892, Amsterdam, The Netherlands, 28 – 30 September, 2010

Ferre, J.; Castells, F. & Flores, J. (1985). Optimization of a Distillation Column with a Direct Vapor Recompression Heat Pump. *Industrial and Engineering Chemistry Process Design and Development*, Vol. 24, pp. 128-132, ISSN 0196-4305

Fonyo, Z. & Benko, N. (1998). Comparison of various heat pump-assisted distillation configurations. *Transactions of the Institution of Chemical Engineers*. Vol. 76, Part A, pp. 348-360, ISSN 0046-9858

Gani, R.; Ruiz, C. & Cameron, I. (1986). A Generalised Model for Distillation Column I: Model Description and Applications. *Computers & Chemical Engineering*, Vol.10, No.3, pp.181-198, ISSN 0098-1354

Gopichand, S.; Devotta, S.; Diggory, P. & Holland, F. (1988). Heat Pump-Assisted Distillation VIII: Design of a System for Separating Ethanol and Water. *International Journal of Energy Research*, Vol.12, No. 1, pp.1-10, ISSN 0363-907X

Halvorsen, I. (2001). Minimum Energy Requirements in Complex Distillation Arrangements. Dr.Ing.Thesis, Department of Chemical Engineering, Norwegian University of Science and Technology. 20.10.2004, Available from http://www.chembio.ntnu.no/users/skoge/publications

Han, M.; Lin, H.; Yuan, Y.; Wang, D. & Jin, Y. (2003). Pressure Drop for Two-phase Counter-current Flow in a Packed Column with a Novel Internal. *Chemical Engineering Journal*. Vol. 94, pp. 179-184, ISSN 1385-8947

Hewitt, G.; Quarini, J. & Morell, M. (1999). More Efficient Distillation. *The Chemical Engineer*, 21st October 1999

Holman, J. (2005). *Heat Transfer*, 9th Edition. McGraw-Hill Book Co., ISBN 0-07-029620-0, New York

Jorapur, R. & Rajvanshi, A. (1991). Alcohol distillation by solar energy, ISES Solar World Congress Proceedings, Vol. I, Part II, pp. 772-777, Denver, Colorado, USA, August 19-21, 1991

Liu, X. & Quian, J. (2001). Modelling, Control and Optimization of Ideal Internal Thermally Coupled Distillation Columns. *Chemical Engineering & Technology*. Vol. 23, No. 3, pp. 235– 241, ISSN 1521-4125

Meili, A .(1990). Heat Pumps for Distillation Columns. *Chemical Engineering Progress*, pp. 60–65, June 1990, ISSN 0360-7275

Mori, H.; Ito, C.; Taguchi, K. & Aragaki, T. (2002). Simplified Heat and Mass Transfer Model for Distillation Column Simulation. *Journal of Chemical Engineering of Japan*. Vol. 35, No.1, pp.100 - 106, ISSN 0021-9592

Muhrer, C.; Collura, M. & Luyben W. (1990). Control of Vapor Recompression Distillation Columns. *Industrial Engineering and Chemistry Research.*, Vol. 29, No. 1, pp. 59-71, ISSN 0888-5885

Null, H. (1976). Heat Pumps in Distillation. *Chemical Engineering Progress*, pp. 58 –64, July 1976, ISSN 0360-7275

Oliveira S.; Marques R. & Parise, J. (2001). Modelling of an Ethanol – Water Distillation Column with Vapor Recompression. *International Journal of Energy Research*. Vol. 25, No. 10, pp. 845–858, ISSN 0363-907X

Oliveira, S.; Marques R. & Parise, J. (2002). Modelling of an Ethanol – Water Distillation Column Assisted by an External Heat Pump. *International Journal of Energy Research*. Vol. 26, pp, 1055 – 1072, ISSN 0363-907X

Olujic, Z.; Fakhri, F.; de Rijke, A.; de Graauw, J. & Jansen, P. (2003). Internal Heat Integration – the key to an energy-conserving distillation column. *Journal of Chemical Technology and Biotechnology*. Vol. 78, pp, 241 – 248, ISSN1097-4660

Pfeffer, M.; Wukovits, W.; Beckmann, G. & Friedl, A. (2007). Analysis and decrease of the energy demand of bioethanol production by process integration. *Applied Thermal Engineering*. Vol.27, No. , pp. 2657-2664, ISSN 1359-4311

Rajput, R. (2002). *Heat and Mass Transfer*. S.Chand and Co. Ltd., New Delhi, ISBN 81-219-1777-8

Sloley, A. (2001). Energy Conservation Seminars for Industry: Texas Energy Conservation. Distillation Column Operations Manual. 16.11.2005, Available from http://www.distillationgroup.com

**4**

# Batch Distillation: Thermodynamic Efficiency

José C. Zavala-Loría* and Asteria Narváez-García
*Universidad Autónoma del Carmen (UNACAR),*
*DES-DAIT: Facultad de Química, Cd. del Carmen, Cam.*
*México*

## 1. Introduction

The batch distillation is a separation processes that requires great amounts of energy. Due to high energy costs, the study of energy consumption is of great interest in this process (Zavala et al. 2007). According to Luyben (1990), the energy consumption increases when operation occurs in a batch manner. Determining how efficient is the heat transfer under specific conditions and to modify them in order to find how efficient the heat is used, is an important task.

The analysis of thermodynamic efficiency in a batch distillation column has been presented by Kim and Diwekar (2000), Zavala-Loría (2004), Zavala et al. (2007), Zavala & Coronado (2008) and Zavala et al. (2011). The first work only developed expressions to calculate thermodynamic efficiency while the rest of the works developed expressions to calculate thermodynamic efficiency and have applied them to a new problem of optimal control: Maximum thermodynamic efficiency.

## 2. Description of the process

For this study we used a conventional batch distillation column consisting of:

- Reboiler
- Tray column
- Condenser
- Reflux drum
- Receiver

Figure 1 Shows a conventional batch column like the one used for this study.

Mass balances resulting from the process shown in Figure 1, allow us to obtain the mathematical model in Table 1. The model considers the following elements:

- Theoretical stages
- Total condenser
- Neglible pressure drop

---

* Corresponding Author

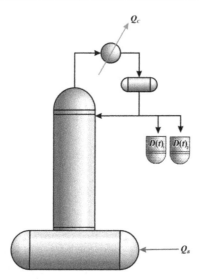

Fig. 1. Conventional batch distillation column.

- Constant flows:
  - Vapor
  - Liquid.
- Constant liquid holdup:
  - Stages
  - Condenser-Reflux drum.
- Adiabatic column

$$\frac{dB}{dt} = -D_t = -\frac{V}{R_t + 1} \tag{1}$$

$$\frac{dx_B^{(i)}}{dt} = \left(\frac{V}{B}\right)\left\{x_B^{(i)} - y_B^{(i)} + \left(\frac{L}{V}\right)\left[x_1^{(i)} - x_B^{(i)}\right]\right\} \quad ; \; i = 1,\ldots,n \tag{2}$$

$$\frac{dx_j^{(i)}}{dt} = \left(\frac{V}{H_j}\right)\left\{y_{j-1}^{(i)} - y_j^{(i)} + \left(\frac{L}{V}\right)\left[x_{j+1}^{(i)} - x_j^{(i)}\right]\right\} \quad ; \; i = 1,\ldots,n \; ; \; j = 1,\ldots,N \tag{3}$$

$$\frac{dx_D^{(i)}}{dt} = \left(\frac{V}{H_D}\right)\left[y_n^{(i)} - x_D^{(i)}\right] \quad ; \; i = 1,\ldots,n \tag{4}$$

$$\sum_{i=1}^{n} y_j^{(i)} = \sum_{i=1}^{n} K_j^{(i)} x_j^{(i)} = 1 \quad ; i = 1,\ldots,n \; ; \; j = 1,\ldots,N \tag{5}$$

Table 1. Mathematical model of a batch distillation column ($n$=number of components and $N$=number of equilibrium stages).

Considering ideal mixtures, the vapor-liquid equilibrium (VLE) can be obtained from Raoult's law equation:

$$K^{(i)} = \frac{y^{(i)}}{x^{(i)}} = \frac{P_i^{sat}}{P} \; ; i = 1, 2,...,n \tag{6}$$

For non-ideal mixtures, we can use the following equation:

$$K^{(i)} = \frac{y^{(i)}}{x^{(i)}} = \frac{\gamma_i P_i^{sat}}{\hat{\phi}_i P} \; ; i = 1, 2,...,n \tag{7}$$

Where $K^{(i)}$ is the constant of the VLE, and $y^{(i)}$ is the mole fraction of the vapor phase, $x^{(i)}$ is the mole fraction of the liquid phase, $\gamma_i$ represents the activity coefficient, $P^{sat}$ is the saturation pressure and $\hat{\phi}_i$ denotes the partial fugacity coefficient, all referring to component $i$, and $P$ represents the system pressure. Activity coefficients can be obtained with a solution model (Wilson, NRTL, etc.) or the Chao-Seader correlation. The fugacity coefficients can be obtained with an equation of state (Redlich-Kwong, Soave, Peng-Robinson, etc.)

At standard or low pressures $P_i^{sat}$ can be obtained with Antoine's equation.

$$\ln\left[P_i^{sat}\right] = A_i - \frac{B_i}{T + C_i} \; ; i = 1, 2,...,n \tag{8}$$

Coefficients $A_i$, $B_i$ and $C_i$ appear in literature related to this area of study.

## 3. Thermodynamic efficiency

By definition, the thermodynamic efficiency ($\eta_t$) of a process based on availability or exergy is defined as:

$$\eta_t = \left(\frac{W_{min}}{W_{total}}\right)_{sep} = \left(\frac{W_{min}}{W_{min} + LW}\right)_{sep} \tag{9}$$

where $W$ is the work and $LW$ is the total loss of work.

Since the minimum work is determined by changes in exergy, they can be determined using the First and Second Law of Thermodynamics. Figure 2 shows the control volume used to obtain the equations that represent thermodynamic efficiency in batch distillation.

Figure 2 shows that the process can exchange energy with the environment but does not perform any mechanical work. The energy balance (enthalpy) given by the First Law of Thermodynamics, considering the reboiler, the column of trays and the condenser-reflux drum, is:

$$\frac{d(ml)_{sist}}{dt} = \frac{d(BI_B)}{dt} + \sum_{i=1}^{n} \frac{d(H_t I_t)_i}{dt} + \frac{d(H_D I_{rt})}{dt} \tag{10}$$

where:

$$\frac{d\left(mI\right)_{sist}}{dt} = Q_0 + Q_B - Q_C - DI_D \tag{11}$$

According to the Second Law of Thermodynamics, the entropy balance is:

$$\frac{d\left(mS\right)_{sist}}{dt} = \frac{d\left(BS_B\right)}{dt} + \sum_{i=1}^{n}\frac{d\left(H_t S_t\right)_i}{dt} + \frac{d\left(H_D S_{rt}\right)}{dt} \tag{12}$$

where:

$$\frac{d\left(mS\right)_{sist}}{dt} = \frac{Q_0}{T_0} + \frac{Q_B}{T_B} - \frac{Q_C}{T_C} - DS_D + d\left(S_{irr}\right) \tag{13}$$

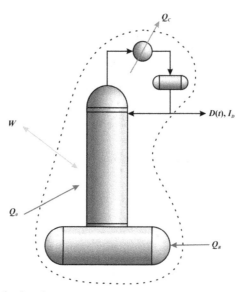

Fig. 2. Control volume for batch process.

$I$ represents enthalpy, $S$ is entropy, $Q$ is the amount of heat, $C$ represents the condenser, $D$ represents the distillate, $B$ represents the bottoms, $m$ is the system mass, $0$ is the reference state, $t$ is the stage or tray and $T$ is the temperature. The available work can be obtained if we combine equations (11) and (13). To do so, equation (13) must be multiplied by the temperature of the reference state $T_0$ (it is considered that the reference state is liquid at 25 °C and 1.0 atm):

$$\frac{d\left(mT_0 S\right)_{sist}}{dt} = Q_0 + \frac{T_0}{T_B}Q_B - \frac{T_0}{T_C}Q_C - DT_0 S_D + T_0 d\left(S_{irr}\right)$$

$$\frac{d\left(mI\right)_{sist}}{dt} - \frac{d\left(mT_0 S\right)_{sist}}{dt} = Q_0 + Q_B - Q_C - DI_D - Q_0 - \frac{T_0}{T_B}Q_B + \frac{T_0}{T_C}Q_C + DT_0 S_D - T_0 d\left(S_{irr}\right)$$

$$\frac{d\left[m\left(I-T_0S\right)\right]}{dt}=\left(1-\frac{T_0}{T_B}\right)Q_B-\left(1-\frac{T_0}{T_C}\right)Q_C-D\left(I_D-T_0S_D\right)-T_0d\left(S_{irr}\right) \tag{14}$$

the availability is represented as $A = I - T_0S$, then:

$$\frac{d\left(mA\right)_{sist}}{dt}=\left(1-\frac{T_0}{T_B}\right)Q_B-\left(1-\frac{T_0}{T_C}\right)Q_C-DA_D-T_0d\left(S_{irr}\right) \tag{15}$$

$$\frac{d\left(mA\right)_{sist}}{dt}=\left(1-\frac{T_0}{T_B}\right)Q_B-\left(1-\frac{T_0}{T_C}\right)Q_C-DA_D-LW \tag{16}$$

Where the loss of work is defined as $LW = T_0dS_{irr}$. The terms of the right hand side of Equation (16) represent the difference between the exergy or the availability of flows entering and leaving the system. The term on the left hand side is the change of exergy in the system. Finding the values in Equation (16), the work loss is:

$$LW=\left(1-\frac{T_0}{T_B}\right)Q_B-\left(1-\frac{T_0}{T_C}\right)Q_C-DA_D-\frac{d\left(mA\right)_{sist}}{dt} \tag{17}$$

$\dfrac{d\left(mA\right)_{sist}}{dt}$ can be calculated, if discretised, by multiplying Equation (12) by $T_0$ minus Equation (10):

$$\frac{\Delta\left(mA\right)_{sist}}{\Delta t}=\frac{\Delta\left(BA_B\right)+\sum\limits_{i=1}^{n}\Delta\left(H_tA_t\right)_i+\Delta\left(H_DA_{rt}\right)}{\Delta t} \tag{18}$$

Another way to estimate the value of $\dfrac{d\left(mA\right)_{sist}}{dt}$ is by applying an exergy balance at the bottom of the column. Applying an exergy balance in the reboiler, we have:

$$\frac{d\left(mA\right)_{sist}}{dt}=\frac{d\left(BA_B\right)}{dt}=B\frac{d\left(A_B\right)}{dt}+A_B\frac{d\left(B\right)}{dt} \tag{19}$$

Introducing Equation (1) from Table (1) in Equation (19), we obtain:

$$\frac{d\left(BA_B\right)}{dt}=B\frac{d\left(A_B\right)}{dt}-DA_B \tag{20}$$

According to Kim and Diwekar (2000) and Zavala-Loría and Coronado-Velasco (2008), the total exergy can be calculated from its physical component ($A_{phis}$) and its chemical component ($A_{chem}$), i.e.:

$$A = A_{phis} + A_{chem} \tag{21}$$

where $A_{phis}$ considers the physical processes that involve thermal interactions with the surroundings and $A_{chem}$ considers the mass and heat transfer with the surroundings. The main contribution to this energy is due to mixing effects and can be estimated from the chemical potential at low pressures (Kim and Diwekar, 2000). $A_{phis}$ is relatively lower than $A_{chem}$. Thus, the physical component can be regarded as constant for all chemical species and the derivative of this term is eliminated. The chemical component of exergy for an ideal mixture can be expressed as:

$$A_{chem} = RT_0 \sum_i x^{(i)} \ln\left[x^{(i)}\right] \tag{22}$$

whereas a non-ideal mixture can be calculated as:

$$A_{chem} = RT_0 \sum_i x^{(i)} \ln\left[\gamma_i x^{(i)}\right] \tag{23}$$

and the exergy exchange in the reboiler for an ideal mixture can be calculated as:

$$A_B = A_{B_{phis}} + A_{B_{chem}} = A_{B_{phis}} + RT_0 \sum_i x_B^{(i)} \ln\left[x_B^{(i)}\right] \tag{24}$$

for a non-ideal mixture:

$$A_B = A_{B_{phis}} + A_{B_{chem}} = A_{B_{phis}} + RT_0 \sum_i x_B^{(i)} \ln\left[\gamma_{B,i} x_B^{(i)}\right] \tag{25}$$

Taking the derivative of Equations (24) and (25) yields:

$$\frac{d(A_B)}{dt} = \frac{d(A_{B_{chem}})}{dt} = RT_0 \frac{d\left\{\sum_i x_B^{(i)} \ln\left[x_B^{(i)}\right]\right\}}{dt} \tag{26}$$

$$\frac{d(A_B)}{dt} = \frac{d(A_{B_{chem}})}{dt} = RT_0 \frac{d\left\{\sum_i x_B^{(i)} \ln\left[\gamma_{B,i} x_B^{(i)}\right]\right\}}{dt} \tag{27}$$

The derivative of the term on the right hand side of Equations (26) and (27) can be represented in terms of the mathematical model of the column; thus, substituting Equation (2) of Table 1 yields:

$$\frac{d\left\{\sum_i x_B^{(i)} \ln\left[x_B^{(i)}\right]\right\}}{dt} = \sum_i \left\{\left[1 + \ln x_B^{(i)}\right] \frac{dx_B^{(i)}}{dt}\right\}$$

$$= \left(\frac{V}{B}\right) \sum_i \left\{x_B^{(i)} - y_B^{(i)} + \left(\frac{L}{V}\right)\left[x_1^{(i)} - x_B^{(i)}\right]\right\} \tag{28}$$

$$
\frac{d\left\{\sum_i x_B^{(i)} \ln\left[\gamma_{i,B} x_B^{(i)}\right]\right\}}{dt} = \sum_i \left\{\left[1 + \ln \gamma_{i,B} x_B^{(i)}\right]\frac{dx_B^{(i)}}{dt} + \frac{x_B^{(i)}}{\gamma_{i,B}}\frac{d\gamma_{i,B}}{dt}\right\}
$$

$$
= \left(\frac{V}{B}\right)\sum_i \left\{x_B^{(i)} - y_B^{(i)} + \left(\frac{L}{V}\right)\left[x_1^{(i)} - x_B^{(i)}\right]\right\}\left[1 + \ln \gamma_{i,B} x_B^{(i)}\right] \tag{29}
$$

$$
+ \sum_i \left[\frac{x_B^{(i)}}{\gamma_{i,B}}\frac{d\gamma_{i,B}}{dt}\right]
$$

Substituting Equation (28) on the right hand side of Equation (20) for an ideal mixture, and Equation (29) on the right hand side of Equation (20) for a non-ideal mixture yields:

**Ideal Mixture:**

$$
\frac{d(B\mathcal{A}_B)}{dt} = VRT_0 \sum_i \left\{x_B^{(i)} - y_B^{(i)} + \left(\frac{L}{V}\right)\left[x_1^{(i)} - x_B^{(i)}\right]\right\}\left\{1 + \ln\left[x_B^{(i)}\right]\right\}
$$

$$
- D\left\{\mathcal{A}_{B,phis} + RT_0 \sum_i x_B^{(i)} \ln\left[x_B^{(i)}\right]\right\} \tag{30}
$$

**Non-ideal Mixture:**

$$
\frac{d(B\mathcal{A}_B)}{dt} = VRT_0 \sum_i \left\{x_B^{(i)} - y_B^{(i)} + \left(\frac{L}{V}\right)\left[x_1^{(i)} - x_B^{(i)}\right]\right\}\left[1 + \ln \gamma_{i,B} x_B^{(i)}\right]
$$

$$
+ BRT_0 \sum_i \left[\frac{x_B^{(i)}}{\gamma_{i,B}}\frac{d\gamma_{i,B}}{dt}\right] \tag{31}
$$

$$
- D_t\left[\mathcal{A}_{B,phis} + RT_0 \sum_i x_B^{(i)} \ln \gamma_{i,B} x_B^{(i)}\right]
$$

The exergy of the current production (dome) that will be used in Equation (6) for an ideal mixture can be calculated as:

$$
\mathcal{A}_D = \mathcal{A}_{D,phis} + RT_0 \sum_i x_D^{(i)} \ln\left[x_D^{(i)}\right] \tag{32}
$$

for a non-ideal mixture:

$$
\mathcal{A}_D = \mathcal{A}_{D,phis} + RT_0 \sum_i x_D^{(i)} \ln\left[\gamma_{i,D} x_D^{(i)}\right] \tag{33}
$$

The exergy transfer associated with the transmission of energy as heat in the process can be calculated by the energy balances in the reboiler and condenser. Considering that $\Delta H^{vap}$ is the same for each component and is not related to the temperature of the process, the Clausius-Clapeyron equation can be used to calculate $\Delta H^{vap}$, and the first two terms on the right hand side of Equation (16) can be calculated by the following equation;

**Ideal Mixture:**

$$\left(1 - \frac{T_0}{T_B}\right)Q_B - \left(1 - \frac{T_0}{T_D}\right)Q_D = V\Delta H^{vap}T_0\left(\frac{1}{T_C} - \frac{1}{T_B}\right) \tag{34}$$

and considering constant relative volatility:

$$\left(1 - \frac{T_0}{T_B}\right)Q_B - \left(1 - \frac{T_0}{T_D}\right)Q_D = VRT_0 \ln\left[\frac{x_D^{(1)}(\alpha_1 - 1) + 1}{x_B^{(1)}(\alpha_1 - 1) + 1}\right] \tag{35}$$

**Non-ideal Mixture:**

$$\left(1 - \frac{T_0}{T_B}\right)Q_B - \left(1 - \frac{T_0}{T_D}\right)Q_D = VRT_0 \ln\left(\frac{\gamma_{1,D}\Phi_{1,B}K_{1,B}}{\gamma_{1,B}\Phi_{1,D}K_{1,D}}\right) \tag{36}$$

where, K is the liquid-vapor equilibrium constant and $\Phi$ is defined as:

$$\Phi_k \equiv \frac{\widehat{\phi}_k}{\phi_k^{sat}}\exp\left[-\frac{V_k\left(P - P_k^{sat}\right)}{RT}\right] ; k = 1, 2, \ldots, n \tag{37}$$

Therefore, the term for exergy loss or work loss, LW in Equation (17) for an ideal mixture, can be calculated with Equation (35) and for a non-ideal mixture with Equation (36):

**Ideal Mixture:**

$$LW = VRT_0 \ln\left[\frac{x_D^{(1)}(\alpha_1 - 1) + 1}{x_B^{(1)}(\alpha_1 - 1) + 1}\right] + DRT_0\left\{\sum_i x_B^{(i)} \ln\left[x_B^{(i)}\right] - \sum_i x_D^{(i)}\left[\ln x_D^{(i)}\right]\right\}$$

$$+ D\left(\mathcal{A}_{B,phis} - \mathcal{A}_{D,phis}\right) \tag{38}$$

$$- VRT_0\sum_i\left\{x_B^{(i)} - y_B^{(i)} + \left(\frac{L}{V}\right)\left[x_1^{(i)} - x_B^{(i)}\right]\right\}\left[1 + \ln x_B^{(i)}\right]$$

**Non-ideal Mixture:**

$$LW = VRT_0 \ln\left(\frac{\gamma_{1,D}\Phi_{1,B}K_{1,B}}{\gamma_{1,B}\Phi_{1,D}K_{1,D}}\right) + D_tRT_0\left\{\sum_i x_B^{(i)} \ln\left[\gamma_{i,B}x_B^{(i)}\right] - \sum_i x_D^{(i)} \ln\left[\gamma_{i,D}x_D^{(i)}\right]\right\}$$

$$+ D_t\left(\mathcal{A}_{B,phis} - \mathcal{A}_{D,phis}\right)$$

$$- VRT_0\sum_i\left\{x_B^{(i)} - y_B^{(i)} + \left(\frac{L}{V}\right)\left[x_1^{(i)} - x_B^{(i)}\right]\right\}\left\{1 + \ln\left[\gamma_{i,B}x_B^{(i)}\right]\right\} \tag{39}$$

$$- BRT_0\sum_i\left[\frac{x_B^{(i)}}{\gamma_{i,B}}\frac{d\gamma_{i,B}}{dt}\right]$$

Considering that $\mathcal{A}_{B,phis} - \mathcal{A}_{D,phis} \approx 0$, then:

**Ideal Mixture:**

$$LW = VRT_0 \ln\left[\frac{x_D^{(1)}(\alpha_1-1)+1}{x_B^{(1)}(\alpha_1-1)+1}\right]$$

$$+DRT_0\left\{\sum_i x_B^{(i)}\ln\left[x_B^{(i)}\right]-\sum_i x_D^{(i)}\left[\ln x_D^{(i)}\right]\right\} \tag{40}$$

$$-VRT_0\sum_i\left\{x_B^{(i)}-y_B^{(i)}+\left(\frac{L}{V}\right)\left[x_1^{(i)}-x_B^{(i)}\right]\right\}\left[1+\ln x_B^{(i)}\right]$$

**Non-ideal Mixture:**

$$LW = VRT_0 \ln\left(\frac{\gamma_{1,D}\Phi_{1,B}K_{1,B}}{\gamma_{1,B}\Phi_{1,D}K_{1,D}}\right)+DRT_0\left\{\sum_i x_B^{(i)}\ln\left[\gamma_{i,B}x_B^{(i)}\right]-\sum_i x_D^{(i)}\ln\left[\gamma_{i,D}x_D^{(i)}\right]\right\}$$

$$-VRT_0\sum_i\left\{x_B^{(i)}-y_B^{(i)}+\left(\frac{L}{V}\right)\left[x_1^{(i)}-x_B^{(i)}\right]\right\}\left\{1+\ln\left[\gamma_{i,B}x_B^{(i)}\right]\right\}-BRT_0\sum_i\left[\frac{x_B^{(i)}}{\gamma_{i,B}}\frac{d\gamma_{i,B}}{dt}\right] \tag{41}$$

the minimum work can be calculated as:

**Ideal Mixture:**

$$W_{\min} = DRT_0\left\{\sum_i x_D^{(i)}\ln\left[x_D^{(i)}\right]-\sum_i x_B^{(i)}\ln\left[x_B^{(i)}\right]\right\}$$

$$+VRT_0\sum_i\left\{x_B^{(i)}-y_B^{(i)}+\left(\frac{L}{V}\right)\left[x_1^{(i)}-x_B^{(i)}\right]\right\}\left\{1+\ln\left[x_B^{(i)}\right]\right\} \tag{42}$$

**Non-ideal Mixture:**

$$W_{\min} = DRT_0\left\{\sum_i x_D^{(i)}\ln\left[\gamma_{i,D}x_D^{(i)}\right]-\sum_i x_B^{(i)}\ln\left[\gamma_{i,B}x_B^{(i)}\right]\right\}$$

$$+VRT_0\sum_i\left\{x_B^{(i)}-y_B^{(i)}+\left(\frac{L}{V}\right)\left[x_1^{(i)}-x_B^{(i)}\right]\right\}\left\{1+\ln\left[\gamma_{i,B}x_B^{(i)}\right]\right\} \tag{43}$$

$$+BRT_0\sum_i\left[\frac{x_B^{(i)}}{\gamma_{i,B}}\frac{d\gamma_{i,B}}{dt}\right]$$

and the thermodynamic efficiency [Equation (9)] results in:

**Ideal Mixture:**

$$\eta_t = \frac{\sum_i\left\{x_D^{(i)}\ln\left[x_D^{(i)}\right]-x_B^{(i)}\ln\left[x_B^{(i)}\right]\right\}+\left\{R\left[x_1^{(i)}-y_B^{(i)}\right]+x_B^{(i)}-y_B^{(i)}\right\}\left\{1+\ln\left[x_B^{(i)}\right]\right\}}{(R+1)\ln\left[\frac{x_D^{(1)}(\alpha_1-1)+1}{x_B^{(1)}(\alpha_1-1)+1}\right]} \tag{44}$$

**Non-ideal Mixture:**

$$\eta_t = \frac{\sum_i \left\{ x_D^{(i)} \ln \gamma_{i,D} x_D^{(i)} - x_B^{(i)} \ln \gamma_{i,B} x_B^{(i)} + \left( R\left[ x_1^{(i)} - y_B^{(i)} \right] + x_B^{(i)} - y_B^{(i)} \right) \left[ 1 + \ln \gamma_{i,B} x_B^{(i)} \right] + \left( \frac{B}{V} \right) \left[ \frac{x_B^{(i)}}{\gamma_{i,B}} \frac{d\gamma_{i,B}}{dt} \right] \right\}}{(R+1) \ln \left( \frac{K_{1,B}}{K_{1,D}} \frac{\Phi_{1,B}}{\Phi_{1,D}} \frac{\gamma_{1,D}}{\gamma_{1,B}} \right)} \tag{45}$$

Equation (45) can be reduced if one considers that the changes of activity coefficients in the reboiler are so small that they can be neglected. Then:

$$\eta_t = \frac{\sum_i \left\{ x_B^{(i)} + x_D^{(i)} \ln \gamma_{i,D} x_D^{(i)} + \left[ R\left( x_1^{(i)} - y_B^{(i)} \right) - y_B^{(i)} \right] \left[ 1 + \ln \gamma_{i,B} x_B^{(i)} \right] \right\}}{(R+1) \ln \left( \frac{K_{1,B}}{K_{1,D}} \frac{\Phi_{1,B}}{\Phi_{1,D}} \frac{\gamma_{1,D}}{\gamma_{1,B}} \right)} \tag{46}$$

If the gas phase is ideal or close to ideality, then Equation (46) can be reduced even more:

$$\eta_t = \frac{\sum_i \left\{ x_B^{(i)} + x_D^{(i)} \ln \gamma_{i,D} x_D^{(i)} + \left[ R\left( x_1^{(i)} - y_B^{(i)} \right) - y_B^{(i)} \right] \left[ 1 + \ln \gamma_{i,B} x_B^{(i)} \right] \right\}}{(R+1) \ln \left( \frac{K_{1,B}}{K_{1,D}} \frac{\gamma_{1,D}}{\gamma_{1,B}} \right)} \tag{47}$$

The solution of Equation (45) or its simplifications [Equation (46) and (47)] require the calculation of activity coefficients of the mixtures, which can be obtained with an appropriate solution model (Wilson, NRTL, UNIQUAC, UNIFAC, etc.). According to Equations (44), (45), (46) and (47), the thermodynamic efficiency not only depends on the reflux ratio but on the concentration of components in both phases and the kind of mixture. A convenient way to express thermodynamic efficiency is the average thermodynamic equation used by Zavala Loría (2004), Zavala et al. (2007), Zavala and Coronado (2008). and Zavala et al. (2011) to solve the maximum thermodynamic efficiency problem:

$$\eta_{average} = \frac{\int_0^t \eta_t dt}{t} \tag{48}$$

## 4. Results and discussion

Using the First and Second Law of Thermodynamics, this work has developed an expression for calculating the thermodynamic efficiency of a batch distillation process [Equations (44) and (45)]. To solve the mathematical model, the feeding was introduced at the top of the column at boiling temperature, neglecting the accumulation of vapor in each stage. The process was performed with total condenser, atmospheric pressure and adiabatic column.

To solve the model, two mixtures were considered. The first one is an ideal mixture of Hexane/Benzene/Chlorobenzene (HBC); the second one is a non-ideal mixture of Ethanol-Water (EW).

## Case of study 1: Hexane/Benzene/Chlorobenzene mixture

The conditions and data used to solve the mathematical model of thermodynamic efficiency are given in Table 2. Relative volatilities were calculated with the Soave-Redlich-Kwong equation of state.

| Variable/Parameter | Amount | Units |
|---|---|---|
| Feed ($F$) | 200.00 | kmol |
| Vapor flow ($V$) | 240.00 | kmol/h |
| Condenser liquid holdup ($H_D$) | 5.00 | kmol |
| Tray liquid holdup ($H_T$) | 1.00 | kmol |
| Batch time | 1.00 | h |
| $Z_{Hexane}$ | 0.20 | |
| $Z_{Benzene}$ | 0.30 | |
| $Z_{Chlorobenzene}$ | 0.50 | |
| $a_{Hexane}$ | 8.54 | |
| $a_{Benzene}$ | 4.71 | |
| Number of trays | 15 | |
| Reflux ratio (Constant) | 5 | |

Table 2. Data used to solve the mathematical model of thermodynamic efficiency

The first step for the resolution of the mathematical model was to reach the steady state to extract the product during one hour. Figure 3, shows the concentrations at the top of the column (including the steady state).

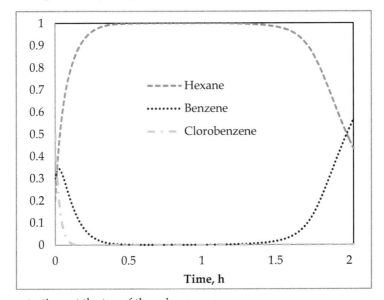

Fig. 3. Concentrations at the top of the column.

The average thermodynamic efficiency is 17.58% for this production time while the average concentration of the most volatile component at the top of the column is 88.55% mol. Table 3, shows the variations observed when the reflux ratio is changed.

| Reflux ratio | $\eta_{average}$ (%) | $x^{(1)}_{D,average}$ (%) |
|:---:|:---:|:---:|
| 2 | 55.60 | 49.50 |
| 3 | 28.61 | 64.34 |
| 4 | 20.60 | 77.66 |
| 5 | 17.58 | 88.55 |
| 6 | 16.16 | 95.55 |
| 7 | 15.08 | 98.51 |
| 8 | 14.02 | 99.40 |

Table 3. Thermodynamic efficiency behavior based on variations in the reflux ratio.

Figure 4 shows the thermodynamic efficiency behavior in the product obtaining.

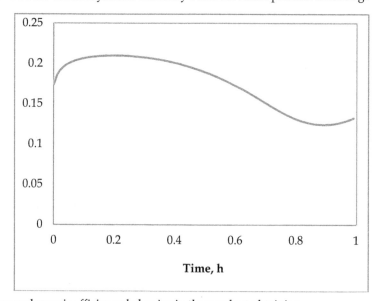

Fig. 4. Thermodynamic efficiency behavior in the product obtaining.

Such variations make evident the influence of the reflux ratio over the thermodynamic efficiency; in other words, the thermodynamic efficiency of the process is smaller when the reflux ratio is higher. We also could observe that the reflux ratio affects the product concentration. If the reflux ratio increases the product concentration increases as well.

## Case of study 2: Ethanol/Water mixture

The conditions used for solving the mathematical model of thermodynamic efficiency for a non ideal mixture (Ethanol/Water) are given in Table 4. Wilson's equation was used to obtain the activity coefficients; the vapor phase is considered to have an ideal behavior. The steady state was reached when Ethanol presented a purity level of 89.61% mol obtained using Wilson's equation.

3.045 hours was the period of time needed to reach the steady state, we obtained an average thermodynamic efficiency of 36.96% and the product presented an average concentration of 86.74% mol of the most volatile component.

| Variable/Parameter | Amount | Units |
|---|---|---|
| Feed (F) | 100.00 | |
| Vapor flow (V) | 120.00 | |
| Condenser liquid holdup ($H_D$) | 5.00 | |
| Tray liquid holdup ($H_T$) | 1.00 | kmol |
| Batch time | 2.00 | kmol/h |
| $Z_{Ethanol}$ | 0.50 | kmol |
| $Z_{Water}$ | 0.50 | kmol |
| $a_{Ethanol}$ | 8.54 | h |
| $a_{Water}$ | 4.71 | |
| Number of trays | 15 | |
| Reflux ratio (Constant) | 2 | |

Table 4. Data used to solve the mathematical model of thermodynamic efficiency.

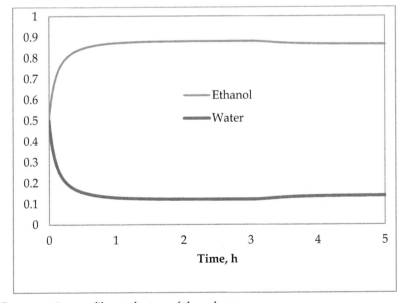

Fig. 5. Concentration profiles at the top of the column.

Figure 5 shows the behavior of the concentrations at the top of the column, while the behavior of punctual thermodynamic efficiency is shown in Figure 6. Table 5 shows the variations observed when different reflux ratios are used.

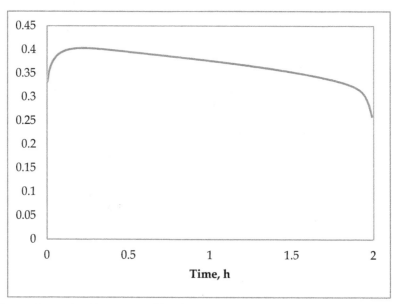

Fig. 6. Thermodynamic efficiency profile (obtaining product).

| Reflux ratio | $\eta_{average}$ (%) | $x^{(1)}_{D,average}$ (%) |
|:---:|:---:|:---:|
| 5 | 36.96 | 86.74 |
| 6 | 32.75 | 86.96 |
| 7 | 29.24 | 87.12 |
| 8 | 26.38 | 87.24 |
| 9 | 24.00 | 87.32 |
| 10 | 22.02 | 87.39 |

Table 5. Thermodynamic efficiency behavior with regards to the reflux ratio variation.

As in the last case, we observe that when the reflux ratio increases, the efficiency decreases; however, the concentration of the most volatile component presents also an increment. We can formulate the following heuristic rule: If the variables of the process are maintained, the reflux ratio will have an inverse effect on the thermodynamic efficiency of the process.

## 5. Concluding remarks

Using the First and Second Law of Thermodynamics (exergy concept), this work has developed an expression for calculating the thermodynamic efficiency of a batch distillation process. The resulting equation was used to find the batch distillation thermodynamic efficiency for an ideal mixture and a non-ideal mixture. The equation obtained is a generalization of the equation developed by Zavala-Loría et al. (2007), Zavala and Coronado (2008) and Zavala et al. (2011).

The results obtained by solving the Equations, allowed us to observe the relationship between reflux and thermodynamic efficiency of the process. Furthermore, variables such as the product purity are affected by the reflux ratio, in other words, the purity of the product requires a greater amount of reflux to obtain a higher concentration.

## 6. Nomenclature

| | |
|---|---|
| $H_j$ | molar hold-up on tray $j$ |
| $H_D$ | molar hold-up on the condenser |
| $I$ | enthalpy (J/mol) |
| $S$ | entropy (J/mol K) |
| $V$ | vapor flow rate (mol/h) |
| $L$ | liquid flow rate (mol/h) |
| $D$ | distillate flow rate (mol/h) |
| $B$ | amount of moles in the reboiler, mol |
| $n$ | number of components |
| $N$ | number of trays in the column |
| $R_t$ | reflux ratio |
| $t$ | time (h) |
| $W$ | work (J) |
| $LW$ | work loss (J) |
| $z$ | mole fraction in the feed |
| $x_j^{(i)}$ | mole fraction in the liquid of component $i$ at plate $j$ |
| $y_j^{(i)}$ | mole fraction in the vapor of component $i$ at plate $j$ |
| $x_B^{(i)}$ | mole fraction in the liquid of component $i$ in the reboiler |
| $x_D^{(i)}$ | mole fraction in the vapor of component $i$ in the distillate |
| $K_j^{(i)}$ | equilibrium constant |
| $T$ | temperature |
| $A$ | availability |
| $\eta_t$ | punctual thermodynamic efficiency |
| $\Delta H^{vap}$ | vaporization heat |

### Subindex

| | |
|---|---|
| 0 | reference |
| $c$ | condenser |
| $f$ | end |
| $D$ | distillate |
| $B$ | boiler or reboiler |
| $t$ | time |

# 7. References

Luyben, William L. (1990); Process modeling, Simulation and Control for Chemical Engineers; McGraw-Hill. Second edition, 725 pp.

Kim, K.J., Diwekar, U.M. (2000); Comparing Batch Column Configurations: Parametric Study Involving Multiple Objetives; Aiche Journal, 46(12), pp. 2475-2488.

Zavala-Loría, J.C. (2004); Optimización del Proceso de Destilación Discontinua; Tesis Doctoral, Departamento de Ingeniería Química, Instituto Tecnológico de Celaya, Celaya, Guanajuato, México.

Zavala, José C., Córdova, Atl., Cerón, Rosa M. and Palí, Ramón J. (2007); Thermodynamic efficiencies of a conventional batch column; Int. J. Exergy, 4(4), pp. 371-383.

Zavala, J. C. and Coronado, Cristina (2008); Optimal Control Problem in Batch Distillation Using Thermodynamic Efficiency; Ind. Eng. Chem. Res., 47, pp. 2788-2793.

Zavala-Loría, J. C., Ruiz-Marín, A. and Coronado-Velásco, Cristina (2011); Maximum Thermodynamic Efficiency Problem in Batch Distillation; Brazilian Journal of Chemical Engineering, 28(2), pp. 333-342.

# Part 2

# Food and Aroma Concentration

# 5

# Changes in the Qualitative and Quantitative Composition of Essential Oils of Clary Sage and Roman Chamomile During Steam Distillation in Pilot Plant Scale

Susanne Wagner, Angela Pfleger, Michael Mandl and Herbert Böchzelt
*Joanneum Research Forschungsgesellschaft mbH Graz, Resources, Institute of Water,*
*Energy and Sustainability, Department for Plant Materials Sciences and Utilisation*
*Austria*

## 1. Introduction

Clary sage (*Salvia sclarea* L.) from the genus Salvia is a biennial or short-living herbaceous perennial plant. Its leaves are united to a basal rosette in the first year, 12 to 25 cm long and 7 to 15 cm wide, ovate, cordate at the base, obtuse and long-stemmed. All the leaves are reticulate-rugose and hairy on both sides. The flowers seem loose to fairly dense, often branched paniculate. They reach approximately 2 cm and the large heart-shaped bracts are long, tapering and purple in early stages, later greenish white. Blossom: June or July to August. The seeds are 2 to 3 mm long nuts. Its native regions are the northern Mediterranean

Fig. 1. Clary sage

along with some other areas in Central Europe, North Africa and Central Asia. It is drought resistant and grows in dry, rocky places at an optimal soil pH of 4.5. Clary sage has a long history as medicinal herb and is currently grown mainly for the production of its essential oil reaching oil yields of 0.1 to 0.3%. (Board, 2003) (Dachler & Pelzmann, 1999) The clear, colourless oil is used in perfumes with a light, warm and sweet fragrance and as muscatel flavouring in food industry for vermouths, wines, and liqueurs. It is also used in aromatherapy for relieving anxiety, stress and fear, menstruation problems such as PMS (premenstrual syndrome) and cramping, and helping with insomnia, amenorrhea, dysmenorrhoea, pre-menopause, depression, fatigue, nervousness, migraine, varicose veins, haemorrhoids, oily skin and hair, spasmodic cough. It shows strong estrogenic, aphrodisiac and regenerative activity. (Lis-Balchin, 2006), (Kintzios, 2000) Sclareol, the essential oil main compound, is used in the perfume industry as a fixative and in the tobacco industry for flavouring. It is part of a large number of amber fragrances with woody note (Heinrich, et al., 2004), (Ferichs & Rimbach, 1992).

Fig. 2. Roman chamomile (Copyright Bernhard Bergmann)

Roman chamomile (*Chamaemelum nobile* L.) is a small (up to 80 cm), perennial plant found in dry fields and around gardens and cultivated grounds. It has daisy-like white flowers and is native to Western Europe, North America, and Argentina. Cultivation areas are mainly found in France (main producer), Belgium, Italy, Czech Republic, Slovakia, India, North America, Brazil and Argentina. Its stem is procumbent with ascending basic axis, the leaves are alternate, twice pinnate, finely dissected and fluffy to almost glabrous. The solitary, terminal flower heads rise eight to twelve inches above the ground and consist of prominent yellow disk flowers and silver-white ray flowers. Seeds are almost triangular, bald and shiny. (Dachler & Pelzmann, 1999) According to (Barnes, et al., 2007) Roman chamomile flower heads contain 0.4 – 1.75%. Blossom time is June and July, and its fragrance is sweet, crisp, fruity and herbaceous. The plant is used in tisanes, perfumes and cosmetics and to flavour foods. It is popular in aromatherapy, whose practitioners believe it to be a calming agent. The pharmacological profile of the Roman chamomile flower heads is similar to German chamomile (*Matricaria recutita* L.). It is used as mild sedative, anti-emetic and antispasmodic (Heinrich , et al., 2004), (Barnes, et al., 2007).

Essential oils are in general a complex mixture of natural compounds (terpene hydrocarbons like mono-, sesqui- and diterpenes - cyclic or non-cyclic – and their corresponding oxygenated isoprenoid compounds alcohols, ketones, esters and aldehydes) with various applications in medicine, pharmacology, perfumery, cosmetics, food preservation or as insect repellent. Essential oils also contain phenolic compounds and alkanes. They are usually produced by distillation, but also by mechanical processes from different plant parts, such as seeds, fruits, fruit peel, roots, leaves, needles and balsams. Unlike fatty oils essential oils are volatile. Several data published in literature show that the quality of essential oils and the oil yield depend on many different parameters e.g. plant development stage, genetic determination, plant organ, drying and storage conditions, soil structure, climate, insolation and weathering, habitat, harvesting methods, date of collection of plant material, analysis conditions used for identification of the compounds, etc. Relative amounts of the compounds and the chemical composition of essential oils may also vary due to the applied distillation process itself (lab or pilot plant scale) and the duration of the distillation.

The essential oil of clary sage is a complex chemical mixture, consisting of up to 100 mostly terpenoid compounds. These are mainly (oxygenated) monoterpenes with small amounts of (oxygenated) sesquiterpenes. The main compounds are linalool (17.2%), linalyl acetate (14.3%), geraniol (6.5%), geranyl acetate (7.5%), terpineol (15.1%), nerol (5.5%), neryl acetate (5.2%) and sclareol (5.2%). Furthermore, α-pinene, β-pinene, camphene, myrcene, limonene, cis-and trans-ocimene, p-cymene, terpinolene, cis-3-hexen-1-ol, caryophyllene, terpinen-4-ol, citronellol, β-gurjunene, caryophyllene oxide, germacrene D, (2R, 5E)-2.12-epoxycaryophyll-5-ene, (2R, 5E)-caryophyllene-5-en-12-al, (2S, 5E)-caryophyllene-5-en-12-al, isospathulenol, (1R, 5R)-1,5-epoxysalvial-4 (14)-ene, salvial-4(14)-en-1-one (Kintzios, 2000).

Unlike chamomile from the Roman chamomile only few pharmacological studies exist. The composition of Roman chamomile oil is very complex and so far app. 100 compounds were identified. Main compounds are the sesquiterpenes and sesquiterpene lactones from the germacranolide type (e.g. nobiline). It is characterized by the presence of terpenoids and saturated and unsaturated fatty acids with four or five carbon atoms, such as butyric, valeric, crotonic, angelic, tiglin and methacrylic acid. These are esterified with C3 to C6-alcohols such as n-butanol, isobutanol, isoamyl alcohol and 3 - methylpentane-1-ol. The relative proportion of total esters in the essential oil of Roman chamomile is known as the highest of all essential oil-producing plants. Furthermore, Roman chamomile flowers contain hydroperoxides (e.g. 1-β-hydroperoxyisonobiline, a sesquiterpene peroxide from germacrane-type and allylic hydroperoxides as well). The content of hydroperoxides in the dried drug varies and is decreased during prolonged storage (Barnes, et al., 2007), (Bajaj, 1996), (Ferichs & Rimbach , 1992).

## 2. Methods

### 2.1 Cultivation and harvest

Both herbs were cultivated in the years 2002 and 2003 and Roman chamomile additionally in 2004 and 2010 by organic farming in three habitats (in 2010 only one habitat) of different altitude (Bad Blumau 285 m, Oberlungitz 400 m, St. Jakob 1000 m above sea level) in the East of Styria (Austria). Fields of the individual sites are divided into parcels a and b to ensure repeatability at each site.

The location of Bad Blumau (I) was located in the thermal spa area Blumau (Austria). Parcels were located within the spa recreation in order to attract interested visitors. The herbal plantation had been converted from a meadow and spa gardener took care of herbal test fields.

Agricultural fields of the family "Ocherbauer" in St. Jakob im Walde were site (II) of the project. As in Blumau a meadow was transformed into a field. The use of machineries for cultivation of herbal test areas was found difficult, because plantation was located on a steep slope.

Agricultural site (III) of the family "Oswald" is located in Oberlungitz. Agricultural areas are laid out in rows for simplifying organic farming cultivation.

Used eight parcels had 10 per 20 m each at the cultivation sites Bad Blumau and St. Jakob, thus covered a total area of 1600 m². In Oberlungitz plants were grown in rows with 4.5 m x 35 m and 4.5 m x 59 m respectively, yielding a total area of 1692 m².

Harvesting techniques differed at the three sites, because available machines and devices were different.

Plant material was harvested in different stages of development. A code for development stages based on the catalogue of codes for grain plants ("BBCH – scale") was established (Meier, 2001). This catalogue was developed by the BBA (Federal Biological Research Centre for Agriculture and Forestry, Germany), the BSA (Protection and National Listing of new plant varieties, Germany), the IVA (Germany) and the IGZ (Institute of Vegetable and Ornamental Crops, Germany) and represents a code for the phenological development stages of mono- and dicotyledonous plants. The used herbs were characterised according to this code.

Clary sage is a biennial plant, which is harvested in the second year of cultivation, when the flowers are blooming. The optimum time for harvesting is June, since at that time the plant reaches the highest oil content. Moreover, in the period of 9 p.m. to 3 a.m., the highest oil yield can be achieved. Roman chamomile is harvested, when the florets wreath in the second third of the domed receptacle is already open. With increasing flowering process, the content of essential oil and chamazulene decrease. The characteristic blue colour is due to the formation of chamazulene in traces, which is formed from matricine by heating the oil. Delayed harvest may also lead to the disintegration of the flower heads. Harvest product should be the blooming chamomile heads (Dachler & Pelzmann, 1999).

For this project the entire herb of both plants was harvested in full bloom using bar- or power mower. For investigational reasons plant parts of clary sage were separated after harvest manually in order to compare the oil yield and composition of the essential oils gained by distillation of different plant parts. The vegetation progressing had been measured and documented regularly by taking several biomass yield samples at various stages; additionally sampling was done in order to document the total content of essential oil of the two herbs and their oil compositions. These data provided important information to determine the optimal harvest time, because the oil content depends on the plant development.

Plant height and inflorescence status of the herbs were noted and the samples were recorded photographically. The timing of sampling was based on the time of harvest.

According to the developed index ("BBCH scale") clary sage was harvested at growth stage 6 (blossom) and the subsidiary stage 66 (beginning of the flowering of the side shoots).

Roman chamomile was harvested at growth stage 6 (blossom) and the subsidiary development stages 65 (full flowering: 50% of flowers open, first ray florets may fall) and 49 (permanent development of young shoots, broadening of the main shoot).

After determining the moisture content in the samples with a moisture analyzer, plants were distilled immediately or dried at 35 ° C in a well ventilated area.

## 2.2 Distillation

Steam distillation was carried out using a hundred litres batch volume distillation plant of the type TWE 250-2000 Herba-Tec produced by the Innotec-Tetkov GmbH in Germany, which in average processes about 10 to 15 kilograms of fresh plant material per batch. For distillation in pilot plant scale plant material was weighed and used fresh or partly dried. Prior to distillation, the plants were cut using a slicer to a size of about 4 to 8 cm.

Main parts of the distillation plant:

1. steam generator
2. steam inlet area
3. pivoted distillation tanks
4. tube cooler
5. cooling water inlet
6. cooling water outlet
7. separation funnel of essential oil
8. control glass
9. lifting device

Fig. 3. Distillation plant in pilot plant scale TWE 250-2000 Herba-Tec

Disintegration of plant material had three main reasons:

1.  Smaller parts of plants are generally easier to handle than the whole herb.
2.  For distillation a homogeneous plant material filling in the column should be achieved avoiding cavities, which lead to shortcut steam flows.
3.  The capacity of the distillation column is for cut plant material in general higher than for uncut material, if stuffing is avoided. A special designed cutter was used to prevent any squeezing and pressing of the herbs during cutting to create a smooth, sharp cutting surface.

A gentle disintegration of plant material is of importance to avoid losses of essential oil.

The steam distillation unit consists of two cylindrical distillation tanks (inner diameter 25 cm, height 2 m), with a 100 litres volumetric capacity. The tanks are rotatable with a swinging sieve to be used separately to allow an alternating filling and distillation process. Thus a continuous batch distillation process can be realized.

The steam production is carried out in a separate, electrically powered steam generator outside the tanks. A constant steam flow of approximately 200 g steam per minute or 12 kg of steam per hour is supplied. This specific amount of steam was claimed by the producer as the optimum amount of operation in connection with the geometry of the tanks.

Fig. 4. Distillation processing steps

After the steam passed through the distillation column the condensation of the water/oil mixture takes place in a heat exchanger (cooler), which is traversed in counter current flow by the cooling water. Separation of the essential oil is carried out in a separation funnel device. By the geometry of the funnel flow conditions are created that support the floating of the oil on the hydrosol (condensed water vapour).The entire distillation process is run under normal atmospheric pressure conditions. All wetted parts are made of stainless steel with an electro-polished surface.

After the cutting process the herbs were weighed and filled into the distillation column manually. A homogeneous bulk body was produced to create comparable conditions and a plugging of plant material was avoided. All distillations were performed using a specific vapour flow of 12 kg of steam per hour. The amount of water content was also measured with a moisture analyzer. The temperature differences of cooling water and steam temperature and any condensate accumulating within the tank were documented. After distillation the used plant material was again weighed and the amount of accumulated flavour water calculated. The yield of essential oil was measured and calculated of 3 to 4 distillations to receive an average result. Prior to chemical analysis the essential oil was stored in a refrigerator.

In some cases every 10 minutes of the distillation process samples were taken in order to investigate the changes of essential oil yield and composition within a steam distillation process. Taken samples represent therefore a mixture of the gained essential oil during the period of 10 minutes. Samples were subjected to capillary gas chromatographic analysis coupled to a mass selective detector (GC/MSD) and coupled to a flame ionization detector (GC/FID).

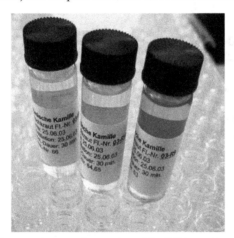

Fig. 5. Essential oils of Roman chamomile produced in different time frames during distillation

## 2.3 Analysis

The essential oils were analysed to state on the essential oil composition and its variation during the steam distillation process.

For quality analysis of essential oils such as the comparison of different samples taken in the distillation process it is necessary, to characterize the high number of compounds qualitatively and quantitatively. This is done by the use of capillary gas chromatographic techniques (GC/MSD and GC/FID). Main principles of these techniques are the separation of a mixture of compounds by transporting the sample through a capillary gas chromatography column. As mobile phase a gas, mainly helium is used. The separated components leave the column due to different exchange mechanisms in the column at different times and are analyzed using special detectors. A frequently used detector is the flame ionization detector (FID), which is characterized by a high dynamic range for the universal detection of a large number of organic compounds. Because of this characteristic, this detector is often used in quantitative determinations. However, only limited information on the structure of the compounds is received. More specific information about the detected substances is provided by mass-selective detectors (MSD). The resulting mass spectra allow identification of chemical structures. By using the values of retention times, mass spectra and comparing results in literature and data bases, the chemical structure of a compound is detectable in most cases unambiguously.

**Gas chromatographic method:**

GC: Hewlett Packard 6890 GC system with integrated auto sampler
MSD: Hewlett Packard 5973 Mass Selective Detector
Column: J&W DB-5MS Capillary Column; Length: 30m; Diameter: 0.25 mm; Film thickness: 0.25 µm (non polar, coated with a phenyl arylene polymer – comparable to a 5% phenyl methyl polysiloxane column)
Mobile phase gas: Helium
Flow rate: 1 ml/min
Split ratio: 20:1
Temperature program: 10 min at 50 °C, with 2 °C/min heated to 220 °C, 10 min hold at 220 °C
Injection temperature: 220 °C
Injection volume: 1 µl
Temperature in Ion source: 250 °C
Quadrupol temperature: 200 °C
MS-Scan area: 20 - 350 m/z
Temperature of FID detector: 250 °C

In this project compositions of the essential oils were determined by comparing the relative retention times of standards and mass spectra from data bases of oil compounds (Adams, NIST, WILEY). Progression and development of the qualitative composition and the relative amounts of the most predominant compounds (>2%) were examined. These are important factors for the pharmaceutical application of essential oils. The results provide the possibility to determine the composition of the oils or the optimised distillation time with the highest relative amount of the main compounds. In addition trends and correlations among these compounds will be shown.

## 3. Results

### 3.1 Clary sage (*Salvia sclarea* L.)

### 3.1.1 Oil yield

Clary sage could only be harvested in the year 2003, because it's a biennial shrub. It shows big differences in essential oil yields in different parts of the plant - highest oil yields were obtained by distillation of the flowers. The oil yield of 7.9 ml/kg dry matter of the distillation of the flowers is 2.5 times higher than by distillation of the whole plant and 4.6 times higher than the distillation of stems (see figure 6).

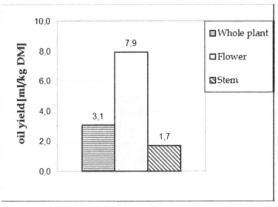

Fig. 6. Oil yield of different plant parts of clary sage (DM…dry matter)

Distillations of the fresh whole plants indicated an oil yield of app. 60% (app. 6.0-7.0 ml/kg dry matter) after 20 minutes distillation. After 30 minutes app. 80% (8.5-9.5 ml/kg dry matter) were obtained. In the last 30 minutes only small oil yields were achieved. Figure 7 indicates the oil yield measured in between specified time frames (9.3 to 20 minutes, 30 to 40 minutes, 50 to 60 minutes).

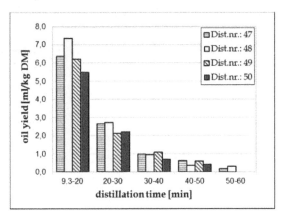

Fig. 7. Oil yields for specific time frames during distillation of the whole plant of clary sage (DM…dry matter)

According to literature data 90% of oil yield was reached after 2 hours distillation time (Kintzios, 2000). The average oil yield of all three agricultural sites reached 7.9 to 9.8 ml/kg dry matter by distillation of the flowers (app. 0.8 to 1.0% oil yield) and by distillation of the whole plant app. 3 ml/kg dry matter (app. 0.3%). These results correspond to the range of the literature values of the essential oil yield of 0.1 to 0.3%.

For economic reasons and according to the results, a distillation time of the fresh plant of 40 minutes and for drug distillations of 30 minutes is recommended by the authors. After that time the oil content increases only slightly and a yield of 90 - 95% is satisfactory. One reason for the extremely high oil content of the distillations of the fresh plant in 2003 could probably have been the weather of spring and summer 2003 with many sunny days (and record heat in Austria).

### 3.1.2 Chemical investigations

75.9 - 91.8% of the essential oil of clary sage, harvested and distilled in pilot plant scale in 2003, was identified. The qualitative and semi-quantitative compositions of the distilled oils differ from each other depending on plant part, location and duration of distillation. The oil consisted mainly of linalyl acetate (16.7 - 69.7%), germacrene D (2.2 - 31.2%), β-caryophyllene (1.2 - 10.7%), bicyclogermacrene (0.6 - 9.1%), linalool (0.6 - 8.9%), α-copaene (0.5 - 6.1%), trans-A/B-sclareol oxide (0.5 - 5.7%) and sclareol (1.6 - 6.9%). Some samples exhibited a relatively high content of the pharmacological important diterpene sclareol. Sclareol is used as starting material for a number of amber fragrances. As the most important compound linalyl acetate was identified (approx. 70%). In most cases, the value of linalyl acetate corresponded to the value of germacrene D - if the relative amount of linalyl acetate decreased the value of germacrene D rose. Based on the data presented (see figures 10 - 12) it can be postulated, that the higher the amount of the flowers in the distilled material, the higher was the amount for linalyl acetate; the higher the amount of stems and leaves the higher was the content of germacrene D (see figures 9 to 11). The highest relative amount of sclareol was detected by distillation of the flowers after a distillation time of 50 minutes.

For economic reasons, only compounds with a higher relative amount than 2% are included in the following graphics.

Figure 10 indicates that linalyl acetate can be seen as the most important compound in the essential oil of the whole clary sage plant, the content does not considerably decrease with

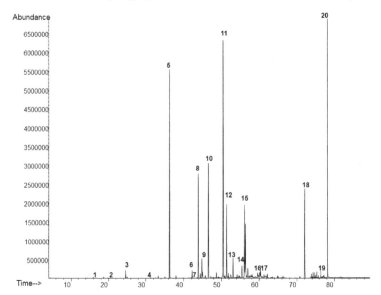

Fig. 8. Chromatogram of the essential oil of clary sage (Peak numbers correspond to compounds in table 1)

| peak number | compound | retention time $t_R$ [min] |
|:---:|:---:|:---:|
| 1 | myrcene | 15.99 |
| 2 | trans – β - ocimene | 20.57 |
| 3 | linalool | 24.46 |
| 4 | α - terpineol | 31.22 |
| 5 | linalyl acetate | 36.39 |
| 6 | neryl acetate | 42.24 |
| 7 | α - cubebene | 43.54 |
| 8 | α - copaene | 43.87 |
| 9 | β - cubebene | 44.84 |
| 10 | β - caryophyllene | 46.57 |
| 11 | germacrene D | 50.48 |
| 12 | bicyclogermacrene | 51.38 |
| 13 | δ - cadinene | 53.08 |
| 14 | 1,5 - epoxysalvial-4(14)-ene | 55.41 |
| 15 | caryophyllene oxide | 56.36 |
| 16 | spathulenol | 59.54 |
| 17 | β - eudesmol | 60.11 |
| 18 | trans – A/B – sclareol oxide | 72.03 |
| 19 | manool | 76.34 |
| 20 | sclareol | 78.16 |

Table 1. Identified compounds from the essential oil of clary sage from the harvest 2003

increasing distillation time. The other compounds are below 10%, linalool and germacrene D decrease with increasing distillation time and α-cubebene and sclareol rise.

In comparison to the whole plant distillation flower distillation shows an increase in the relative amount of linalyl acetate in the final minutes of the distillation. The same trend

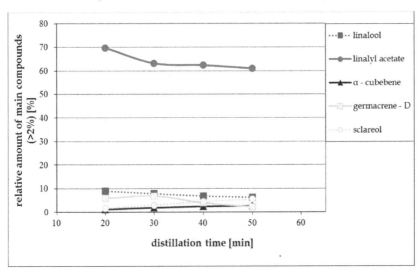

Fig. 9. Change of essential oil composition of main compounds during steam distillation of clary sage (habitat Bad Blumau, first cut, whole plant, beginning of flowering of the side shoots)

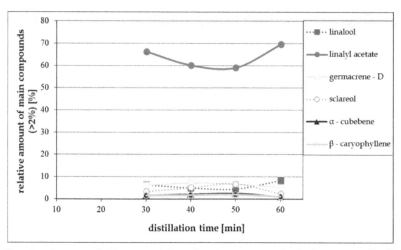

Fig. 10. Change of essential oil composition of main compounds during steam distillation of clary sage (habitat Bad Blumau, first cut, flowers only, beginning of flowering of the side shoots)

(increasing or decreasing with changes of direction in the final minutes respectively) show the graphs for linalool, sclareol and germacrene D.

The content of linalyl acetate in the essential oil produced by distillation of the stems and leaves is lower. It can be concluded that linalyl acetate is predominantly present in the flowers. The content of linalyl acetate decreases in the oil of the stems and leaves material with increasing distillation time, the levels of germacrene D, α-copaene, β-caryophyllene and bicyclogermacrene increase; sclareol content remains the same.

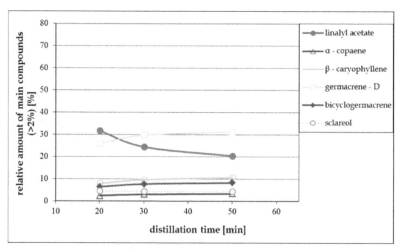

Fig. 11. Change of essential oil composition of main compounds during steam distillation of clary sage (habitat Bad Blumau, first cut, stems and leaves, beginning of flowering of the side shoots)

## 3.2 Roman chamomile

### 3.2.1 Oil yield

In 2003 by distillation of fresh Roman chamomile plants an average of 4.6 ml oil/kg dry matter is obtained after 20 minutes distillation time (equivalent to about 80% of oil yield - 3.5 ml oil/kg dry matter) and after another 10 minutes, 0.8 ml oil/kg dry matter. A distillation time of 30 minutes, thus proves to be useful in order to achieve an oil yield of about 95%.

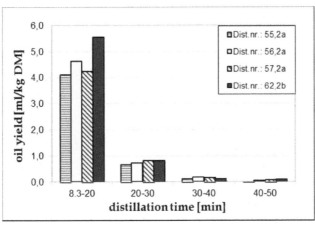

Fig. 12. Oil yields for specific time frames during distillation of the whole plant of Roman chamomile in 2003 (DM...dry matter)

Basically, Roman chamomile can be distilled easily. The colour of the essential oil is light to medium blue (see Figure 5), clear and low viscous. From an economic point of view a distillation time of 30 minutes of the fresh, and distillations of 20 minutes for dried plant material makes sense. The average yield of essential oil by distillation of the fresh Roman chamomile plant material reached values of 3.5 ml/kg dry matter (0.4%) and the dried 2.4 ml/kg dry matter (0.2%). The achieved oil content of Roman chamomile distillations in literature refers to the distillation of the flower heads and is not comparable with the results of whole plant distillations.

Harvest at the full flowering stage of the plants and the high number of sunshine hours in 2003 provided very good conditions for the increase of the content of the essential oil in the plants. By distillation of fresh Roman chamomile plants higher oil content with less effort (drying process) was achieved compared to distillation of drugs. A distillation of the drug is therefore not useful, if the capacity of steam distillation is capable for fresh processing.

Apart from a lower oil yield in 2004 compared to 2003, the curves of the distillations were not related. While in 2003, already after 25 minutes 90% of the oil yield was achieved, in 2004 this required a period of about 40 minutes. The oil yield was approaching its maximum more slowly. The reason for this fact could possibly have been the weather conditions, since, as already mentioned, the summer was exceptionally hot and dry in 2003. The oil cells of the plants in 2003 were better developed and thus distillation has been accelerated.

Due to the variability of results from different years, a distillation time of 50 minutes is advised.

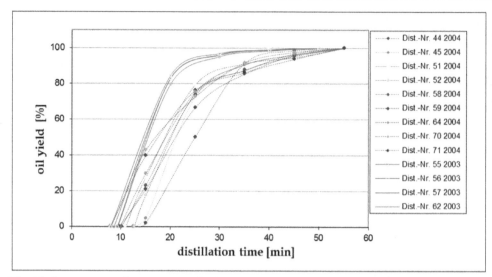

Fig. 13. Comparison of approximated oil recovery curves for the whole plant distillation of Roman chamomile in 2003 and 2004

By comparing the relative oil yields for the last two years the recovery curve is significantly different. One reason for the inhomogeneous curves from 2004 may have been the defective water dosing pump. The water supply had to be operated manually in this year, which might have led to fluctuations in the amount of steam. Apart from the already described different distillation maxima of 2004 and 2003, different heights of the curves can be recognized.

In 2003 Roman chamomile plants were harvested in full bloom and without any weeds, while in the cultivation year 2004 weeds proportion was high and the plants were harvested partially wet and prostrate. In summary, not only the weather, but also inflorescence, moisture and weeds are involved in the formation of the oil.

The distillation in 2010 differs from the distillation of the years 2003 and 2004. In the first cut 90% of the total oil yield was achieved after 20 minutes. The same distillation behaviour can be described for the second cut.

### 3.2.2 Chemical investigations

As the number of literature surveys on the Roman chamomile is scarce, few comparisons could be drawn to literature regarding the composition of the essential oil. In addition identification of compounds proved to be rather difficult, since the spectra exhibited many very similar esters, which are difficult to distinguish. As the composition also differed in the investigated years the following table is an example for the results of one investigation year (2003).

In both test series (fresh plant and drug material), the relative amounts of angelic acid 3-methylpentyl ester, 3-methyl-2-butenoic acid 3-methylbutyl ester, myrtenal and trans-pinocarveol increase, while the values for 3-methyl-2-butenoic acid pentadecyl ester, 3-methyl-2-butenoic acid cyclobutyl ester, 3-methyl 2-butenoic acid pentyl ester, α-pinene,

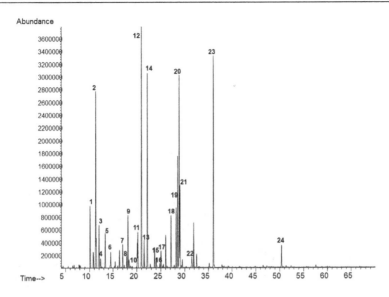

Fig. 14. Chromatogram of the essential oil of Roman chamomile (Peak numbers correspond
to compounds in table 2)

| peak number | compound | retention time $t_R$ [min] |
|---|---|---|
| 1 | isobutyl isobutyrate | 10.13 |
| 2 | α - pinene | 11.29 |
| 3 | (E)-2-butenoic acid 2-methylpropyl ester | 11.98 |
| 4 | camphene | 12.32 |
| 5 | 3-methacrylic acid-1-butene-4-yl-ester | 13.27 |
| 6 | β - pinene | 14.47 |
| 7 | 2 - methyl butanoic acid 2-isobutyl ester | 17.02 |
| 8 | isoamyl butyrate | 17.85 |
| 9 | isobutyric acid 3-methylbutyl ester | 18.12 |
| 10 | p - cymene | 18.37 |
| 11 | cyclopropane carboxylic acid 3-methylbutyl ester | 20.08 |
| 12 | 3-methyl-2-butenoic acid pentadecyl ester | 20.97 |
| 13 | isobutyric acid 3-methyl-2-butenyl ester | 21.54 |
| 14 | 3-methyl-2-butenoic acid cyclobutyl ester | 22.28 |
| 15 | butyl angelate | 24.00 |
| 16 | isoamyl 2-methylbutyrate | 24.75 |
| 17 | 2-methyl-butanoic acid 2-methylbutyl ester | 25.04 |
| 18 | trans - pinocarveol | 27.15 |
| 19 | 3-methyl-2-butenoic acid pentyl ester | 28.54 |
| 20 | 3-methyl-2-butenoic acid 3-methylbutyl ester | 28.85 |
| 21 | 6,6-dimethyl-2-methylene bicyclo[2,2,1]heptane-3-one | 29.00 |
| 22 | myrtenal | 31.47 |
| 23 | angelic acid 3-methylpentyl ester | 36,07 |
| 24 | germacrene-D | 50,39 |

Table 2. Identified compounds from the essential oil of roman chamomile harvested in 2003

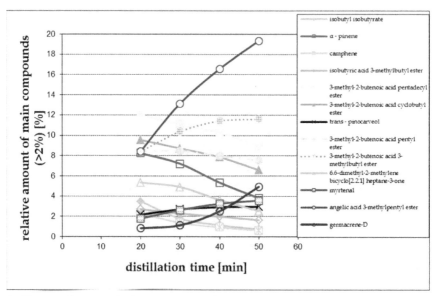

Fig. 15. Change of essential oil composition of main compounds during steam distillation of Roman chamomile (habitat Oberlungitz, first cut, fresh plant)

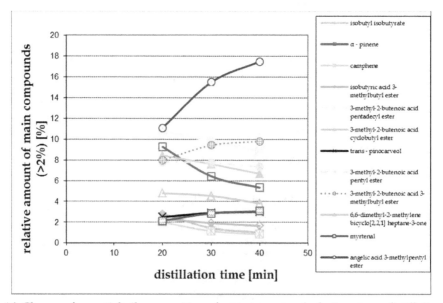

Fig. 16. Change of essential oil composition of main compounds during steam distillation of Roman chamomile (habitat Oberlungitz, first cut, drug)

6,6-dimethyl-2-methylene-bicyclo[2.2.1]heptane-3-one, isobutyl butyrate, isobutyric acid 3-methylbutyl ester and camphene decrease. 78.0 – 87.6% of the total Roman chamomile oil was identified and angelic acid 3-methylpentyl ester (8.4 - 19.3%), 3-methyl-2-butenoic acid pentadecyl ester (8.0 - 13.2%), α-pinene (3.8 - 12.2%), 3-methyl-2-butenoic acid 3-methylbutyl ester (7.0 - 11.6%), 3-methyl-2-butenoic acid cyclobutyl ester (6.6 - 10.2%), 3-methyl 2-butenoic acid pentyl ester (7.3 - 9.0%) and 6.6-dimethyl-2-methylene-bicyclo[2.2.1]heptane-3-one (2.5 - 6.1%) were identified as main compounds.

# 4. Conclusions

Based on the sampling during the steam distillation process of the two essential oils changes in the composition of the oils were observed during distillation. Thus, the achievable relative content of individual compounds of the essential oil depending on the plant (plant part) can be optimized. Additionally other plants and their essential oils were investigated. This might be of importance for future product developments, if specific essential oil compositions are required. In most cases the pilot plant scale distillation must not exceed 60 minutes distillation time, because after this period only very small amounts of essential oils are generated. That fact would not justify a longer distillation process for economic reasons.

# 5. Acknowledgment

Following organizations have provided financial support which is thankfully acknowledged:

- REG plus – Funding Regional Innovation (a Funding Programme of FFG - Austrian Research Promotion Agency)
- EUREGIO Styria – co financed by the INTERREG IIIa Programme

# 6. References

Bajaj, Y., 1996. *Biotechnology in Agriculture and Forestry - Medicinal and aromatic plants.* 37 ed. s.l.:Springer.

Barnes, J., Anderson, L. & Phillipson, J., 2007. *Herbal Medicines.* Third edition ed. s.l.:Pharmaceutical Press.

Board, N., 2003. *The complete technology book of essential oils (aromatic chemicals).* s.l.:Asia Pacific Business Press Inc..

Dachler, M. & Pelzmann, H., 1999. *Arznei- und Gewürzpflanzenanbau, - ernte und -aufbereitung.* s.l.:Österreichischer Agrarverlag.

Ferichs, G. & Rimbach , E., 1992. *Hagers Handbuch der pharmazeutischen Praxis: für Apotheker, Arzneimittelhersteller, Drogisten, Ärzte und Medizinalbeamte.* Berlin: Springer.

Heinrich , M., Barnes , J., Gibbons, S. & Williamson, E., 2004. *Fundamentals of Pharmacognosy and Phytotherapy.* s.l.:Elsevier Science.

Kintzios, S., 2000. *SAGE - The genus Salvia.* s.l.:harwood academic publishers.

Lis-Balchin, M., 2006. *Aromatherapy science: a guide for healthcare professionals.* London: Pharmaceutical Press.

Meier, U., 2001. *Entwicklungsstadien mono- und dikotyler Pflanzen: BBCH Monografie.* 2. Auflage ed. Berlin: Biologische Bundesanstalt für Land- und Forstwirtschaft.

# 6

# Distillation of Natural Fatty Acids and Their Chemical Derivatives

Steven C. Cermak, Roque L. Evangelista and James A. Kenar

*National Center for Agricultural Utilization Research,
Agricultural Research Service, United States Department of Agriculture
USA*

## 1. Introduction

Well over 1,000 different fatty acids are known which are natural components of fats, oils (triacylglycerols), and other related compounds (Gunstone & Norris, 1983). These fatty acids can have different alkyl chain lengths (typically ten or more carbon atoms), 0-6 carbon-carbon double bonds posessing *cis-* or *trans*-geometry, and can contain a variety of functional groups along the alkyl chain (Gunstone et al., 2007b). Of these, there are approximately 20-25 fatty acids that occur widely in nature, are produced from commodity oils and fats, and find

| Fat/Oil | Fatty acid length and unsaturation | | | | | | | | | | | | |
|---|---|---|---|---|---|---|---|---|---|---|---|---|---|
| | 8:0 | 10:0 | 12:0 | 14:0 | 16:0 | 18:0 | 18:1 | 18:2 | 18:3 | 20:0 | 20:1 | 22:0 | 22:1 |
| Canola | | | | 0.1 | 4.1 | 1.8 | 60.9 | 21.0 | | 0.7 | | 0.3 | |
| Coconut | 7.8 | 6.7 | 47.5 | 18.1 | 8.8 | 2.6 | 6.2 | 1.6 | | 0.1 | | | |
| Cottonseed | | | | 0.7 | 21.6 | 2.6 | 18.6 | 54.4 | 0.7 | 0.3 | | 0.2 | |
| Crambe | | | | | 1.7 | 0.8 | 16.1 | 8.2 | 2.9 | 3.3 | | 2.2 | 59.5 |
| Cuphea (PSR-23) | 0.8 | 81.9 | 3.2 | 4.3 | 3.7 | 0.3 | 3.6 | 2.0 | 0.3 | | | | |
| Palm | | | 0.2 | 1.1 | 44.0 | 4.5 | 39.1 | 10.1 | 0.4 | 0.4 | | | |
| Palm kernel | 3.3 | 3.4 | 48.2 | 16.2 | 8.4 | 2.5 | 15.3 | 2.3 | | 0.1 | 0.1 | | |
| Rapeseed | | | | 2.7 | 1.1 | | 14.9 | 10.1 | 5.1 | 10.9 | | 0.7 | 49.8 |
| Soybean | | | 0.1 | 0.2 | 10.7 | 3.9 | 22.8 | 50.8 | 6.8 | 0.2 | | | |
| Sunflower | | | | | 3.7 | 5.4 | 81.3 | 9.0 | | 0.4 | | | |
| Lard | | 0.1 | 0.1 | 1.5 | 26.0 | 13.5 | 43.9 | 9.5 | 0.4 | 0.2 | 0.7 | | |
| Tallow | | | 0.1 | 3.2 | 23.4 | 18.6 | 42.6 | 2.6 | 0.7 | 0.2 | 0.3 | | |

Table 1. Fatty acid composition of selected fats and oils (Evangelista & Cermak, 2007; Knapp, 1993, O'Brien, 2004; Stauffer, 1996)

major use for food and nutrition applications with the remainder being used by the oleochemical industry to produce soaps, detergents, personal care products, lubricants, paints, and more recently, biodiesel. Approximately 17 commodity fats and oils are obtained from various domesticated plants and animals. The largest vegetable oil sources are the oilseed crops (soybean, rapeseed, sunflower, and cottonseed) grown in relatively temperate climates. Another major oil source are oil-bearing trees (palm, coconut, and olive) grown in tropical or warm climates (O'Brien et al., 2000). The triglyceride-containing oils are extracted from oilseeds by mechanical pressing or by using solvent extraction ($n$-hexane). Seeds containing high oil contents are usually mechanically extracted first to reduce the oil content in the seed by 60% before solvent extraction. Animal fats are obtained by rendering inedible animal by-products like fat trim, meat, viscera, bone, and blood, generated by slaughter houses and meat processing industry and mortalities on farms (Dijkstra & Segers, 2007; Hamilton et al., 2006). World fat and oil production in 1998 was 101 million tons, of which 14.2% (14.3 million tons) was used as basic oleochemicals (Hill, 2000). In 2009, the global production of fats and oils increased to 137.5 million tons with 21.2% (29.3 million tons) used for non-food industrial purposes (Gunstone, 2011). This growth was driven by the high petroleum prices as well as the growing demand for natural or renewable products (de Guzman, 2009).

| Symbol | Systematic Name | Trivial Name | Melting Point[a,b] (°C) | | Boiling Point[c] (°C/(10 mm Hg)) | |
|---|---|---|---|---|---|---|
| Saturated fatty acids | | | Acid | Methyl Ester | Acid | Methyl Ester |
| 10:0 | decanoic | capric | 31.0 | -13.5 | 150 | 108 |
| 12:0 | dodecanoic | lauric | 44.8 | 4.3 | 173 | 133 |
| 14:0 | tetradecanoic | myristic | 54.4 | 18.1 | 193 | 161 |
| 16:0 | hexadecanoic | palmitic | 62.9 | 28.5 | 212 | 184 |
| 18:0 | octadecanoic | stearic | 70.1 | 37.7 | 227 | 205 |
| 20:0 | eicosanoic | arachidic | 76.1 | 46.4 | 248[d] | 223[d] |
| 22:0 | docosanoic | behenic | 80.0 | 53.2 | 263 | 240 |
| 24:0 | tetracosanoic | lignoceric | 84.2 | 58.6 | --- | 198(0.2)[e] |
| Unsaturated fatty acids[f] | | | | | | |
| 16:1 | 9-hexadecenoic | palmitoleic | 0.5 | -34.1 | 180(1)[a] | 182 |
| 18:1 | 9-octadecenoic | oleic | 16.3 | -20.2 | 223 | 201 |
| 18:2 | 9,12-octadecadienoic | linoleic | -6.5 | -43.1 | 224 | 200 |
| 18:3 | 9,12,15-octadecatrienoic | linolenic | -12.8 | -52.4 | 225 | 202 |
| 20:1 | 9-eicosenoic | gadoleic | 23.0 | --- | 170(0.1)[a] | 154(0.1)[e] |
| 20:4 | 5,8,11,14-eicosatetraenoic | arachidonic | -49.5 | --- | 163(1)[a] | 194(0.7)[a] |
| 22:1 | 13-docosenoic | erucic | 33.5 | -3.5 | 255 | 242 |

Table 2. Nomenclature of selected fatty acids and their respective melting and boiling points. [a]Gunstone et al., 2007b. [b]Knothe & Dunn, 2009. [c]Budde, 1968. [d]Farris, 1979. [e]Ethyl ester. [f]Double bonds in the all *cis*- geometry.

The fatty acid composition of fats and oils varies widely depending on the source (Table 1). Coconut and palm kernel oils contain high amounts of medium chain saturated fatty acids like lauric and myristic acids (Table 2). Palm, tallow and lard oils are high in longer saturated fatty acids (palmitic and stearic acids) and monounsaturated oleic acid. Canola, and sunflower oils are high in oleic acid while soybean oil has more linoleic acid. Rapeseed and crambe are good sources of long chain fatty acids like erucic acid.

The first step in fatty acid production (Fig. 1) is the splitting or hydrolysis of the triglyceride molecules of fats and oils in the presence of water to yield glycerine (10% yield) and a mixture of fatty acids (96% yield), (Gunstone et al., 2007a).

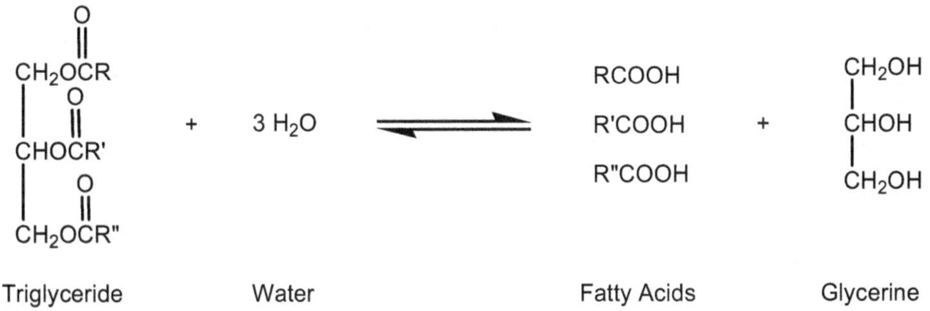

$$
\begin{array}{ccccccc}
\underset{\text{Triglyceride}}{
\begin{array}{l}
CH_2OCR \\
\quad\; \overset{O}{\overset{\|}{\phantom{l}}} \\
CHOCR' \\
\quad\; \overset{O}{\overset{\|}{\phantom{l}}} \\
CH_2OCR''
\end{array}}
& + &
\underset{\text{Water}}{3\,H_2O}
& \rightleftharpoons &
\underset{\text{Fatty Acids}}{
\begin{array}{l}
RCOOH \\
R'COOH \\
R''COOH
\end{array}}
& + &
\underset{\text{Glycerine}}{
\begin{array}{l}
CH_2OH \\
CHOH \\
CH_2OH
\end{array}}
\end{array}
$$

Fig. 1. Splitting or hydrolysis of fat or oil triglycerides to fatty acids and glycerine

This can be done batch-wise using the Twitchell process (Ackelsberg, 1958; Twitchell, 1898) or continuously at high pressure and temperature like the Colgate-Emery process (Barnebey & Brown, 1948). Typically, the crude fatty acids obtained by the Colgate-Emery process are considerably lighter in color in comparison to those obtained by the Twitchell process. The degree of triglyceride hydrolysis is important as residual mono-, di-, triglycerides and free glycerol in the fatty acid prior to distillation will result in more distillation pot residue (Potts, 1956). The fatty acids from the fat splitting process are relatively dark in color and contain various impurities. The fatty acids are subsequently purified or separated into fractions by distillation and fractionation.

## 2. Distillation methods used in fatty acid industry

Purification of fatty acids by distillation has been practiced for well over a hundred years and is still the most common and most efficient means of producing high purity fatty acids. Distillation removes both the low and high boiling impurities as well as odor substances. Distillation of fatty acids may be either batch or continous process, at atmospheric pressure or under reduced pressure. It may be simple distillation involving purification of mixed fatty acids or fractional distillation consisting of both purification and separation of fatty acids according to chain length (Gervajio, 2005; Muckerheide, 1952; Potts & White, 1953). Because of the inherent sensitivity of fatty acids toward heat, the distillation methods employed should be conducted at as low a temperature as practically and economically feasible while maintaining the shortest residence time of the fatty acid in the distillation unit. Today's, modern distillation units rely upon high vacuum, effective heating, short contact times, effective mass transfer between vapor and condensate, and steam economy (Lausberg et al., 2008).

## 2.1 Batch distillation

Batch distillation at atmospheric pressure is probably the oldest of the commercial processes used in fatty acid distillation. It uses a direct-fired still pot fitted with a steam sparger. The pot is charged with fatty acids and heated to 260° to 316°C and sparged with saturated steam at 149°C. The ratio of steam to fatty acid vapor is typically 5 to 1. The steam and fatty acid vapor are condensed separately. The economics of this type of distillation is poor due to the large amount of steam used. Considerable amounts of fatty acids are also entrained in the steam condensate. Distillation is further complicated because of the prolonged heating of the fatty acids at high temperatures and the inherent thermal instability of the fatty acids. This combination often results in considerable decarboxylation and polymerization with consequently large amounts of viscous residue and pitch. Tall fatty acids of about 95% hydrolysis when distilled in this manner yield 15 to 20% entrained fatty acids and 10 to 15% residue. Re-splitting the residue and distillation yields low quality fatty acids and a final pitch residue of 5 to 8% (Muckerheide, 1952). Later improvements in this distillation technique included working at reduced pressure (5-50 mm Hg) and lowering the amount of injected steam. The water from the steam is desirable as it suppresses anhydride formation (Potts, 1956).

## 2.2 Continuous distillation

Probably the first fatty acid still to use continuous distillation was developed by Wecker (1927). A simplified diagram of this process is illustrated in Fig. 2. Preheated fatty acid feed enters through pipe c and flows through a series of reaction chambers a interconnected by pipe b. The reaction chambers are heated at the bottom by gas or oil burners. Superheated steam is introduced through pipe l and injected into the feed in each chamber by a sparger (m and n). The low pressure imposed in the reaction chamber and the high temperature of the feed caused the superheated steam to evaporate vigorously resulting in an instantaneous distillation of the fatty acids. The vapor are led to a pipe header g, condensed by a water-cooled condenser h, and collected in i. The steam passes on to the barometric condenser through k and the non-condensable gases are removed by a vacuum pump. The residue leaving the last reaction chamber is cooled in d and into collector e. Vacuum on the still is maintained at 30-35 mm Hg and the temperature in the still chambers ranged from 196° to 260°F. Residence time of fatty acids is about 30 min.

One disadvantage of steam distillation of fatty acids is the formation of emulsions in the last stage of condensation where a water spray is used. The calcium and magnesium salts in the water spray react with the fatty acids forming soaps. To recover the fatty acids, the soap is acidified and redistilled if desired. This can be avoided by employing dry distillation, i.e., distillation without using steam or any gaseous medium as carrier of the fatty acids. Such process was developed by Mills (1942) who employed a combination of dry and flash distillation to recover fatty acids from hydrolyzed fats and oils (Fig. 3). The fatty acid to be distilled is rapidly heated using a heat exchanger (4) to the boiling point corresponding to the operating pressure (≤ 12.7 mm Hg absolute) in the still (10). When the heated feed is introduced to the bottom of the tube (13) and exposed to the lower pressure in the still, the fatty acids vaporizes immediately. The vapors lift the undistilled residue (11) from the bottom of the still up the tube and splashes against the bottom of the baffle (15) creating a continuous curtain of liquid undistilled material. The vapor proceeds to the condensers (17

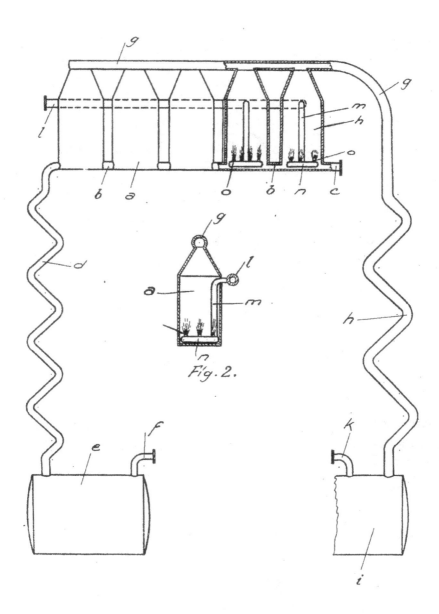

Fig. 2. Continuous distillation (Wecker, 1927)

and 18) and the fatty acid condensates are collected in the receivers (20 and 21) which can be withdrawn continuously or intermittently. The undistilled material is withdrawn continuously through pipe (30) which can be directed by valve (35) back to the heat exchanger or by valve (34) to the residue collector (31).

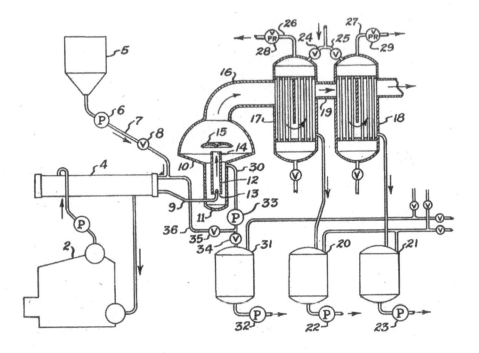

Fig. 3. Continuous dry distillation (Mills, 1942)

## 2.3 Fractional distillation

Because fatty acids are derived from natural sources, their initial and distilled compositional mixtures tend to vary even when the same type of fat or oil is used. Users generally prefer pure fatty acids or mixtures of fatty acids of consistent composition and known properties. Fractional distillation separates fatty acids based on their boiling points. Fatty acids which differ in chain length by two carbons are easily separated, thus; fatty acid fractions of 90% or better purity are obtained (Potts & White, 1953; Ruston, 1952). Fundamentally, fractional distillation is carried out in the same manner as continuous distillation. The main difference is in the design of the main fractionating column which is fitted with several bubble cap trays, means for removal of side stream distillates of fatty acids and return part of these streams as reflux (Muckerheide, 1952; Stage, 1984).

In the fractionating column, vapors move upwards through the column and condensed at the top. A portion of the condensate is returned as reflux downwards through the column where it is brought into more or less intimate contact with the ascending vapors. Heat is exchanged between the rising warmer vapor and the cooler descending condensate. The more volatile fraction in the condensate is vaporized and the easily condensable fraction in the vapor is condensed. Under ideal conditions, the heat lost by the rising vapor is gained by the descending condensate, with no heat loss or gain from the outside. The net result is

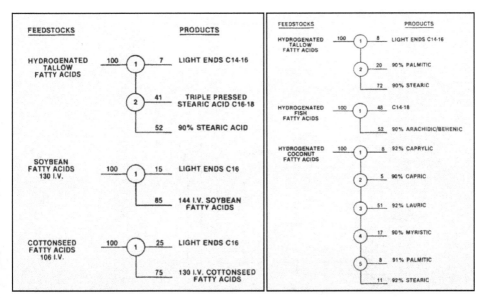

Fig. 4. Partially fractionated hydrogenated tallow, soybean and cottonseed fatty acids and fully fractionated hydrogenated tallow, fish, and coconut fatty acids (Berger & McPherson, 1979)

the concentration of more volatile fractions on top of the column and the increasing concentration of less volatile fraction at the bottom of the column (Norris & Terry, 1945). Fractionating stills are custom designed to suit the feedstock and product requirements. With lauric type fatty acids from coconut and palm kernel oils, up to 30 fractionating trays can be used for highest purity fraction because of the higher volatility and greater stability of the shorter chain fatty acids. Long chain fatty acids like erucic (C22:1) in rapeseed oil have much lower vapor pressure and would need a limited number of fractionating trays to keep the reboiler below the decomposition temperature (Berger & McPherson, 1979). Commercial fatty acid products that can be obtained by fractional distillation are shown in Fig. 4.

The first continuous fractional distillation unit for the separation of a fatty acid mixture was installed by Armour and Company in 1933 (Fig. 5). The system consisted of the main fractionating tower, two smaller side stripping towers, conventional air ejectors and boosters, condensers, coolers, and a direct-fired fatty acid heater. The direct-fired fatty acid heater was susceptible to coking and corrosion from the fatty acids which resulted in operation downtime. Shell and tube heaters using condensing Dowtherm vapor as source of heat replaced the direct-fired heater in subsequent installations (Potts & White, 1953).

Fractional distillation was also employed by General Mills in their fat and oil processing plant which started operation in 1948. The feed stock is introduced into the first distillation tower and heated by the rising vapours from the base of the tower (Fig. 6). This eliminated the problem of fouling in heating tubes when preheating incoming feed. Also, to conserve space and construction cost, the second distillation tower was superimposed on the third. Fractionated fatty acids, fatty acid esters, and their derivatives were produced from low grade fats, oils, acid oils, and tall oil.

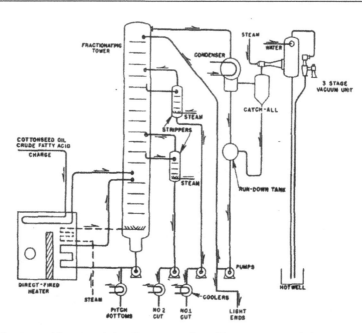

Fig. 5. Flow diagram of fractional distillation employed by Armour and Company in 1933 (Potts & White, 1953)

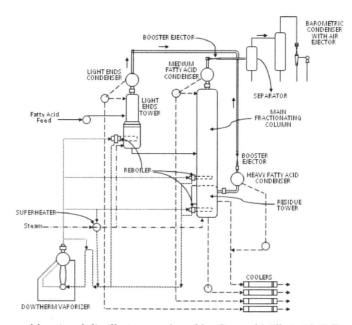

Fig. 6. Flow diagram of fractional distillation employed by General Mills in 1948 (Potts & White, 1953)

## 2.4 Molecular distillation

Molecular distillation is industrially useful in the purification of unstable or highly oxidatively unstable fats, oils and their derivatives. Molecular distillation consists of at least the following types: wiped film molecular distillation unit (Fig. 7) and centrifugal molecular distillation unit (Fig. 8).

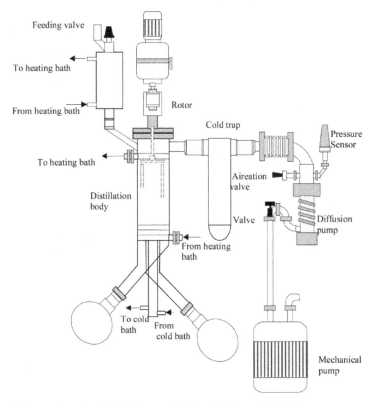

Fig. 7. Wiped film molecular distillation unit (Marttinello et al., 2008)

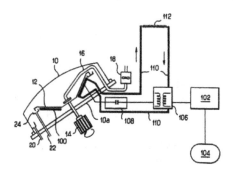

Fig. 8. Centrifugal film molecular distillation unit (Nuns et al., 1994)

Molecular distillations are conducted under vacuum conditions, which allows for reduced distillation temperatures compared to conventional distillation techniques thus reducing the risk of oxidative damage. Separating the oil's components by weight allows contaminates to be reduced far below industry standards. Current industrial applications include cosmetic applications, the concentration of omega-3 fatty acids (EPA and DHA) and corresponding esters in fish oil (Rossi et al., 2011), and contaminant removal. Additionally, Vitamin E (Pramparo et al., 2005), Vitamin A, cocoa butter, dimer acids, epoxy resins, lubricants, monoglycerides, insecticides, pharmaceuticals, perfumery and flavours, essential oils, Azadirachtin (Neem based pesticides) and its formulations as well as many other natural & herbal products have been distilled on an industrial scale using this process. Additional applications and information on fatty acid molecular distillation will be reported in Section 3.3 Distillation.

## 3. New crops and products

Development of new crops in the United States is based on a history of screening and identifying plants for novel chemicals, germplasm development of a select few unique plants, isolation of raw materials (processing, refining and distilling), product development, evaluation, and scale-up to commercial production and, finally, the transfer of knowledge to industrial partners and farmers. In most cases, these new crop identifications initially begin with the individual selection based on oil composition and unique fatty acid profiles, followed by continued evaluation and development based on novel raw materials, industrial applications of the raw materials and agronomic potential. From this extensive survey, two crops were selected to focus on (Cuphea and Meadowfoam). Additionally, these crops have had been converted into industrial products which require some degree of purification. Thus separations of both the new crop fatty acids and new products have undergone distillations.

### 3.1 New crop examples

### 3.1.1 Meadowfoam

One example of a successful new crop is meadowfoam (*Limnanthes alba*). In the 1960's, USDA scientists identified this plant as a potential new crop out of hundreds of others that needed further research. In the 1970's, a meadowfoam breeding program began the daunting task of domesticating this plant. By the 1990's, numerous varieties had been released and meadowfoam reached commercial scale (approximately 8000 acres in 1997) in the Willamette Valley of Oregon located in the Northwestern US. The unique long chain fatty acids of meadowfoam [5-eicosenoic acid (62%); 5,13-docosadienoic acid (19%); 5-docosenoic acid (3%); and 13-docosenoic acid (10%)] helped advance its development (Phillips et al., 1971). Finally, the $\Delta 5$ unsaturation itself has enhanced resistance to oxidative degradation as evidenced by the high OSI of the oil (246.9 h @ 110°C) which makes meadowfoam oil one of the most stable oils available on the market (Isbell et al., 1999). Meadowfoam oil is found in numerous cosmetic applications, such as hair shampoos and conditioners, skin creams, hair setting aids, permanents, hair relaxers, and hair colors.

## 3.1.2 Cuphea

Cuphea (*Lythraceae*) is a large genus of over 200 species of herbs and shrubs that produces a small seed with oil that is rich in saturated medium-chain triacylglycerols (which can be converted to medium chain fatty acids, MCFAs) (Graham et al., 1981; Knapp, 1993; Miller et al., 1964; Wolf et al., 1983). MCFAs (C8:0-C12:0) are used primarily in soaps (Nandi et al., 2004), detergents (Molly & Bruggeman, 2004), cosmetics (Brown et al., 2008), lubricants (Nagaoka & Ibuki, 2000), and food applications (Tholstrup et al., 2004). One-half of the MCFAs used by the US soap and detergent industry are obtained from coconut and palm kernel oils, while the other half is from petroleum (Hardin, 1991).

With the need for higher seed yields, oil content, and less seed shattering, Steve Knapp (1993) at Oregon State University began developing promising cuphea crosses. Cuphea PSR-23 is a hybrid between *Cuphea viscosissima* (a species native to the US) and *Cuphea lanceolata* (a species native to Mexico). One of these new germplasm lines, PSR-23, with partial seed retention and high in C10:0, decanoic acid, was developed and has been planted in the Midwest and mechanically harvested and dried (Cermak et al., 2005) by researchers at USDA since 2000. The seeds weighed 538 g/L (3.3 g/1,000 seeds) and contain up to 35% oil (Table 1). The oil typically contains around 82% capric, 4% oleic, 4% palmitic, 2% linoleic, 4% myristic, and 3% lauric acids (Evangelista & Manthey, 2004; Gesch et al., 2005; Kim et al., 2011).

Recent seed increases in Cuphea provided sufficient amounts of seed to conduct oil extraction studies and, at the same time, produce much needed oil for product development and applications testing. The first report on full press oil extraction from Cuphea seeds was also optimized in a pilot-scale study (Evangelista & Cermak, 2007). Oil extracted by pilot plant screw pressing of whole Cuphea seeds produced dark green colored oil. Chlorophyll content of the oil, ranging from 200-260 ppm, had been reported, but levels up to 326 ppm have recently been observed (Evangelista & Cermak, 2007). About 6.5-8% bleaching clay had been used in the bleaching step to bring the chlorophyll level in the refined oil to 0.5 ppm. Aside from the added cost of the bleaching clay, more oil is also lost as these adsorbents retain between 50-75% their weight of oil. With a supply of cuphea oil a series of new products could be developed from this new oil or corresponding fatty acids ranging from biodiesel (Geller et al., 1999), to cosmetics (Brown et al., 2008) to lubricants (Cermak & Isbell, 2004a,b; Cermak et al., 2008).

## 3.2 Estolides

Estolides have been used to help develop new products from industrial crops (Cermak & Isbell, 2001b). Estolides are formed by the formation of a carbocation that can undergo nucleophilic addition with or without carbocation migration along the length of the chain. The carboxylic acid functionality of one fatty acid links to the site of unsaturation of another fatty acid to form oligomeric esters. The extent of oligomerization is reported by estolide number (EN) which is defined as the average number of fatty acids added to the base fatty acid. The estolide carboxylic acid functionality can be converted *in situ* under esterification conditions with the addition of an alcohol to yield the corresponding estolide ester (Fig. 9).

Estolides from a number of fatty acids have been shown to have desirable low-temperature properties (Cermak & Isbell, 2001a,b; Isbell et al., 2000a) making them suitable for lubricants. However, the current market price of meadowfoam precludes its use in these markets.

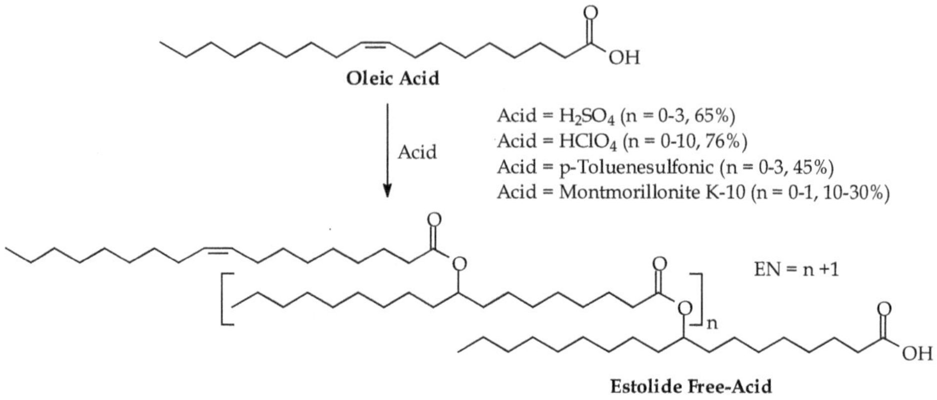

Fig. 9. General Oleic Estolide Free-Acid Synthesis

Therefore, meadowfoam has been marketed into cosmetics where the estolide has shown good properties for use in hair conditioners (Isbell et al., 2000b), but dark colored products have limited the estolides use. Frykman and Isbell (1999) have shown that some color reduction is possible by bleaching with sodium borohydride, but further color reduction may be required. A second method for estolide purification was reported based on chromatography, but this method only provided trace quantities of individually purified oligomers for qualitative characterization (Isbell & Kleiman, 1994 and 1996) and would not serve as a means for industrial-scale estolide production.

### 3.3 Distillation

Coconut and palm kernel oils alone provided more than 450 thousand tons of decanoic acid to the global market in 2006, where it is used as a wood preservative (Yoshida & Iinuma, 2004) cosmetic agents for hair (Hoppe & Engel, 1989), activity against termites (Goettsche & Borck, 1994), dental compositions (Velamakanni et al., 2006), lubricants (Cermak & Isbell, 2004a), fabric softeners (Hohener & Frick, 2003) and cosmetics (Ishii & Mikami, 1995).

Cuts of saturated fatty acids can be successfully distilled from coconut and palm kernel fatty acids at relatively low temperatures to produce nearly colorless fatty acids. There are many compounds that are sensitive to heat such as high vacuum oils (Rees, 1980), vegetable oils (Cermak & Isbell, 2002; Isbell & Cermak, 2004), pharmaceuticals, and cosmetics (Batistella & Marciel, 1996) which prohibit the use of conventional distillation techniques. Molecular or short-path distillation, which has been known for some time (Biehler et al., 1949), uses a high vacuum to achieve distillation of thermally unstable materials, and is often the most economically feasible method of purification. Centrifugal and falling films are two basic types of molecular distillation units which use a short exposure of the distilled liquid to the evaporating cylinder. The high temperature exposure time in these stills is on the order of a few seconds to tenths of seconds as the liquid is spread evenly in the form of a film (Micov et al., 1997). These types of distillation units have been used successfully to demonstrate and compare the distillation of many different compounds, such as carotenoids from palm oil (Batistella & Marciel, 1998), cuphea fatty acids (Cermak et al., 2007), and estolides (Isbell & Cermak, 2004).

At the USDA laboratory, estolides and new crop fatty acids have been distilled on both the Myers Lab 3 and Myers Pilot 15 molecular distillation units (Myers Vacuum, Kittanning, PA), the simpler of the two being the Myers Lab 3 (diagram illustrating the operational features of this unit is presented in Fig. 10). The condenser temperature can be set to the desired temperature, rotor speed was constant at 1725 rpm, and cold tap water was used to cool the diffusion pump and rotor bearing. Vacuum pressure is usually maintained at $6x10^{-4}$ – $3x10^{-3}$ mm Hg at both the chamber and foreline pressure sensors.

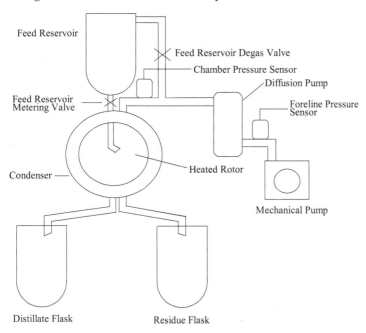

Fig. 10. Schematic diagram of Myers Lab 3 short path molecular distillation unit

The more complex unit, a Myers Pilot-15 distillation unit (Fig. 11), was used for pilot scale distillations (Cermak & Isbell, 2002). It is a continuous, centrifugal 38.1 cm molecular still that contains all of the components needed for distilling raw feedstock. Raw feedstock is delivered with a metering valve and a gear pump. It first enters a degasser unit which is maintained at pressures between $5.4x10^{-3}$ – $6.8x10^{-2}$ mm Hg. It then enters the heated evaporator cone or distillation chamber, where the molecular distillation takes place. The distillate and residue are continuously removed by transfer pumps. The material is passed to and from each station through stainless steel transfer lines, which are traced with heating tapes. This unit was designed for use in extended or large scale distillations of ~50 gals/day.

The major component of the Myers Pilot-15 distillation unit is a concave, heated evaporator cone, which rotates within the distillation chamber and is maintained at pressures as low as $<6.8x10^{-5}$ mm Hg (Fig. 12). Degassed and/or stripped feedstock is metered by the feed pump into the center of the spinning evaporator cone (1700 rpm) and is spread rapidly and evenly outward in a thin film over the entire surface by centrifugal forces. As the film spreads and is heated, part of the feedstock reaches a temperature at which it vaporizes and

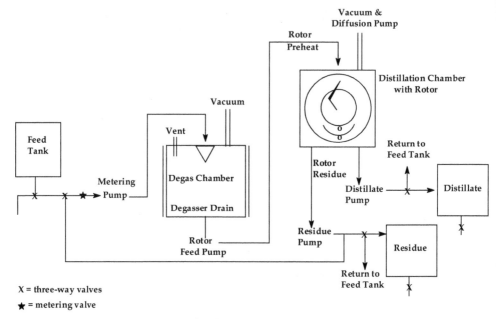

Fig. 11. Operational Flow Diagram for the Myers Pilot-15 Distillation Unit

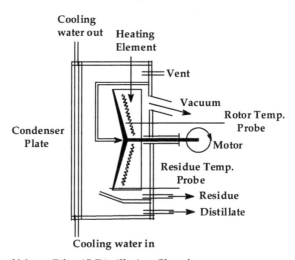

Fig. 12. Side View of Myers Pilot-15 Distillation Chamber

leaves the rotor surface to condense on the cooler surface of the condenser plate. With the help of centrifugal force the unvaporized feedstock, the residue, is spun into a gutter, where it is removed from the still by a constant-speed transfer pump. The condensed vapor, the distillate, flows by gravity into a removal pipe located at the bottom of the distillation chamber and then pumped from the still by another constant speed transfer pump. The operational distillation unit has two temperature sensors in the distillation chamber. The first sensor, which monitors the temperature of the rotor, is located by the electric heating elements just beneath the 38.1 cm rotor. This first temperature corresponds to the temperature of the heating element, not the temperature of the rotor. The actual temperature of the rotor can be varied by varying the flow of feed material without affecting the element temperature. The second and most important temperature sensor measures the temperature of the residue. This sensor is located in the gutter where the residue is collected during distillation. The residue sensor provides the closest temperature of the rotor, since it is a measurement of the temperature of the material that was just in contact with the rotor.

All the transfer lines and pumps are wrapped with heating tapes. The heating tape and electrical elements are controlled with digital controllers either based on percent power demand or actual measured temperature. Each region of the transfer line and pump has internal temperature probes that measure the temperature of the material as it passes the sensors.

These two distillation units common to industrial processes have allowed new crop fatty acids and products to be distilled at levels acceptable for industrial purposes. These materials have than been used in commercial products such as cosmetics (Brown et al., 2008; Isbell et al., 2000b) and lubricants (Cermak & Isbell, 2004a,b; Cermak et al., 2008).

### 3.3.1 Cuphea

Estolides have been previously synthesized by USDA's National Laboratory in Peoria, IL, from C10:0 decanoic acid, cuphea oil, Table 1 (Cermak & Isbell, 2004a). These estolides have useful properties as lubricants based on the physical properties exhibited by these C10:0 materials. An issue associated with the decanoic estolides is the color of the final product. Commercial decanoic acid is not as refined as other mid-chain fatty acids, so removal of the color bodies is necessary for light colored estolides. The measurement of the color of a material is designated as the Gardner color. The Gardner color scale is from 1 to 18 with 1 containing the least amount of color and 18 with the maximum amount of color. Lubricant manufacturers and consumers would prefer lubricants with colors similar to current petroleum oils, lower numbers i.e. Gardner 1 to 3. To obtain these properties, the starting material must be distilled, which is costly and can lead to undesired additional color bodies with certain distillation techniques. The ability to achieve a fast and mild distillation of short-chain fatty acids could be used to obtain oil with low Gardner color while the residue would not be darkened by the distillation process. These improvements with decanoic acid would help with the commercialization of cuphea as a new oil seed and give US farmers a valuable rotation crop. This studies objective was to investigate the general conditions necessary for laboratory molecular distillation of decanoic acid or enrichment from cuphea fatty acids. Fatty acid profiles, Gardner colors, and flow rate requirements were examined to determine the best set of operating conditions.

The basic experimental conditions for the two different flow rates, high and low, over eight different rotor temperature settings, their effects on split ratio determined by mass, Gardner colors, and percent short saturated fatty acids are reported in Table 3. The split ratios of the high and low flow rates across the rotor are similar (Fig. 13A). However, as the rotor temperature was increased to 110°C and greater, the split ratios increased with the lower flow rate. The increased mass in the distillate fraction would be expected at the lower flow

| Trial | Rotor temp[a] (°C) | Flow (g/min) | Split flow | C8 and C10 FAs (%)[b] | | Gardner color | |
|-------|--------------------|--------------|------------|------------|---------|------------|---------|
| | | | | Distillate | Residue | Distillate | Residue |
| 1 | 40 | 1.49 | 0.03 | 98.9 | 81.9 | 1- | 1 |
| 2 | 40 | 0.49 | 0.03 | 98.9 | 81.5 | 1- | 1 |
| 3 | 50 | 1.36 | 0.30 | 98.5 | 77.4 | 1- | 2 |
| 4 | 50 | 0.54 | 0.12 | 98.8 | 80.3 | 1- | 3 |
| 5 | 55 | 1.65 | 0.93 | 97.0 | 69.5 | 1- | 1 |
| 6 | 55 | 0.49 | 0.37 | 98.3 | 76.9 | 1- | 1 |
| 7 | 60 | 1.99 | 1.30 | 96.5 | 63.4 | 1- | 1 |
| 8 | 60 | 0.52 | 0.85 | 96.2 | 69.2 | 1- | 2 |
| 9 | 65 | 2.23 | 1.47 | 95.5 | 62.7 | 1- | 1 |
| 10 | 70 | 1.36 | 1.44 | 95.1 | 62.3 | 1- | 3- |
| 11 | 70 | 0.46 | 0.98 | 96.1 | 67.9 | 1- | 2 |
| 12 | 90 | 1.72 | 1.79 | 88.2 | 58.9 | 1- | 1 |
| 13 | 90 | 0.50 | 1.97 | 89.8 | 66.6 | 1- | 1 |
| 14 | 110 | 1.87 | 4.63 | 83.8 | 69.1 | 1- | 6- |
| 15 | 110 | 0.54 | 4.92 | 83.3 | 66.9 | 1- | 7 |

[a]Heated rotor spinning at 28.75 Hz under a high vacuum ($6 \times 10^{-4}$ – $3 \times 10^{-3}$ mm Hg)
[b]Saturated fatty acids determined by GC (SP-2380, 30 m x 0.25 mm i.d.)

Table 3. Effects of molecular distillation parameters on fatty acids, split ratio, and color

rate because heat transfer to a smaller amount of material passing across the rotor would be more efficient (Isbell & Cermak, 2004). Data points are not reported past 110°C because the saturated fatty acids in this study had distilled before that temperature.

As the rotor temperature was increased the split ratios of distillate to residue increased and affected what was being distilled. At increased temperatures, more unsaturates and higher chain saturated fatty acids were distilled affecting the purity of the sample. The purity of the distillate as defined by the percent of octanoic (C8:0) and decanoic (C10:0) fatty acids is shown in Fig. 13B. Both the high and low flow rates gave similar results as the temp was increased. All temperatures 70°C and less provided materials that were greater than 95% enriched in C8:0 and C10:0 FAs. With a lower rotor temperature a greater quantity of short saturated fatty acids was present in the distillate faction. As the rotor temperature was increased, both the low and high flow rate distillates contained lower percentages of the short saturated fatty acids, but the amounts of distillate increased over the amount of residue.

As the rotor temperature was increased, the distillate to residue split ratio increased, but the percent of short-chain fatty acids decreased in the distillate samples. The split ratios of the multiple-pass distillations increased with each consecutive distillation from 2.50 (Table 4, Trial 16) to 16.59 (Trial 19). The distillate percents of C8:0 and C10:0 and overall saturates all increased as expected as the distillate was further purified to obtain high percents of total

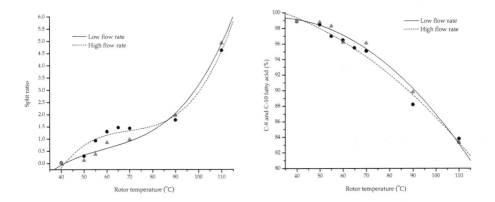

Fig. 13. A) Effect of rotor temperature on split ratio or B) Effect of rotor temperature on C8:0 and C10:0 fatty acids

| Trial | Distillation | Split Flow Ratio | Fatty acid (%)[a] | | | | | | | |
|-------|-------------|------------------|------|------|------|------|------|------|------|------|
| | (pass) | | 8:0 | 10:0 | 12:0 | 14:0 | 16:0 | 18:0 | 18:1 | 18:2 |
| 16 | First | 2.50 | 0.8 | 89.8 | 3.1 | 3.0 | 1.6 | 0.0 | 1.0 | 0.6 |
| 17 | Second | 11.33 | 0.7 | 91.4 | 3.0 | 2.6 | 1.2 | 0.0 | 0.6 | 0.4 |
| 18 | Third | 15.50 | 0.7 | 92.4 | 2.9 | 2.4 | 0.9 | 0.0 | 0.4 | 0.3 |
| 19 | Fourth | 16.59 | 0.6 | 92.9 | 2.9 | 2.2 | 0.8 | 0.0 | 0.4 | 0.2 |
| 20 | Fifth | 11.79 | 0.6 | 93.8 | 2.7 | 2.0 | 0.6 | 0.0 | 0.2 | 0.0 |

[a]Determined by GC (SP-2380, 30 m x 0.25 mm i.d.) and standard deviation < + 0.10

Table 4. Fatty acid profile of distillate distillations at 70°C

saturates (99.8 %, Trial 19). The fatty acid profile for the multiple-pass distillation is shown in Table 4. After the first pass (Trial 16), stearic and linolenic fatty acids were removed and each additional pass started to remove the linoleic and oleic fatty acids. The final pass yielded a material that contained only 0.2% oleic as the only unsaturation in the fraction.

Single-pass distillation is the simplest and least expensive mode of distillation, but some applications require additional distillations to achieve difficult separations. Separating materials that are very close in molecular weight or have very similar boiling points can require multiple-pass distillations. One of the main reasons for performing single or multiple-pass distillations is to aid in the removal of color bodies. All the cuphea FA distillates had very excellent Gardner colors of 1-, which is the lowest color rating (Table 3). The residues varied in the amount of color bodies in the samples depending on whether the residue was a single or multiple-pass distillation. A multiple-pass distillation of a distillate is prime example to explore the removal of color bodies. The first distillation gave a distillate with a Gardner color of 1-, however, when that material was re-distilled, the residue led to a Gardner color of 6+ (Cermak et al., 2007). Each consecutive distillation had higher split flow rates and showed that most of the material was being distilled. This increase was expected as only small amounts of saturates and/or color bodies were left to be removed with each distillation. This multiple-pass distillate would be very desirable for very color sensitive applications.

With the Myers 3 being primarily a laboratory molecular distillation unit, one of the main goals was to have a high throughput while collecting quality saturated material. This equipment would meet these high throughput conditions when the rotor temperature was set to 65°C with a high flow rate of 2.2 g/min while maintaining a distillate to residue split ratio of about 1.5 as demonstrated in Table 3 (Trial 9). The distillate produced would contain high percents of C8:0 and C10:0 as well as no unsaturates. The Gardner color values were excellent under these conditions as well. Thus, cuphea fatty acids were effectively separated with the Myers 3 lab-scale centrifugal molecular distillation unit.

### 3.3.2 Meadowfoam

Previous studies of meadowfoam oil and the corresponding fatty acids have resulted in several novel compounds, i.e., estolides (Isbell & Kleiman, 1994 and 1996) and lactones (Isbell & Cermak, 2001). Meadowfoam estolides have cosmetic applications as they improve conditioning, shine, and comb-out compared with existing conditioners (Isbell et al., 2000b). One of the biggest problems with meadowfoam estolides is their color. Ideally, cosmetic ingredients should be colorless (Gardner=0). Crude meadowfoam fatty acids are very dark (Gardner=15+), which leads to very dark estolides (Gardner=18+). If crude meadowfoam fatty acids could be purified to a low Gardner color at a low cost, this would lead to a lower colored estolide and eliminate post-distillation decolorization (Frykman & Isbell, 1999). Color improvements in meadowfoam estolide and fatty acids would help the commercialization of this new oil seed and give US farmers a valuable rotation crop.

Crude mixtures of meadowfoam fatty acids were separated using a Myers Pilot-15 molecular centrifugal distillation unit. A series of conditions were examined to identify the optimum operating conditions including: rotor temperature, degas temperature, rotor preheat, and flow to the rotor (Cermak & Isbell, 2002). The main heating source for the distillation unit is the rotor element, which is located beneath the rotor and can range anywhere from room temperature to 800°C (Fig. 12). Thus, the easiest place to make a significant impact on heat available for distillation is in the distillation chamber. A set of conditions where the rotor temperature was varied while other system conditions remained constant is shown in Table 5. The flow rate of the feed stock was set at 100 g/min. The rotor temperature was changed from 275 to 475°C as the data points were collected. As the temperature increased, the distillate to residue ratio also increased, which was expected. At a rotor temperature of 475 °C, most of the material was distilled and only a small fraction was collected as residue. With the increased temperature, color bodies also distilled. The Gardner color for the distillation products with varying rotor temperature are reported in Table 5. The 475°C distillation had a Gardner color of 4, whereas the ideal Gardner color is 1.

The increased rotor temperature, unfortunately, maximized the amount of distilled material at the sacrifice of color. The fatty acid profile of both the residue and distillate are reported in Table 5. The crude meadowfoam main components were 5-eicosenoic acid (59.9%), 5,13-docosadienoic acid (17.3%), 5-docosenoic acid (4.3%), and 13-docosenoic acid (11.9%). The distillation at a rotor temperature of 425°C resulted in the closest percent composition to the

| Rotor Temp | C20:1 (%) | | C22:1 (%) | | | | C22:2 (%) | | Gardner color |
|---|---|---|---|---|---|---|---|---|---|
| °C | Δ5-d | Δ5-r | Δ5-d | Δ5-r | Δ13-d | Δ13-r | Δ5,13-d | Δ5,13-r | Distillate |
| S.M. | 59.9 | | 4.3 | | 17.3 | | 11.9 | | 15 |
| 275 | 72.0 | 56.9 | 2.0 | 4.9 | 5.8 | 13.4 | 10.1 | 18.9 | 3 |
| 325 | 69.2 | 54.1 | 2.6 | 5.6 | 7.3 | 15.3 | 11.9 | 20.2 | 1 |
| 375 | 69.6 | 40.1 | 2.8 | 7.4 | 7.9 | 20.2 | 12.9 | 26.1 | 1 |
| 400 | 63.2 | 38.2 | 3.4 | 8.9 | 9.4 | 23.7 | 14.6 | 27.1 | 1 |
| 425 | 59.7 | 35.0 | 4.3 | 7.1 | 12.0 | 19.4 | 17.4 | 23.8 | 1 |
| 475 | 59.2 | 35.2 | 4.4 | 6.6 | 18.0 | 18.0 | 17.5 | 21.9 | 4 |

Table 5. Fatty acid profile - varying rotor temperature

crude meadowfoam fatty acids. Different ratios of fatty acids were distilled depending on the rotor temperature. For the two rotor temperature extremes of 275 and 475°C, the amount of 5-eicosenoic acid was enriched in the distillate to 72.0% at 275°C from 59.2%, whereas the 5,13-docosadienoic acid was enriched in the distillate to 17.5% at the upper temperature of 475°C from 10.1%.

Meadowfoam fatty acids were effectively separated by a Myers Pilot-15 molecular distillation unit. The precise distillation conditions were determined by varying conditions to obtain material that was light in color, Gardner color of 1. Conditions were then chosen to minimize high energy demands on any one element of the system. The varying conditions described were used to determine the ideal distillation conditions (Cermak & Isbell, 2002). All of the conditions played a vital role in conducting a successful distillation. The rotor temperature and flow rate had the greatest impact on the Gardner color and the fatty acid composition of the distillate. After the ideal conditions were determined, an additional larger volume (95 L) of meadowfoam fatty acids was distilled to verify the conditions.

Additionally, meadowfoam estolides can be distilled on the Myers Lab 3 (Fig. 10) to separate the monoestolides (EN=1) and polyestolides (EN>1) (Fig. 9). Table 6 shows experimental conditions for two flow rates over seven different temperature settings and their effect on split ratio, which was determined by mass. The split ratio for both high and low flow rates across the rotor are similar. However, at the low flow rate (0.32 g/min) and high temperatures (300°C), the split ratio is nearly twice that of the high flow rate (1.79 g/min). The increased mass in the distillate fraction would be expected at the lower flow rate because heat transfer to a smaller amount of material passing across the rotor would be more efficient. However, these higher split ratios lowers monoestolide purity which decreases beyond 275°C (Fig. 14). The highest distillate purity occurs for both the low and high flow rates at 200°C (Fig. 14). At 325°C, splatter from the feed stock bumping as it comes into contact with the hot rotor became a significant problem and contaminated the distillate. Therefore, the sample at high flow rate at 325°C was not collected.

Myers Lab 3 distillation unit proved effective. Color from the distillate was effectively removed from the monoestolide fraction with a Gardner color of 1 for those fractions below 250°C (Table 6). This corresponds well with the optimum distillation temperature of 225°C.

| Set point (°C) | Temperature (°C) | | Isolated mass (g) | | Rate (g/min) | Split ratio | Gardner color | |
|---|---|---|---|---|---|---|---|---|
| | Rotor | Distillate | Residue | Distillate | | | Residue | Distillate |
| Low flow rate | | | | | | | | |
| 175 | 174 | 29.3 | 0.9 | 19.2 | 1.01 | 0.05 | 12 | 1 |
| 200 | 201 | 29.4 | 2.4 | 15.9 | 0.92 | 0.15 | 12 | 1 |
| 225 | 226 | 29.8 | 2.7 | 12.9 | 0.78 | 0.21 | 12 | 1 |
| 250 | 249 | 29.9 | 3.6 | 9.5 | 0.66 | 0.38 | 12 | 1 |
| 275 | 274 | 29.8 | 5.2 | 9.5 | 0.74 | 0.55 | 12 | 4 |
| 300 | 299 | 30.1 | 3.0 | 3.3 | 0.32 | 0.91 | 12 | 5 |
| 325 | 322 | 30.2 | 8.1 | 6.2 | 0.72 | 1.31 | 12 | 6 |
| High flow rate | | | | | | | | |
| 175 | 174 | 29.7 | 4.0 | 39.9 | 2.20 | 0.10 | 12 | 3 |
| 200 | 198 | 29.5 | 7.6 | 34.2 | 2.09 | 0.22 | 13 | 1 |
| 225 | 225 | 30.2 | 7.5 | 31.0 | 1.93 | 0.24 | 14 | 3 |
| 250 | 248 | 30.5 | 8.7 | 18.7 | 1.37 | 0.47 | 14 | 4 |
| 275 | 274 | 29.7 | 8.2 | 20.2 | 1.42 | 0.41 | 14 | 4 |
| 300 | 299 | 30.4 | 13.8 | 21.9 | 1.79 | 0.63 | 14 | 6 |

Table 6. Effect of Myer Lab 3 distillation parameters on split and color

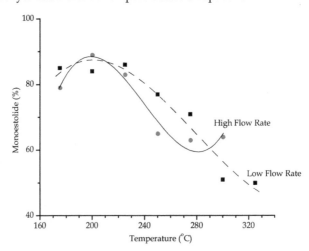

Fig. 14. Effect of rotor temperature on monoestolide composition in distillate fraction

Higher temperatures show degradation in color values as a result of co-distillation and splatter from the color bodies found in the polyestolide fraction. This color improvement from the starting material (Gardner color of 12) should greatly enhance the value of this material for cosmetic applications where product color is an important factor.

# 4. Fatty acid alkyl ester derivatives

Industrially, mono-alkyl fatty acid esters can be prepared by reacting fat and oil triglycerides with an alcohol using akaline-catalyzed interesterification (alcoholysis) or from the direct esterification of fatty acids (Farris, 1979; Sonntag, 1982). Although a variety of alcohols can be utilized, fatty acid methyl esters (FAME) prepared using methanol are most common based on price and availability. These fatty acid esters not only serve as specialty chemicals but are also used extensively in various oleochemical processes as intermediates to produce fatty alcohols, alkanolamides, and α-sulfonated methyl esters (Gervajio, 2005; Gunstone et al., 2007a). Additionally, fatty acid esters, particularly fatty acid methyl esters, are used extensively for the burgeoning biodiesel industry (mono-alkyl esters of long-chain fatty acids derived from vegetable oils or animal fats).

## 4.1 Distillation

With regards to distillation, mono-alkyl fatty acid esters have several significant advantages when compared to the distillation of fatty acids (Budde, 1968; Farris, 1979). Because the acid moiety is in the ester form, the fatty acid esters are less corrosive. Therefore, expensive corrosion resistant equipment may not be necessary. Since the ester group cannot participate in hydrogen bonding the esters have lower boiling points, are oftentimes easier to fractionate, and require less energy for their fractionation, Table 2. As can be seen from Table 2, the esters tend to boil approximately 30°C below their corresponding fatty acids. As a result of the two aforementioned properties, the esters are also less susceptible to color formation, decarboxylation and degradation during the distillation. Finally, the fatty acid methyl esters are more ameneable to fractional distillation and separation since they follow Raoult's law more closely than their corresponding fatty acids, which show significant deviation from Raoult's law making it difficult to fractionally distill and separate fatty acids differing by two carbon atoms in chain length (Budde, 1968; Markley, 1964).

Careful control of the distillation parameters allows the advantages inherent to the fatty acid esters to be exploited and used advantageously especially when attempting to fractionate thermolabile materials such as highly unsaturated esters derived from marine oils. Fractional distillation has been performed for some time on a variety of saturated and highly unsaturated fatty acid ester mixtures at reduced pressures on the laboratory scale as a method to purify fatty acid esters. Early reports examined the separation of various fatty acid methyl esters of corn (Baughman & Jamieson, 1921), sunflower (Baughman & Jamieson, 1922b), soybean (Baughman & Jamieson, 1922a), and menhaden oils (Brown & Beal, 1923) by fractional distillation into their various components. Later spinning band columns improved distillation fractionation; for example, Weitkamp (1945) reported the isolation of 32 compounds into four main compound classes (fatty acids, 2-hydroxy acids, and two types of branched iso and anteiso acids) from wool wax by distilling fatty acid methyl esters. Privett and coworkers (1959) distilled several fractions of methyl esters derived from pork liver lipids. Fractional distillation of unsaturated C20 esters from rapeseed (Haeffner, 1970) and herring oil methyl esters (Ackman et al., 1973) have also been reported. Methyl docosahexaenoate (C22:6) was fractionated from a fatty acid methyl ester mixture derived from tuna oil using spinning band distillation at a pressure of 0.025-0.030 mm Hg (Privett et al., 1969). Unique to this method, an amplified distillation process was used, wherein a mixture of carrier components based on long chain acetates (3 g myristyl, 7 g palmitoleyl, 7 g oleyl, 5 g of 11-eicosenoyl, and 3 g erucyl acetates) was employed to facilitate the fractionation of minor components and minimize artifacts arising from C22:6 methyl ester degradation by keeping the distillation temperature from rising sharply during fractionation. The carrier acetates were present in two-to three-fold excess over the fatty acid methyl esters and were chosen to have a range of boiling points covering that of the sample, not form azeotropes and could be separated easily with any sample components. Distillation gave fractions containing mixtures of the enriched methyl esters and acetates. These separate fractions containing the desired methyl esters and carrier acetates were then saponified to give long chain alcohols (from carriers) which were extracted from the soaps as nonsaponifiable matter. The sodium acetate was converted to acetic acid by acidification and separated from the desired fatty acids by extraction with distilled water. By this method a cut of tuna oil containing 84% C22:6 was obtained.

The development of molecular distillation techniques characterized by short exposure of the sample to high temperatures, short path length for the distillate, and high vacuum led to better column efficiencies. Many reports on the molecular distillation of fatty acid esters from sunflower and soybean esters (Pramparo et al., 2005), rapeseed fatty acid esters and tocopherols (Jiang et al., 2006), eicosapentaenoic (EPA) and docosahexaenoic acid ethyl esters (Brevik et al., 1997), and squid visceral oil ethyl esters (Liang & Hwang, 2000; Rossi et al., 2011) have been reported. Vázquez and Akoh (2010) reported a detailed study on the fractionation of short and medium chain fatty acid ethyl esters via short-path distillation that were obtained from a blend of coconut oil and dairy fat. They examined feed rates, temperatures, and single or multiple steps. They were able to obtain fractions containing a high purity of specific fatty acid ethyl esters with a desired composition and yield.

On an industrial scale, a recent set of patents describe a continuous process (~2,000 kg/h) whereby fractional distillation of palm kernel fatty acid methyl esters is done to produce technical grade oleic acid methyl esters (~75 wt%) containing low levels of saturated methyl palmitate (< 5 wt%) and methyl stearate (<2 wt%) (Heck et al., 2000; Heck et al., 2005). This method avoids additional crystallization steps currently used by other processes. The resulting enriched methyl oleate fraction could then be reduced to obtain the corresponding unsaturated fatty alcohols. In this process, the starting palm kernel methyl esters are first fractionally distilled to separate the C8-C14 esters from the C16-C18 esters. The bottom C16-C18 ester fraction is subsequently fractionally distilled to separate the C18 esters (saturated and unsaturated) from the C16 esters. Finally, the C18 esters are fractionally distilled to obtain an enriched C18 unsaturated fraction (methyl oleate) while reducing the saturated C18 ester (methyl stearate) to less than 2 wt%.

A 1999 patent describes the large scale fractional distillation of rapeseed methyl esters, among other fats and oil esters, to obtain a colorless fraction enriched in C22 (behenic) methyl esters with low acid value and purity of at least 86 wt% (Kenneally et al., 1999). The starting rapeseed methyl esters composition was C16-3.5%, C18-38.0%, C20-9.7%, C22-47.4%, C24-1.4%. Batch fractional distillation at 232-274°C of the rapeseed methyl esters using a packed column with 10 theoretical stages, overhead condenser, receiver, and vacuum pump operating between 5-25 mm Hg gave a 32% yield of the C22 methyl ester cut containing 92.6% C22 and 1.8% C18.

The explosive growth of the biodiesel industry has focused attention on mono-alkyl esters of long-chain fatty acids. Typically, biodiesel is comprised of fatty acid methyl esters, although ethyl, propyl, butyl esters, etc. might be used for either B100 or blending applications. Because various esters may be used and these higher esters likely have boiling points higher than the methyl esters, focus on the distillation characteristics of these fatty acid esters with regards to biodiesel specifications are important. Accordingly, Schober and coworkers (2010) have examined the distillation characteristics for a series of fatty acid ethyl esters derived from various feedstock following ASTM D1160 specifications. They found the ethyl esters exceeded the maximum limit set for in the ASTM specifications which has important implications for higher esters to be used as biodiesel.

## 4.2 Reactive distillation

Reactive distillation has been known since the 1920's but has gotten renewed interest in recent years. The complexities of reactive distillation have been thoroughly reviewed (Malone & Doherty, 2000; Sharma & Singh, 2010; Taylor & Krishna, 2000). Reactive distillation is the simultaneous implementation of reaction and distillation within a single column unit, whereby the reactants are converted to products in a reaction zone in the presence of catalyst with simultaneous distillation of the products and recycling of unused reactants to the reaction zone (Doherty & Buzad, 1992; Kiss et al. 2008, Malone & Doherty, 2000; Omota et al., 2001; Omota et al., 2003ab; Sharma & Singh, 2010; Taylor & Krishna, 2000). A diagram comparing a conventional reaction and distillation process to a reactive distillation process is shown in Fig. 15.

Briefly, as seen in Fig 15a., the reaction in the presence of catalyst takes place in a separate reactor. The crude products are transferred to a sequence of distillation columns are required to produce pure C and D by the conventional process. The unreacted components, A and B, are recycled back to the reactor. In contrast, Fig 15b shows that by combining the reaction and distillation operations, A and B continuously react in the reaction zone while products C and D are removed from A and B in the rectifying and stripping sections at the top and bottom, respectively (Taylor, & Krishna, 2000). This technique can reach 100% conversion and potentially minimize operational and equipment costs and decrease waste energy. Reactive distillation is especially suited for the chemical reactions limited by thermodynamic and equilibrium constraints, since one or more of the products of the reaction are continuously separated from the reactants (Kiss, 2011; Nguyen & Demirel, 2011).

With regards to fatty acid ester production and purification, and more specifically to large scale production of biodiesel, it would appear that reactive distillation could provide an efficient and integrated approach to obtain the desired fatty acid esters. However, to date most studies have focused on computer aided modeling, simulation and economics of reactive distillation to produce fatty acid alkyl esters (Bhatia et al., 2007; Kiss, 2011; Kiss et al., 2006; Nguyen & Demirel, 2011; Omota et al., 2001; Omota et al., 2003ab) while more recent experimentally based studies are not as prevalent (Bhatia et al., 2006; He et al., 2006; Silva et al., 2010; Steinigeweg & Gmehling, 2003).

Steinigeweg and Gmehling (2003) reported the development of a reactive distillation process for the pilot plant production of decanoic acid methyl esters. The reaction was catalyzed heterogeneously using a strong acidic ion-exchange resin and supported by structured corrugated wire mesh sheets (Katapak-S) in the columns and reaction parameters such as

Reaction Sequence: A + B ⇌ C + D

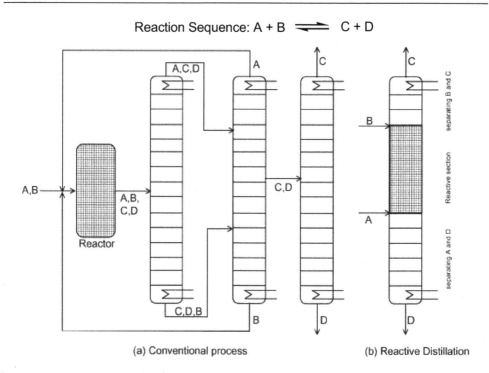

(a) Conventional process                    (b) Reactive Distillation

Fig. 15. Scheme comparing the reaction sequence where A and B are reactants and C and D are the desired products. (a) conventional reactor and distillation process; (b) reactive distillation column. Adapted from Taylor & Krishna (2000)

reflux ratio and reactant ratios were examined (Steinigeweg & Gmehling, 2003). They found that a low reflux ratio of 0.01, a (1:2) decanoic acid:methanol in feed stream, distillate to feed ratio of (1:2), 393K and 3 bar resulted in good fatty acid conversion to the corresponding methyl decanoate. Recently, computer simulation for the esterification of decanoic acid with methanol using reactive distillation has been carried out by Machado and coworkers (2011) using the experimental data obtained Steinigeweg and Gmehling (2003) to validate their models. Furthermore, Machado and coworkers modeled experimental results obtained for the esterification of oleic acid with methanol (Silva et al., 2010; Kiss et al., 2006) and lauric acid with ethanol (De Pietre et al., 2010). They found good agreement between their modeling and the experimentally determined esters found in the literature and predict conversions above 98%.

Bhatia and coworkers performed a thorough study on the esterification of palmitic acid with isopropanol in a reactive distillation column using zinc acetate supported on functionalized silica gel as a catalyst in Katapak-SP structured packing (Bhatia et al., 2006). The column performance was evaluated by varying the operating parameters such as feed flow rate, reboiler temperature, palmitic acid feed composition, palmitic acid feed temperature, molar ratio of isopropanol feed to palmitic acid feed and reflux ratio. From their work they proposed a technically optimized reactive distillation process for the production of isopropyl palmitate.

Using a laboratory scale continuous flow reactive distillation apparatus, He, Singh, & Thompson, (2006) examined a combined process which used a conventional pre-reactor coupled to a reactive distillation column to convert canola oil into its corresponding fatty acid methyl esters, Fig. 16.

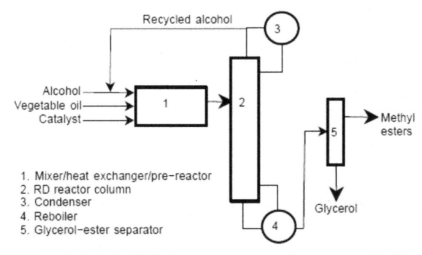

Fig. 16. Schematic of reactive distillation reactor system used to convert canola oil into its corresponding fatty acid methyl esters (He et al., 2006)

Reaction parameters such as methanol:canola oil, feed rate of canola oil and potassium hydroxide catalyst, and temperature were examined for runs with and without the pre-reactor. A 95% conversion with a 94% yield (esters contained 1.1%, 2.0%, and 2.0% mono-, di-, and triglyceride, respectively) was obtained using a column temperature of 65°C and a 4:1 methanol:canola oil molar ratio with the pre-reactor. They were able to reduce reaction times 10 to 15 times while also reducing methanol consumption by 66% over conventional biodiesel processes.

Recently, Brazilian researchers used a pre-reactor (with sodium hydroxide catalyst) coupled to a reactive distillation column in a semi-batch system to examine the preparation of fatty acid methyl esters from soybean oil and bioethanol (Silva et al., 2010). They used experimental design to optimize the catalyst concentration (from 0.5 wt% to 1.5 wt %) and the ethanol/soybean oil molar ratio (from 3:1 to 9:1). The reactive column reflux rate was 83 ml/min, and the reaction time was 6 min. Their best conversion to esters was 98.18 wt% with 0.65 wt% of sodium hydroxide, ethanol/soybean oil molar ratio of 8:1, and reaction time of 6 min. The reaction time of 6 min means: 1 min in the pre-reactor and 5 min in the reactive distillation column.

A recently patent published application demonstrated a process to produce mixed fatty acid esters with a variety of heterogeneous catalysts. Again, a pre-reactor coupled to a reactive distillation column was used (Asthana et al., 2009). Crude fatty acid methyl esters are first prepared in the pre-reactor much like a conventional biodiesel process. After a crude separation of glycerol and other components the fatty acid methyl esters are transferred to the reactive distillation column where they undergo further reaction with different alcohols from 2-8 carbon atoms to give a fatty acid alkyl ester mixture upon exiting the reactive distillation column.

Although the results described above are promising and represent good first steps toward implementation of reactive distillation for fatty acid ester production, as concluded by Taylor and Krishna in their 2000 review on reactive distillation, and with particular focus towards production and separation of fatty acid esters, they state:

> ...There is a crying need for research in this area. It is perhaps worth noting here that modern tools of computational fluid dynamics could be invaluable in developing better insights into hydrodynamics and mass transfer in RD columns. Besides more research on hydrodynamics and mass transfer, there is need for more experimental work with the express purpose of model validation...

There remains more work to be done in the use of reactive distillation to produce and purify fatty acid alkyl esters.

## 5. References

Ackelsberg, O.J. (1958). Fat splitting. *J. Am. Oil Chem. Soc.* Vol.35, pp. 635-640, ISSN 0003-021X

Ackman, R.G., Ke, P.J., & Jangaard, P.M. (1973). Fractional vacuum distillation of herring oil methyl esters. *J. Am. Oil Chem. Soc.* Vol.50, pp. 1-8, ISSN 0003-021X

Asthana, N.S., Miller, D.J., Lira, C.T., & Bittner, E. (2009). Process for producing mixed esters of fatty acids as biofuels. U.S. Patent Application 2009/0126262

Barnebey, H. & Brown, A.C. (1948). Continuous fat splitting plants using the Colgate-Emery process. *J. Am. Oil Chem. Soc.* Vol.25, pp. 95-99, ISSN 0003-021X

Batistella, C.B. & Maciel, M.R.W. (1996). Modeling, simulation and analysis of molecular distillators: centrifugal and falling film. *Computers & Chem. Eng.* Vol.20, pp. S19-S24, ISSN 0098-1354

Batistella, C.B. & Maciel, M.R.W. (1998). Recovery of carotenoids from palm oil by molecular distillation. *Computers & Chem. Eng.* Vol.22, pp. S53-S60, ISSN 0098-1354

Baughman, W.F. & Jamieson, G.S. (1921). The chemical composition of corn oil. *J. Am. Chem. Soc.* Vol.43, pp. 2696-2702, ISSN 1520-5126

Baughman, W.F. & Jamieson, G.S. (1922a). The chemical composition of soya bean oil. *J. Am. Chem. Soc.* Vol.44, pp. 2947-2952, ISSN 1520-5126

Baughman, W.F. & Jamieson, G.S. (1922b). The chemical composition of sunflower-seed oil. *J. Am. Chem. Soc.* Vol.44, pp. 2952-2957, ISSN 1520-5126

Berger, R. & McPherson, W. (1979). Fractional distillation. *J. Am. Oil Chem. Soc.* Vol.56, pp. 743-744, ISSN 0003-021X

Bhatia, S., Ahmad, A.L., Mohamed, A.R., & Chin, S.Y. (2006). Production of isopropyl palmitate in a catalytic distillation column: Experimental studies. *Chem. Eng. Sci.* Vol.61, pp. 7436-7447, ISSN 0009-2509

Bhatia, S., Mohamed, A.R., Ahmad, A.L., & Chin, S.Y. (2007). Production of isopropyl palmitate in a catalytic distillation column: Comparison between experimental and simulation studies, *Computers & Chem. Eng.* Vol.31, pp. 1187-1198, ISSN 0098-1354

Biehler, R.M., Hickman, K.C.D., & Perry, E.S. (1949). Small laboratory centrifugal molecular still. *Anal. Chem.* Vol.21, pp. 638-640, ISSN 0003-2700

Brevik, H., Haraldsson, G.G., & Kristinsson, B. (1997). Preparation of highly purified concentrates of eicosapentaenoic acid and docosahexaenoic acid. *J. Am. Oil Chem. Soc.* Vol.74, pp. 1425-1429, ISSN 0003-021X

Brown, J.B. & Beal G.D. (1923). The highly unsaturated fatty acids of fish oils. *J. Am. Chem.* Vol.45, pp. 1289-1303, ISSN 1520-5126

Brown, J.H., Kleiman, R., Hill, J., Koritala, S., & Lotts, K. (2008). Cosmetic and topical compositions comprising cuphea oil and derivatives thereof. May 7, 2008. European Patent Application EP 1,916,988

Budde, W.M., Jr. (1968) General physical and chemical properties of fatty acids, In: *Fatty Acids and Their Industrial Applications*. Pattison, E.S. pp. 47-76, Marcel Dekker, Inc. ISBN 0318165236, New York, New York

Cermak, S.C., Biresaw, G., & Isbell, T.A. (2008). Comparison of a new estolide oxidative stability package. *J. Am. Oil Chem. Soc.* Vol.85, pp. 879-885, ISSN 0003-021X

Cermak, S.C. & Isbell, T.A. (2001a). Synthesis of estolides from oleic and saturated fatty acids. *J. Am. Oil Chem. Soc.* Vol.78, pp. 557-565, ISSN 0003-021X

Cermak, S.C. & Isbell, T.A. (2001b). Biodegradable oleic estolide ester having saturated fatty acid end group useful as lubricant base stock. November 13, 2001. U.S. Patent 6,316,649 B1

Cermak, S.C. & Isbell, T.A. (2002). Pilot-plant distillation of meadowfoam fatty acids. *Ind. Crops Prod.* Vol.15, pp. 145-154, ISSN 0926-6690

Cermak, S.C. & Isbell, T.A. (2004a). Synthesis and physical properties of Cuphea-oleic estolides and esters. *J. Am. Oil Chem. Soc.* Vol.81, pp. 297-303, ISSN 0003-021X

Cermak, S.C. & Isbell, T.A. (2004b). Estolides - The next bio-based functional fluid. *Inform.* Vol.15, pp. 515-517, ISSN 0897-8026

Cermak, S.C., Isbell, T.A., Isbell, J.E., Akerman, G.G., Lowery, B.A., & Deppe, A.B. (2005). Batch drying of cuphea seeds. *Ind. Crops Prod.* Vol.21, pp. 353-359, ISSN 0926-6690

Cermak, S.C., John, A.L., & Evangelista, R.L. (2007). Enrichment of decanoic acid in Cuphea fatty acids by molecular distillation. *Ind. Crops Prod.* Vol.26, pp. 93-99, ISSN 0926-6690

De Pietre, M.K., Almeida, L.C.P., Landers, R., Vinhas, R.C.G., & Luna, F. J. (2010). $H_3PO_4$ and $H_2SO_4$ treated niobic acid as heterogeneous catalyst for methyl ester production. *React. Kinet., Mech. Catal.* Vol.99, pp. 269-280, ISSN 1878-5204

Dijkstra, A.J. & Segers, J.C. (2007). Production and refining of oils and fats, In: *The Lipid Handbook with CD-ROM, Third Edition*, Gunstone, F., Harwood, J.L., & Dijkstra, A.J. pp. 143-262, CRC Press, ISBN 0-8496-9688-3, Boca Raton, Florida

Doherty, M.F. & Buzad, G. (1992). Reactive distillation by design. *Trans. IChemE* Vol.70, Part A, pp. 448-458, ISSN 0957-5820

Evangelista, R.L. & Cermak, S.C. (2007). Full-press oil extraction of Cuphea (PSR23) seeds. *J. Am. Oil Chem. Soc.* Vol.84, pp. 1169-1175, ISSN 0003-021X

Evangelista, R.L. & Manthey, L.K. (2004). Protein and oil contents and fatty acid compositions of Cuphea PSR23 seeds. The American Chemical Society 36th Annual Great Lakes Regional Meeting Program and Abstracts. October 17-20, Peoria, IL, pp. 188

Farris, R.D. (1979). Methyl esters in the fatty acid industry. *J. Am. Oil Chem. Soc.* Vol.56, pp. 770A-773A, ISSN 0003-021X

Frykman, H.B. & Isbell, T.A. (1999). Decolorization of meadowfoam estolides using sodium borohydride. *J. Am. Oil Chem. Soc.* Vol.76, pp. 765-767, ISSN 0003-021X

Geller, D.P., Goodruma, J.W., & Knapp S.J. (1999). Fuel properties of oil from genetically altered Cuphea viscosissima. *Ind. Crops Prod.* Vol.9, pp. 85-91, ISSN 0926-6690

Gervajio, G.C. (2005). Fatty acids and derivatives from coconut oil, In: *Bailey's Industrial Oil and Fat Products, six volume set, sixth Edition*, Shahidi, F. Vol. 6. pp. 1-56, John Wiley & Sons Inc., ISBN 978-0-471-38546-2, New York, New York

Gesch, R.W., Cermak, S.C., Isbell, T.A., & Forcella, F. (2005). Seed yield and oil content of Cuphea as affected by harvest date. *Agron. J.* Vol.97, pp. 817-822, ISSN 0926-6690

Goettsche, R. & Borck, H-A. (1994). Wood preservative compositions having activity against termites and fungi. U.S. Patent 5,276,029

Graham, S.A., Hirsinger, F., & Röbbelen, G. (1981). Fatty acids of Cuphea (*lythraceae*) seed lipids and their systematic significance. *Am. J. Bot.* Vol.68, pp. 908-917, ISSN 0002-9122

Gunstone, F.D. (2011). Non-food uses of vegetable oils. *Lipid Technol.* Vol.23, pp. 24, ISSN 1863-5377

Gunstone, F.D. & Norris, F. (1983). *Lipids in Foods: Chemistry, Biochemistry and Technology*, Pergamon Press, ISBN 0-08-025499-3, Oxford, England

Gunstone, F.D., Alander, J., Erhan, S.Z., Sharma, B.K., McKeon, T.A., & Lin, J.-T. (2007a). Nonfood uses of oils and fats, In: *The Lipid Handbook with CD-ROM, Third Edition*, Gunstone, F., Harwood, J.L., & Dijkstra, A.J. pp. 591-636, CRC Press, ISBN 0-8496-9688-3, Boca Raton, Florida

Gunstone, F.D., Harwood, J.L., & Dijkstra, A.J. (Eds.) (2007b). *The Lipid Handbook with CD-ROM, Third Edition*, CRC Press, ISBN 0-8496-9688-3, Boca Raton, Florida

de Guzman, D. (April 2009). Sizzling Fats and Oils, In: *ICIS Chemical Business, 19.10.2011*, Available from http://www.icis.com/Articles/2009/04/27/9209502/development-for-fats-and-oils-based-chemicals-rises.html

Haeffner, E.W. (1970). Separation of the C2O unsaturated fatty acids from rapeseed oil by countercurrent distribution. *Lipids* Vol.5, pp. 489-491, ISSN 1558-9307

Hamilton, C.R., Kirstein, D., & Breitmeyer, R.E. (2006). The Rendering industry's biosecurity contribution to public and animal health, In: *Essential Rendering: All About the Animal By-Products Industry*, Meeker, D., pp. 82-94, Kirby Lithographic Company, Inc., ISBN 0-9654660-3-5, Arlington, VA, USA

He, B.B., Singh, A.P., & Thompson, J.C. (2006). A novel continuous-flow reactor using reactive distillation for biodiesel production. *Trans. Am. Soc. Agric. Bio. Eng.* Vol. 49, pp. 107-112 ISSN 0001-2351

Hardin, B. (1991). Cuphea - Plants with beautiful future - Industrial applications of Cuphea: *Agricultural Research*. September, pp 16-17, ISBN 0-16-009204-3

Heck, S., Winterhoff, V., Gutsche, B., Fieg, G., Mueller, U., Rigal, J., & Kapala, T. (2005). Method for producing technical oleic acid methyl esters. U.S. Patent Application 2005/0090676 A1

Heck, S., Klein, N., Komp, H.-D., Boehr, C., Huebner, N., & Westfechtel, A. (2000). Production of unsaturated coconut and/or palm kernel fatty alcohols, useful as chemical intermediates, e.g. for surfactants and skin care products, involves fractionating lower alkyl ester twice and hydrogenating unsaturated fraction. Germany patent application DE 19912684 A1

Hill, K. (2000). Fats and oils as oleochemical raw materials. *Pure Appl. Chem.* Vol.72, pp. 1255-1264, ISSN 1365-3075

Hohener, A. & Frick, R. (2003). Fabric softener composition. U.S. Patent 6,583,105 B1

Hoppe, U. & Engel, W. (1989). Cosmetic agents for hair. U.S. Patent 4,839,165

Isbell, T.A., Abbott, T.P., Asadauskas, S., & Lohr, J.E. (2000a). Biodegradable oleic estolide ester base stocks and lubricants. U.S. Patent 6,018,063

Isbell, T.A., Abbott, T.P., & Carlson, K.D. (1999). Oxidative stability index of vegetable oils in binary mixtures with Meadowfoam oil. *Ind. Crops Prod.* Vol.9, pp. 115-123, ISSN 0926-6690

Isbell, T.A., Abbott, T.A., & Dworak, J.A. (2000b). Shampoos and conditioners containing estolides. U.S. Patent 6,051,214

Isbell, T.A. & Cermak, S.C. (2004). Purification of meadowfoam monoestolide from polyestolide. *Ind. Crops Prod.* Vol.19, pp. 113–118, ISSN 0926-6690

Isbell, T.A. & Cermak, S.C. (2001). Synthesis of delta-eicosanolactone and delta-docosanolactone directly from meadowfoam oil. *J. Am. Oil Chem. Soc.* Vol.78, pp. 527-531, ISSN 0003-021X

Isbell, T.A. & Kleiman, R. (1994). Characterization of estolides produced from the acid-catalyzed condensation of oleic acid. *J. Am. Oil Chem. Soc.* Vol.71, pp. 379–383, ISSN 0003-021X

Isbell, T.A. & Kleiman, R. (1996). Mineral acid-catalyzed condensation of Meadowfoam fatty acids into estolides. *J. Am. Oil Chem. Soc.* Vol.73, pp. 1097-1107, ISSN 0003-021X

Ishii, H. & Mikami, N. (1995). Preparation of *n*-(long-chain acyl)amino acid esters as moisture-holding agents and emulsifiers for cosmetics and skin medicaments. Japan Patent 93-263900

Jiang, S.T., Shao, P., Pan, L.J., & Zhao, Y.Y. (2006). Molecular distillation for recovering tocopherol and fatty acid methyl esters from rapeseed oil deodoriser distillate, *Biosystems Engineering* Vol.93, pp. 383-391, ISSN 1537-5110

Kenneally, C.J., Busch, G.A., & Gansmuller, E.W. (1999). A process for making high purity fatty acid lower alkyl esters. World Patent WO 9924387 A1

Kim, K-I., Gesch, R.G., Cermak, S.C., Phippen, W.B., Berti, M.T., Johnson, B.L., & Marek, L. (2011). Cuphea growth, yield, and oil characteristics as influence by climate and soil environments across the upper Midwest USA. *Ind. Crops Prod.* Vol.33, pp. 99-107, ISSN 0926-6690

Kiss, A.A. Dimian, A.C., & Rothenberg, G. (2008). Biodiesel by catalytic reactive distillation powered by metal oxides. *Energy & Fuels* Vol.22, pp. 598–604, ISSN 1520-5029

Kiss, A.A., Omota, F., Dimian, A.C., & Rothenberg, G. (2006). The heterogeneous advantage: biodiesel by catalytic reactive distillation. *Top. Catal.* Vol.40, pp. 141-150, ISSN 1022-5528

Kiss, A.A. (2011). Heat-integrated reactive distillation process for synthesis of fatty esters, *Fuel Processing Technology* Vol.92, pp. 1288-1296, ISSN 0378-3820

Knapp, S.J. (1993). Breakthrough towards the domestication of Cuphea, in new crops. J. Janick and J.E. Simon (eds), Wiley, New York, pp 372-379, ISBN 0–471–59374–5

Knothe, G. & Dunn, R.O. (2009). A comprehensive evaluation of the melting points of fatty acids and esters determined by differential scanning calorimetry. *J. Am. Oil Chem. Soc.* Vol.86, pp. 843-856, ISSN 0003-021X

Liang J.-H. & Hwang L. (2000). Fractionation of squid visceral oil ethyl esters by short-path distillation. *J. Am. Oil Chem. Soc.* Vol.77, pp. 773-777, ISSN 0003-021X

Lausberg, N., Josten, H., Fieg, G., Kapala, T., Christoph, R., Suessenbach, A., Heidbreder, A., Mrozek, I., & Schwerin, A. (2008). Process for obtaining fatty acids with improved odor, color and heat stability. U.S. Patent 7,449,088

Machado, G.D., Aranda, D.A.G., Castier, M., Cabral, V.F., & Cardozo-Filho, L. (2011). Computer simulation of fatty acid esterification in reactive distillation columns, *Ind.Eng. Chem. Res.* Vol.50, pp. 10176-10184, ISSN 1520-5045

Markley, K. S. (1964). Techniques of separation, In: Fatty Acids, Their Chemistry, Properties, Production and Uses, , 2nd edition, Part 3, Markley, K.S., pp. 2016-2071, Interscience Publishers, New York, New York, ISBN 60-10162

Malone, M.F. & Doherty, M.F. (2000). Reactive distillation, Ind. Eng. Chem. Res. Vol.39, pp. 3953-3957, ISSN 1520-5045

Marttinello, M.A., Leone, I., & Prámparo, M. (2008). Simulation of deacidification process by molecular distillation of deodorizer distillate. *Lat. Am. appl. res.* Vol.38, pp. 299-304, ISSN 0327-0793

Micov, M., Lutisan, J., & Cvengros, J. (1997). Balance equations for molecular distillation. *Sep. Sci. Technol.* Vol.32, pp. 3051-3066, ISSN 0149-6395

Miller, R.W., Earle, F.R., & Wolff, L.A. (1964). Search for new industrial oils. IX. Cuphea, a versatile source of fatty acids. *J. Am. Oil Chem. Soc.* Vol.41, pp. 279-280, ISSN 0003-021X

Mills, V. (1942). Process of distilling higher fatty acids. U.S. Patent 2,274,801

Molly, K. & Bruggeman, G. (2004). Medium chain fatty acids applicable as antimicrobial agents December 15, 2004. European Patent EP 1,294,371

Muckerheide, V. (1952). Fat splitting and distillation. *J. Am. Oil Chem. Soc.* Vol.29, pp. 490-495, ISSN 0003-021X

Nagaoka, S. & Ibuki, M. (2000). Biodegradable lubricant base oil and its manufacturing process. U.S. Patent 6,117,827

Nandi, S., Gangopadhyay, S., & Ghosh, S. (2004). Production of medium chain glycerides and monolaurin from coconut acid oil by lipase-catalyzed reactions. *J. Oleo. Sci.* Vol.53, pp. 497-501, ISSN 1345-8957

Nguyen, N. & Demirel, Y. (2011). Using thermally coupled reactive distillation columns in biodiesel production. *Energy* Vol.36, pp. 4838-4847, ISSN 0360-5442

Norris, F. & Terry, D. (1945). Precise laboratory fractional distillation of fatty acid esters. *J. Am. Oil Chem. Soc.* Vol.22, pp. 41-46, ISSN 0003-021X

Nuns, J., Girault, A., & Rancurel, A. (1994). Molecular distillation apparatus having induction heating. U.S. Patent 5,334,290

O'Brien, R. (2004). *Fats and Oils: Formulating and Processing for Applications* (second edition), CRC Press, ISBN 0-8493-1599-9, Boca Raton, Florida, USA.

O'Brien, R., Farr, W., & Wan, P. (Eds.). (2000) *Introduction to Fats and Oils Technology* (second edition), AOCS Press, ISBN 1-893997-13-8, Champaign, Illinois, USA.

Omota F., Dimian A.C., & Bliek A. (2001). Design of reactive distillation process for fatty acid esterification. *Computer Aided Chemical Engineering* Vol.9, pp. 463-468, ISSN 0009-2509.

Omota F., Dimian A.C., & Bliek A. (2003a). Fatty acid esterification by reactive distillation: Part 2 - kinetics-based design for sulphated zirconia catalysts, *Chem. Eng. Sci.* Vol.58, pp. 3175-3185, ISSN 0009-2509

Omota F., Dimian A.C., & Bliek A. (2003b). Fatty acid esterification by reactive distillation. Part 1: Equilibrium-based design, *Chem. Eng. Sci.* Vol.58, pp. 3159-3174, ISSN 0009-2509

Phillips, B.E., Smith, C.R., & Tallent, W.H. (1971). Glycerides of Limnanthes douglasii seed oil. *Lipids.* Vol.6, pp. 93-99, ISSN 0024-4201

Potts, R.H (1956). Distillation of fatty acid. *J. Am. Oil Chem. Soc.* Vol.33, pp. 545-548, ISSN 0003-021X

Potts, R.H & White, F.B. (1953). Fractional distillation of fatty acids. *J. Am. Oil Chem. Soc.* Vol.30, pp. 49-53, ISSN 0003-021X

Pramparo, M., Prizzon, S., & Martinello, M.A. (2005). Estudio de la purificación de ácidos grasos, tocoferoles y esteroles a partir del destilado de desodorización. *Grasas y Aceites* Vol.56, pp. 228-234, ISSN 0017-3495

Privett, O.S., Weber, R.P. & Nickell, E.C. (1959). Preparation and properties of methyl pork liver. *J. Am. Oil Chem. Soc.* Vol.36, pp. 443-449, ISSN 0003-021X

Privett, O.S., Nadenicek, J.D., Pusch, F.J., & Nickell, E.C. (1969). Application of high vacuum fractional distillation to complex mixtures of methyl esters of polyunsaturated fatty acids. *J. Am. Oil Chem. Soc.* Vol.46, pp. 13-17, ISSN 0003-021X

Rees, G.J. (1980). Centrifugal molecular distillation - II. *Chem. Eng. Sci.* Vol.35, pp. 841-845, ISSN 0009-2509

Rossi, P.C., Pramparo, M., Gaich, M.C., Grosso, N.R., & Nepote, V. (2011). Optimization of molecular distillation to concentrate ethyl esters of eicosapentaenoic (20:5 ω-3) and docosahexaenoic acids (22:6 ω-3) using simplified phenomenological modeling. *J. Sci. food Agric.* Vol.91, pp. 1452-1458, ISSN 1097-0010

Ruston, N. (1952). Commercial uses of fatty acids. *J. Am. Oil Chem. Soc.* Vol.29, pp. 495-498, ISSN 0003-021X

Schober, S., Wolf, M., & Mittelbach, M. (2010). Distillation characteristics of fatty acid ethyl esters. *Energy Fuels,* Vol.24, pp. 6693-6695, ISSN 1520-5029

Sharma, N. & Singh, K. (2010). Control of reactive distillation column: a review. *IJCRE* Vol. 8, pp. R5,1-55, ISSN 1542-6580

Silva, N. L., Santander, C. M. G., Batistella, C. B., Maciel Filho, R., & Maciel, M. R. W. (2010). Biodiesel production from integration between reaction and separation system: reactive distillation process. *Appl. Biochem. Biotechnol.* Vol.161, pp. 245-254, ISSN 1559-0291

Sonntag, N.O. (1982). Fat splitting, esterification, and interesterification, In: *Bailey's Industrial Oil and Fat Products, Fourth Edition,* Swern, D. Vol. 2. pp. 97-174, John Wiley & Sons Inc., ISBN 0-471-83958-2, New York, New York

Stage, H. (1984). Fatty acid fractionation by column distillation: Purity, energy consumption and operating conditions. *J. Am. Oil Chem. Soc.* Vol.61, pp. 204-214, ISSN 0003-021X

Steinigeweg, S. & Gmehling, J. (2003). Esterification of a fatty acid by reactive distillation. *Ind. Eng. Chem. Res.* Vol.42, pp. 3612-2319, ISSN 1520-5045

Stauffer, C. (1996). *Fats & Oils*. Eagan Press, ISBN 0-913250-90-2. St. Paul, Minnesota, USA.

Taylor, R. & Krishna, R. (2000). Modelling reactive distillation. *Chem. Eng. Sci.* Vol.55, pp. 5183-5229, ISSN 0009-2509

Tholstrup, T., Ehnholm, C., Jauhiainen, M., Petersen, M., Høy, C., Lund, P., & Sandström, B. (2004). Effects of medium-chain fatty acids and oleic acid on blood lipids, lipoproteins, glucose, insulin, and lipid transfer protein activities. *Am J Clin Nutr.* Vol.79, pp. 564-569, ISSN 0002-9165

Twitchell, E. (1898). Process of decomposing fats or oils into fatty acids and glycerine. U.S. Patent 601,603

Vázquez, L. & Akoh, C. (2010). Fractionation of short and medium chain fatty acid ethyl esters from a blend of oils via ethanolysis and short-path distillation. *J. Am. Oil Chem. Soc.* Vol.87, pp. 917-928, ISSN 0003-021X

Velamakanni, B.V., Mitra, S.B., Wang, D., Scholz, M.T., & Aasen, S.M. (2006). Hardenable antimicrobial dental compositions and methods. U.S. Patent Appl. 2006/0205838

Wecker, H. (1927). Process for separation of volatile substances. U.S. Patent 1,622,126

Weitkamp, A.W. (1945). The acidic constituents of degras. A new method of structure elucidation, *J. Am. Chem. Soc.* Vol.67, pp. 447-454, ISSN 1520-5126

Wolf, R.B., Graham, S.A., & Kleiman, R. (1983). Fatty acid composition of Cuphea seed oils. *J. Am. Oil Chem. Soc.* Vol.60, pp. 103-104, ISSN 0003-021X

Yoshida, S. & Iinuma, M. (2004). Wood preservative. U.S. Patent 6,827,949 B2

# 7

# Distillation of Brazilian Sugar Cane Spirits (Cachaças)

Sergio Nicolau Freire Bruno
*Laboratório Nacional Agropecuário,*
*RJ-MG, Ministério da Agricultura*
*Brazil*

## 1. Introduction

*Cachaça* is the sugar cane spirit typical of and exclusively produced in Brazil which has alcohol content between 38 and 48% in volume, at 20ºC. It is obtained from the sugar cane fermented wort and has peculiar sensory characteristics. It may also include the addition of up to 6g/L of various sugars, expressed as sucrose (Brasil, 2005).

In the production of *cachaças*, as of other distilled beverages, a distillation process is used to isolate, select and concentrate specific volatile components of the "liquid mixture" by heating it (Boza & Horii, 1988; Léauté, 1990). After the fermentation, the distillation is the most important step for the quality of distilled beverages (Janzantti, 2004; Boza & Horii, 1988).

Distillation also promotes some heat induced chemical reactions such as the synthesis of acrolein through a Maillard reaction (Boza & Horii, 1988 ; Nykänen & Nykänen, 1991) and of heterocyclic aromatic compounds, such as furans (furfural, etc.), pyrazines and pyridines (Janzantti, 2004 ; Léauté, 1990). Besides that, distillation causes the extraction of certain long chain esters retained in the yeast cells at the end of the fermentation step, transferring them to the distillates (Nykänen & Nykänen, 1991).

The *cachaças* "wine" composition is quite complex and contains liquid, solid and gaseous substances. Ethanol is the main liquid component, with 5 to 8% v/v, and water is the substance present in the greatest amount - about 89 to 92% v/v. Other liquid components present in smaller amounts are glycerol, lactic acid and butyric acid; volatile components such as esters, acetic acid, propanoic acid, aldehydes and higher alcohols, among others (Novaes, 1999; Bruno, 2006 ; Nascimento et al., 1998a,b).

At every production of a distilled beverage, each volatile component will be more easily distilled according to three criteria: solubility in alcohol or water, boiling point and alcoholic content variation in the vapor phase during distillation (Janzantti, 2004 ; Léauté, 1990).

According to the components' volatility, it is possible to isolate the volatile (water, ethanol and others) components from the non-volatile ones (suspended solids, minerals, yeast cells, non-fermentable sugars, proteins, etc.) obtaining two fractions, phlegm and vinasses, the residue from the distillation of the "wine". The phlegm, main distillation product, is an impure hydroalcoholic mixture, and its content depends on the type of equipment used (Mutton & Mutton, 2005).

## 2. Distillation system and practices

*Cachaças* can be produced by two very different systems: "continuous" and "by batches". In the traditional continuous system the distillation column used is continuously fed with the "wine", while in the "by batches" system, typical of alembics, the whole wine volume to be distilled is transferred to a pan before the distillation starts (Bruno, 2006).

The different distillation systems are frequently used to distinguish the type of *cachaça* and the production method. Therefore, *industrial cachaça* would be the one produced from distillation columns of any size, through a continuous process, and *alembic cachaça*, the one prepared in copper alembics, with a limited volume. Among the practices that do not fit in the description of the last process, we can mention the sugar cane burning and the use of chemical adjuncts in the fermentation (Tonéis & Cia, 2005; Ampaq, 2006).

As the distillation process is modified, samples of the same "wines", distilled in the same type of equipment, produced distillates with different quality. A slower distillation results in higher yield and in an increase of the aromatic compounds content, such as of aromatic esters. A fast distillation reduces the yield and produces distillates that are more acidic and contain an excess of higher alcohols (Almeida Lima, 2001; Mutton and Mutton, 2005 ).

Reche et al, (2007), use Principal component analysis (PCA), linear discriminant analysis (LDA) and ethanal, ethyl carbamate, dimethyl sulfide, isobutyl alcohol, n-propanal, copper, ethyl acetate,and phenylmethanal as chemical descriptors, to develop a model that showed 95.1% accuracy in distinguishing between *cachaças* distilled in copper pot stills and *cachaças* distilled in stainless steel columns. First, an exploratory analysis was carried out by PCA using analytical data for 82 samples (55 samples of *cachaças* distilled in copper pot stills and 27 samples distilled in stainless steel columns) to verify group formation and data structure. Afterwards, LDA was used for classification purposes. The training set used in the LDA was composed of the 82 samples used in the PCA. The self-consistency of the LDA model was examined by cross-validation using 33 unknown samples. During cross-validation, one sample at a time (of *n* samples) is left out, and the prediction ability is tested on the sample omitted. This procedure is repeated *n* times, resulting in *n* models, and will give an estimate of the average prediction ability for the *n* models.

### 2.1 Distillation by batch or intermittent: Alembics

As seen in Figure 1, the common alembic has only one pan, usually made of copper or stainless steel, in which the "wine" is heated to boiling then distilled. This equipment, also called distillation apparatus, can include columns with various external geometric characteristics and internal rectifiers (use of plates or trays, bubble traps, etc.), various extension tubes and cooling systems (Bruno, 2006; Maia, 2000; Pinto, 1986).

The distillation kettle (or pan) heating process may be direct or by steam. This process should be slow and gradual, since abrupt heating of the "wine" may cause the apparatus to overflow.

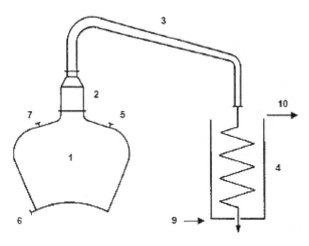

1. Curcubit or kettle
2. Column head, dome or helmet
3. Extension tube or condensation tube
4. Condenser
5. Wine inlet

6. Stillage discharge
7. Pressure equalizing valves
8. Distillate outlet
9. Water inlet
10. Water outlet

Fig. 1. Common Alembic (Source: Nogueira & Venturuni Filho, 2005)

Besides that, a gradual temperature increase allows the production of vapors which, upon reaching the column head, will partially condense and return to the kettle. The porting of uncondensed vapor reaches the extension tube and partially condenses when becomes in contact with a colder surface. Then it reaches the cooler, where the condensation is completed.

The distillate shows a high alcoholic content (65 - 70% vol) at the beginning of the process and a separation of 5 -10% of the total theoretical volume of spirit by the initial distillation is recommended. This is known as the "head" distillate, rich in aldehydes, ethyl acetate, fatty acids, ethyl caprate and ethyl caprylate (Léauté, 1990; , Mutton & Mutton, 2002; Bruno, 2007), dymethil sulphide (Nicol; Faria et al., 2003) and other volatile compounds that have greater affinity for ethanol than for water, a factor of greater significance than the individual boiling temperatures (Maia, 2000; Leauté, 1990). The fraction known as "heart distillate", the core distillate, is isolated next and contains a smaller proportion of the "head fraction" components, such as esters, aldehydes, higher alcohols, besides ethyl lactate and the fraction of long chain volatile acids and other undesirable secondary products in concentrations higher than those recommended, formed during the fermentation or inside the alembic itself (Janzantti, 2004 ; Léauté, 1990 ; Mutton & Mutton , 2005). The "heart fraction" represents about 80% of the distillate volume. Since it contains the smallest amount of undesirable substances, this is the best fraction. In practice, usually the phlegm content is controlled around 45 - 50% vol in the receiving box and then the "cut" is made.

Finally, the components with higher boiling points and greater affinity for water are removed. This fraction ("tail") has a high content of phurphural and of other less desirable components, such as acetic acid, and of the "heavier" fraction of higher alcohols, known as "fusel oil" (Mutton & Mutton , 2005; Nykänen & Nykänen, 1991). The "tail fraction", also called "weak water", corresponds to 10% of the distillate total volume, and is collected from an alcohol content of 38% vol up to 10% vol, approximately.

The practices of mixing the "head" and/or "tail" fraction with the new wine (Mutton & Mutton, 2005; Novaes, 1999) are extensively used, either for the recovery of alcohol or to allow the reactions among the "head" fraction compounds (acetaldehyde, ethyl acetate) and the "tail" (acetic and lactic acids) compounds with the wine alcohol, producing aromatic components important for the quality of the *cachaça*. In fact, the richer aroma of the cachaças distilled in alembics is related not only to the concentration changes that occur during the distillation processes, but also to the chemical reactions that occur among the components in contact with the hot alembics walls (Faria et al, 2003). These reactions are also favored by the presence of copper and by direct heating (Faria et al., 2003, Léauté, 1990; Nicol, 2003). For such, the distillation should be well conducted, allowing the isolation of compounds that are undesired and harmful to human health (Mutton & Mutton, 2002).

Besides the alembic shown in Figure 1, there are also two or three bodies alembics, which are common alembics adjusted accordingly, in order to save fuel, facilitate and accelerate the distillation process (Maia, 2000; Bruno, 2007).

Figure 2 shows a version of an alembic (2) with two pans. The pans on the right side function as heat exchangers, both to pre-heat the "wine" (up to 60°C) that will be distilled in the second pans at the left, and to pre-cool the emerging vapors condensed on the extension tube, simultaneously.

Fig. 2. Two two-body alembics with direct heating (CQE A and CQE B, Paraty-RJ)

The system may also include one more kettle or pan, making a three-body alembic, as shown in Figure 3. The upper pan works as a heat exchanger, such as in the two-body alembic. The apparatus is fed through the "wine" heater (1). The three bodies are interconnected by pipelines and valves and are at different levels, the "wine" falls by gravity and accumulates in the lowest body, either the exhaustion pan or kettle (2). When the working level is reached (75% of the total volume), the wine starts to accumulate in the distillation pan (3). Once the operative load is completed (75%), the "wine" heater is equally loaded.

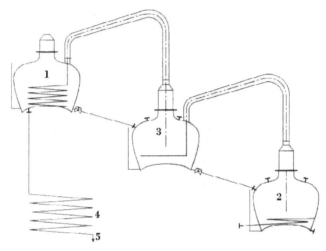

Fig. 3. Three-body alembic (Source: Mutton & Mutton, 2005)

When the valves of the three bodies are closed, the wine in the exhaustion pan is heated by steam or direct heat. The hydroalcoholic vapors containing the most volatile components produced during the progressive heating are, then, injected into the "wine" in the central pan. This enriched and heated wine will produce hydroalcoholic vapors, whose alcohol composition is richer than in those vapors received in the exhaustion kettle (Novaes, 1999). Part of the vapors, richer in water, condenses in the column and returns to the "wine" from where it came, while the fraction richer in alcohol passes through this condensation zone as vapor, reaching the serpentine coil in the "wine" heater, where it condenses. The recently condensed distillate then goes to the cooler serpentine coil (4), and is collected in the reception box (5).

When the alcohol content of the distillate collected in the reception box reaches values between 45 to 48% vol, the heat source and the distillation cycle are interrupted. The vinasses in the third body, which will not contain alcohol anymore, runs out and it will be completed with the distillation kettle "wine". The kettle is filled with the pre-heater "wine", which will then be fed with the new "wine". A new distillation cycle starts with the introduction of steam in the exhaustion kettle (Novaes, 1999 ; Bruno, 2006).

As the distillation progresses, both the uncondensed gases and the future "head" fraction resulting from the "wine" contained in the heater may be removed. Therefore, it will not be necessary to remove them in the next cycle. The "tail" fraction does not have to be removed either, since the vapors that would generate it are those produced in the exhaustion kettle, which will join the "wine" in the distillation pan (Novaes, 1999).

## 2.2 Systematic or continuous distillation

Many distillation columns have simply been adapted from those used in the production of ethyl alcohol and produced phlegms with high alcohol content, resulting in products poor in flavor. The distillation apparatus that were made of copper started to be made of stainless steel, at a lower cost. The distillates started to have strange odors, which disappeared inside the vat. However, storage time in the vats was not always enough for that.

In order to eliminate these problems, the distillation columns were redesigned to produce low degree, less rectified phlegm, with enough congeners' content to produce a more favorable *flavor* for marketing purposes (Almeida Lima, 2001). As shown in Figure 4, the column consists of a series of overlapping "plates" or "trays", which make its core. Two

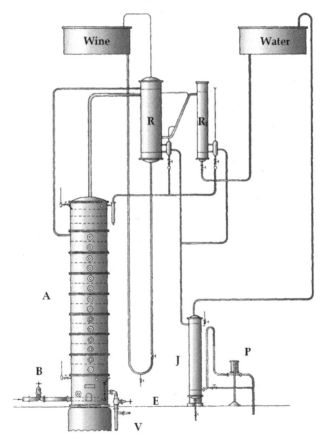

A- Distillation Column; R - "Wine" heater; R₁ – Auxiliary Condenser; J - Cooler; V – Vinasses; P – Graduate Cylinder; B – Vapor valve, E – Drainage Test.

Fig. 4. Distillation column for *cachaça* (Source: Stupiello, 1992)

plates together results in a "segment". The two trays communicate through a syphon. According to Figure 5, the upper part of this siphon is prominent and conditions the formation of a layer of liquid over the upper tray.

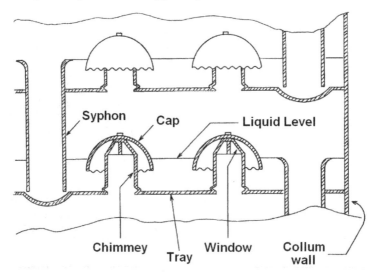

Fig. 5. Scheme of two trays stacked in a distillation column (Source: Novaes, 1993)

Besides the siphon, the tray has a certain number of chimneys stacks with side windows or hatches over which the caps or hats are. The edge of these caps stay immersed in the liquid, offering resistance to the steam passage through the siphon.

The "wine" is introduced in the column (15 to 20 plates) through the upper segment and the steam serpentine coil, for heating purposed, is introduced from the lower segment. To start the process, the "wine" is introduced in the column until it reaches the desired level at the lower segment. The system is cooled with water and the steam is introduced in the serpentine coil. The vapor formed in the lower plate reaches the chimneys of the plate immediately above it and is accumulated in their caps.

Once the pressure increases, vapor starts to bubble in the liquid in the plate, heating and enriching the "wine" on this plate. After some time, the "wine" on the plate above also starts to boil and heats the next plate, and then the "wine" in there and so on until the last plate is reached.

As the phlegm is removed, there is a depletion of the alcoholic content on the plates, with a subsequent increase of the upper plate temperature. When the temperature reaches 92°C, the column starts to be fed again with the new "wine", setting a continuous flow. The vapor formed in this plate will then be condensed.

System stability is maintained by controlling the vapor inlet, the "wine" flow and the vinasses removal. The use of low alcoholic content "wine" can be worked out efficiently by controlling the condensers downgrading and determining the most adequate selective condition for the final condenser, in order to obtain a better quality *cachaça*. Some compact distillation columns with a "wine" heater installed at the top of the column make it harder to control the selectivity, and cause the "wine" to be carried to the distillate (Novaes, 1993; Bruno, 2006).

## 2.3 Bidestillation or double distillation

Bidistillation was effectively introduced in Brazil in 1991, by Novaes (1994), with the purpose of suppressing the ethyl carbamate in the *cachaça* produced at a plant in city of Nova Friburgo (Rio de Janeiro state, Brazil).

The double distillation process in *cachaça* is based on the processes used in the production of whisky and cognac, through the distillation in copper alembics (Novaes, 1994). The first step ends when the distillate inside the collecting box reaches 27% vol, similar to the "weak water". The vinasses is discarded. This "wine" distillation will be done three or four times until the total "weak water" volume obtained is equal to that of the work load of the alembic.

All fractions from these distillates with low alcoholic content are stored in a flask that has at least the volume of the alembic load, and stay there until the next distillation.

At the end of the last "wine" distillation, the alembic will receive the "weak water" to be distilled. The sample should be heated slowly and, initially, the undesirable gases that do not condense will be released through channeling in the cooler and will be lost into the air. Right after that, a distillate with 77% vol will be collected when the cooling water valve is opened, and kept like that until the end of the process.

Three fractions will be obtained from this distillation: "head", 1.5% of the "weak water" volume distilled (72 to 75% vol); "heart", collected in another flask until the alcoholic content reached 66-68% vol; and "tail", produced until its volume in the collecting box was equal to 30% of the initial "weak water" load.

"Head" and "tail" fractions will be mixed and the volume divided in three parts. Each part will be mixed with a new "wine" load, and so on. The "heart" fraction will be stored in flask according to the type of beverage desired. It can be aged, left to "rest" in adequate vats for a certain period of time, or, simply be consumed *in natura*.

Ethyl carbamate is not too volatile in alcoholic solutions, due to its affinity with water and alcohol. Since it is formed by the distillation process, when a new distillation is done, the product obtained will be poor in ethyl carbamate. Most of the ethyl carbamate will remain in the residue from the first distillation (Riffkin et al., 1989; Novaes, 1994). Similarly, the amount of compounds responsible for carrying copper from inside the serpentine coils (organic acids) is very small, which also results in the production of *cachaças* with low levels of copper and of volatile acidity (Bizelli et al., 2000).

## 3. Reflux, reflux rate, column geometry and the effect over the *cachaça* composition

Distillation efficiency is controlled by the appropriate column design. The column design is a determining factor for the reflux adjustment, which consists on successive recondensation and revaporization occurring throughout the alembic column. Each time the vapors condense, a liquid with higher alcoholic content is generated and this liquid is capable of producing new vapors, even richer in ethanol.

At sea level, water boils at 100°C and ethanol at 78.5°C. The boiling temperature for a mixture of these liquids will be closer to 100°C as the ethanol concentration decreases and closer to 78.5°C as the ethanol content increases. A "wine" containing 8% v/v of ethanol starts to boil at a temperature around 94°C and, since ethanol is more volatile than water, the vapors formed will be richer in ethanol (45% v/v) than the mixture that produced them, according to the ethanol-water phase diagram shown in Figure 6. Once the vapors return to liquid state, the ethanol rate will be the same (45% v/v), but the temperature required to vaporize it again will decrease to about 83.5°C.

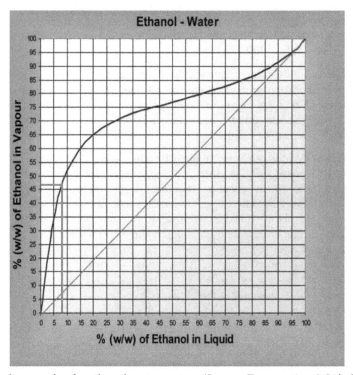

Fig. 6. Phase diagram for the ethanol-water system (Source: Evaporation & Life Science, 2002)

This phenomenon allows the vapors condensed in the column to return to gas phase, even if they stay at a position in the column in which the temperature is permanently lower than that of the liquid in the pan. This is the useful reflux. If the vapors condensed in the column return to the pan, the reflux would have been useless: it will only delay the distillation (Maia, 2000).

The efficiency of the still is a function of its design and of the operation of the column as well as of the reflux apparatus associated with the still head (Maia, 2000). This part of the still divides the condensate so that a portion (D) is taken off as a fresh distillate (P) and the remainder (L) is returned to the column as a reflux liquid. The ratio L/P is called the reflux ratio. In addition to other factors, the columns give better performance at lower throughputs (Peters et al., 1974) and higher temperature gradients. Under these conditions, the residence time in the column is longer, which results in an equilibrium improvement, distillations with lower temperatures for the ethanol-water vapor phase, which is richer in ethanol than the liquid phase, and in an increased reflux rate (Maia, 2000; Peters et al., 1974).

In the distillation in alembic, the reflux reflects the combined effect of various factors: a) temperature of vapors, from the moment they reach boiling temperature until they reach the access to the extension tube (temperature gradient); b) column geometry and length; c) number and geometry of column plates (Maia, 2000 , Claus & Berglund, 2005; Tham, 2006); d) type of bubble traps (Tham, 2006); e) presence and geometry of cooling system on the top of the column (dephlegmator or hood); f) diameter and geometry of extension tube; g) conduction of process (Maia, 2000).

The efficiency of a distillation column separation includes a set of variables, whose complexity makes the appropriate process control less accessible to most of the average *cachaça* producers. Very high vapor or liquid rates can cause problems such as: liquid entrainment, flooding, formation of vapor cones or excess of foaming, among others (Kalid, 2002; Tham, 2006).

The vapor inlet flow (Vin) is generated in the exhaustion section. It is possible to manipulate the reflux rate, equivalent to L, to control the quality of the distillate, because with an increase in reflux, the distillate becomes richer in the lighter products. Likewise, the bottom product composition may become richer in heavier substances, if the rate of heating fluid (water vapor) increases. However, the rate of reflux is a manipulated variable that controls the quality of the distillate (controlled variable), while Vin is another manipulated variable associated to the quality of the bottom product. Yet, making measurement in the mixture line is a complex and expensive task. Alternatively, temperature is used to infer the composition. In this case, the controlled variable is the temperature or the temperature difference between two column plates (Kalid, 2002).

## 4. Influence of distillation systems and processes on the ethyl carbamate content in *cachaças*

Distillated beverages containing cyanogenic glycosides, characteristic of the raw material used, are those in which ethyl carbamate (EC) contents reach the highest values, such as in *brandies* of stone fruits (peach, plum, cherry, etc.) (Schehl, 2005), and *tiquira* (Cagnon et al., 2002), a manioc spirit. The control of the distillation processes for these beverages is not enough to reduce the EC contents to acceptable levels, due to the large cyanide concentration formed. Cyanide is the immediate precursor of EC and this large cyanide concentration cannot be suppressed in the distillate, avoiding the precursors less volatile than cyanides from transferring to the distillates (Bruno, 2006; Lachenmeier et al., 2005). Although sugar cane is considered a cyanogenic plant, cyanide formation in *cachaças* is not very well defined. This is also true for other nitrogen-based EC precursors. However, the main role of the cyanate formed from the oxidation of cyanide, no matter its source (aminoacids, urea, cyanide or carbamylic compounds), has been suggested since the beginning of the last decade and supported by recent studies and reports (Bruno, 2006; Aresta et al., 2001, Riffkin et al.,1999).

One chemical pathway proposed for the EC formation involves the oxidation of cyanide (CN-) to cyanate (NCO-) catalyzed by Cu (II) ions (Aresta et al., 2001; Beattie & Polyblank,1995; Mackenzie et al., 1990) followed by reaction with ethanol (Aresta et al., 2001; Taki et al., 1992) according to the reaction scheme below:

$$2Cu(II) + 4\ CN^- \rightarrow 2Cu(CN)_2$$

$$2Cu(CN)_2 \rightarrow 2CuCN + C_2N_2$$

$$C_2N_2 + 2\ OH^- \rightarrow NCO^- + CN^- + H_2O$$

$$NCO^- + C_2H_5OH \rightarrow C_2H_5OCONH_2$$

Another possible mechanism involves the formation of isocyanic acid, which is released directly from the thermal decomposition of urea present in the wort. Aresta et al., (2001), have reported other possible mechanisms, most of them involving cyanide, copper and cyanate.

According to Riffkin et al., (1989), who used an experimental copper alembic to produce whisky "low wines", the amount of EC formed in the first two hours after distillation, in the presence of 0.8 mg/L of copper, represents about 20% of the EC's final concentration. Approximately 80% of EC were formed within 48 hours in the fresh distillate. By using an Amberlite IRC ion exchange resin a complete inhibition of the EC formation was also verified in fresh distillates when copper ions were suppressed.

Double distillation has been the most common procedure for removing ECs from distilled spirits. Nevertheless, this procedure generally leads to losses in ethanol yield and aroma. Boscolo, (2001), analyzed 84 samples of *cachaças* from various Brazilian regions, with EC levels varying from of 42 to 5689 µg/L and an average value of 904 µg/L. Only 13% presented EC levels below the maximum international limit established (150 µg/L). Those authors reported the occurrence of smaller amounts of EC in Brazilian sugar cane spirits obtained from distillation systems in which the descendent parts (end part of adapter and serpentine coil) are made of stainless steel. However, these systems produce spirits with sensorial defects, mainly due to the presence of sulfur compounds (Faria et al., 2003; Andrade-Sobrinho et al., 2002)., In alembics made entirely of copper, these off-odors usually react more effectively with this metal producing an odorless salt (Andrade-Sobrinho et al., 2002).

A preventive action commonly reported for the reduction of the EC levels in distilled spirits consists in fitting the upper portion of the columns with either a bubble cap tray or a section packed with copper rings (Bujake, 1992) or other copper devices (Andrade-Sobrinho et al., 2002). However, Andrade-Sobrinho et al. (2002) reported the presence of high levels of EC when the distillation system has a small area made of copper in the ascendant parts. They used 126 samples of commercial *cachaças* from several states and reported an average value of 770 µg/L. The average value of the *cachaça* samples produced using distillation columns was of 930 µg/L, whereas those from alembics showed an average of 630 µg/L.

Bruno et al., (2007), studied the influence of the distillation systems and of the distillation process on the EC levels. They analyzed 34 sugar cane spirit samples from 28 main producers (total of 30 different distillation systems) from various regions of the Rio de Janeiro State, Brazil. The first 17 samples consisted of freshly distilled fractions collected in glass flasks from 13 different distillation systems (2 producers have 2 alembics) and four (4) samples of them were recollected afterwards. Samples from alembics were collected from the heart fractions. Seventeen (17) samples were bottled *cachaças* acquired directly at the distilleries. The selection criteria of the 28 producing locations were their different distillation systems, their legal condition, economic importance and the quality of the facilities. These systems were divided in 7 continuous distillation columns (DC); 16 alembics made entirely of copper (CA, pot still); 5 stainless steel alembics with a copper serpentine coil (SSC); 1 alembic made entirely of stainless steel (SSS); and 1 double distillation copper alembic (DDC).

As shown in Table 1, only one out of the seven medium sized continuous distillation columns produced EC levels below the maximum limit of 150 µg/L. Ten out of the 16 copper alembics complied with this limit. The results also show that the average of EC levels in the *cachaças* from alembics was lower than the average for *cachaças* from distillation columns (Figure 7).

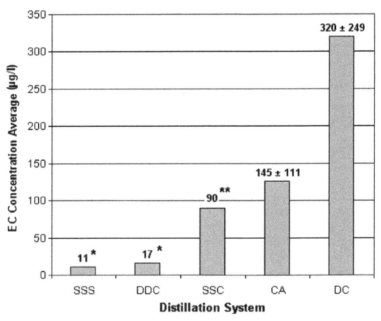

Fig. 7. EC concentration averages for each type of distillation system used for *cachaças* (± confidence intervals; P = 0.05). *One sample; **non parametric distribution (Source: Bruno et al., 2007)

The results are partially in agreement with those reported by Andrade-Sobrinho *et al.*, (2002), for *cachaças* from various states in Brazil. However, the average for *cachaças* from Rio de Janeiro was 5 to 6 times lower than that described by Boscolo, (2001) and Andrade-

| Producer/ Sample | | Distillation System/ heating system | EC ± SD (µg/L) | Confidence limits (± µ) | Apparatus coupled (alembics) | Dephlegmator |
|---|---|---|---|---|---|---|
| EDF | D | CA/DF | <LOQ | - | * | * |
| PTJ | D | CA/DF | 45 ± 3 | 8 | * | * |
| PLK | C | CA/DF | 214 ± 4 | 9 | * | * |
| JLG (1st) | D | CA/DF | 73 ± 1 | 2 | Pre-heater | Tubular (Cu) |
| JLG (2sc) | D | CA/DF | 298 ± 4 | 10 | Pre-heater | Tubular (Cu) |
| CQE (A) | D | CA/DF | 142 ± 1 | 2 | Pre-heater | Bowl jacket |
| CQE (B) | D | CA/DF | 156 ± 3 | 7 | Pre-heater | Bowl jacket |
| CRD (A) | D | CA/DF | 95 ± 6 | 15 | Pre-heater | Bowl jacket |
| CRD (B) | D | CA/DF | 195 ± 6 | 15 | Pre-heater | Bowl jacket |
| PHM | C | CA/SG | 89 ± 2 | 5 | Pre-heater | Tubular (Cu) |
| SLS | C | CA/SG | 50 ± 1 | 2 | Pre-heater | Tubular (Cu) |
| BVS | C | CA/SG | <LOQ | - | Pre-heater | Tubular (Cu) |
| SRV (1st) | C | CA/SG | <LOQ | - | Pre-heater | Tubular (Cu) |
| BAV | C | CA/SG | 30 ± 1 | 5 | Pre-heater | Tubular (Cu) |
| RCL | C | CA/SG | 27 ± 2 | 2 | Pre-heater | Tubular (Cu) |
| CHV | C | CA/SG | 599 ± 4 | 10 | Pre-heater + WSA | Bowl jacket |
| MAG | C | CA/SG | 232 ± 5 | 12 | Pre-heater + WSA | * |
| SBH | D | SSC/SG | 79 ± 2 | 5 | * | Tubular (SS) |
| PRI | D | SSC/SG | 35 ± 1 | 2 | * | Tubular (SS) |
| CTS | C | SSC/SG | 17 ± 2 | 5 | * | Tubular (SS) |
| DNC | C | SSC/SG | 215 ± 1 | 2 | * | Tubular (SS) |
| CPT | C | SSC/SG | 106 ± 3 | 7 | * | Tubular (SS) |
| PCA(1st) | D | DC/SG | 714 ± 7 | 17 | * | * |
| PCA(2sc) | D | DC/SG | 456 ± 5 | 12 | * | * |
| VMC(1st) | D | DC/SG | 262 ± 7 | 5 | * | * |
| VMC(2sc) | D | DC/SG | 155 ± 2 | 5 | * | * |
| BTB(1st) | D | DC/SG | 40 ± 3 | 7 | * | * |
| BTB(2sc) | D | DC/SG | 61 ± 2 | 5 | * | * |
| CAC | C | DC/SG | 323 ± 4 | 9 | * | * |
| SFC | C | DC/SG | 252 ± 2 | 7 | * | * |
| VRI | C | DC/SG | 216 ± 6 | 15 | * | * |
| CAM | C | DC/SG | 607 ± 2 | 5 | * | * |
| SRR | D | SSS | 11 ± 2 | 5 | * | * |
| NFB | C | DDC | 17 ± 3 | 8 | * | * |

CA: copper alembic; SSC: alembic with stainless steel pot and copper serpentine; DC: continuous distillation column; SSS: alembic entirely in stainless steel; DDC: double distillation in a big pot still used for distillation of whisky; C: commercial cachaças; D: distillates; SG: steam generator; DF: direct fire; WSA = wine stripping alembic; *- absent; (1st), (2nd) - first and second sampling; A, B - alembics A and B of the same producer; SD: standard deviation.

Table 1. Main characteristics and EC levels of the distillation systems evaluated (Source: Bruno et al., 2007)

Sobrinho et al., (2002). About 45% of the products and distilled fractions exceeded the maximum allowed limit of 150 µg/L, with an average of 160 ± 68 µg/L.

During the evaluation of the EC concentrations of 25 brands of pot still *cachaças* produced in the Paraíba State, Brazil, Nóbrega et al. (2009) reported an EC concentration range and average value of 55–700 and 221 µg/L, respectively. EC levels in 70% of the brands exceeded the international limit for spirits (150 µg/L).The average EC levels found in pot still *cachaças* from Paraíba State are considerably lower than the mean values reported previously for 34 samples of pot still *cachaças* from different parts of Brazil (630 µg/L, Andrade-Sobrinho et al., 2002) and for several commercial *cachaças* produced in Minas Gerais State (1206 µg/L ,Baffa Júnior et al., 2007; 893 µg/L, Labanca et al., 2004). Although not clearly stated, the samples investigated by Baffa Júnior et al. (2007) and Labanca et al. (2004) are likely to be pot still types, because Minas Gerais State, in Southeastern Brazil, is the national leader in pot still production of *cachaça*.

However, the average level reported here are higher than the values reported for 13 commercial samples of pot still *cachaças* produced in Rio de Janeiro State (123 µg/L, Bruno et al., 2007).

At first, the lower EC levels found in pot still alembics could be explained by the reactions of cyanide with copper in the ascending parts, which produce non-volatile complexes (Andrade-Sobrinho et al., 2002; Aresta et al., 2001; Boscolo, 2001). However, 2 alembics, without any copper-made part or device inside the columns, produced low EC levels (< 10 and 45 µg/L). These alembics are characterized by a distillation process with low flow rates and temperatures, and suggest that they are favored by their appropriate design (Bruno et al., 2007).

For pot still systems, as the alcoholic content decreases, the temperature of the wine increases. However, it is possible to keep the vapor temperature lower than 80°C in the dephlegmator by increasing the water flow rate inside it during the distillation process. Figures 8.A and 8.B show that it was possible to drastically reduce the EC levels in the freshly distillate fractions from a copper alembic labeled as JLG through this control method. On the other hand, the other 5 alembics that produced lower EC levels (< LQ to 50 µg/L), with the same copper tubular dephlegmator inside the column head of JLG, have two lower sections with stainless steel or copper bubble cap trays (Figure 5 and 9), which leads to high separation efficiencies.

Aiming at reducing the levels of EC in alembic *cachaças*, today, many distillation systems in Brazil are already made with those two lower sections. Therefore, alembic SRV stands out as the one with greater chance of rectification, with three plates with four bubble traps in each plate, made of stainless steel, and a reasonable reflux control system in the dephlegmator. In this system, EC formation was not detected either (< LQ), even when two samples of different batches were analyzed (see Table ), which indicated the absence of nitrogenated precursors in more rectifier systems.

In the alembics with a condensing bowl jacket around the still top, it is not possible to properly control the reflux, which, sometimes, can result in distillation temperatures higher than 90°C and accelerate the distillation process. This is the case for the CQE (Figure 2), CRD (Figure 12), and CHV (Figure 13) alembics. The highest EC level among all alembics was found in the CHV unit (595 µg/L), which has in the bottom a wine stripping alembic producing alcohol vapors which will warm it up through the bowl jacket and make the final distillation fraction.

Bruno et al. (2007) also observed that the presence of cloudiness was very common when the temperatures in the column head were above 80-82°C. As a result of entrainment (Tham, 2006; APV,1998), there is an enrichment of less volatile components, such as nitrogenous compounds, into the lower parts of the distillation system (serpentine coil, etc.), which in the presence of Cu (II) ions and increasing temperatures could eventually promote the decomposition or oxidation reactions responsible for the highest EC levels formed within these systems. These reasons could explain the high EC levels found in the heart and tail samples from the copper alembics (with a pre-heater) CQE and CRD, respectively (Bruno et al., 2007).

As shown in Figure 8.A, the high EC production in the beginning of the first distillation in the JLG (1) could be related to the high initial distillation temperatures (>90°C), which

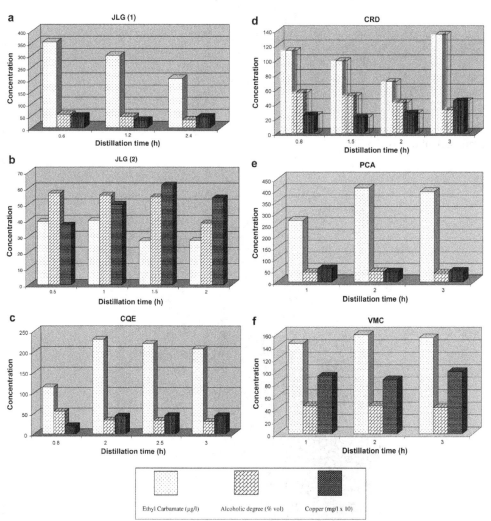

Fig. 8. Ethyl carbamate formation throughout different distillation processes. Cu levels and alcoholic content were measured in the fractions collected at various times during the distillation process. (a) Distillation in JLG with uncontrolled reflux; (b) distillation in JLG with controlled reflux; (c)–(f)distillations in the CQE, CRD, PCA and VMC systems. Each fraction in the histogram was analyzed for ethyl carbamate, ethanol and copper.

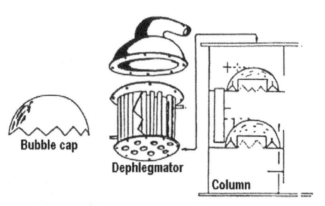

Fig. 9. Scheme of the parts from an alembic column designed to afford an efficient reflux (Source: Unknown).

Fig. 10. Alembic SRV (Valença-RJ).

Fig. 11. Column (at the left) with temperature control by a tubular dephlegmator (alembic JLG ; Bemposta-RJ).

Fig. 12. Details of a cooling system on top of a hood-like column (CRD B, Paraty-RJ).

Fig. 13. Alembic of three bodies (CHV) with a hood-type cooling system (Vassouras/RJ).

promoted the early entrainment previously mentioned and subsequent depletion of the nitrogen precursors. An expressive reduction in the EC levels was observed in the second distillation in the JLG (2) as a result of an adjustment on the reflux, which was obtained by controlling the water flow circulating in the tubular dephlegmator. In this case, the distillation temperatures stayed around 78°C (Figure 8.B).

The lower EC levels seem to come from alembics operating with high reflux ratios, low throughputs and low distillation temperatures (<80 °C). The highest EC levels were measured for those alembics with improper reflux ratios, emphasizing that the reflux ratio plays an important role in the EC formation (Bruno et al., 2007).

In the results obtained by Nóbrega et al (2009), shown in Table 2, with regard to the cooling system, a clear concentration of hot head systems was observed in the most heavily contaminated range (200–700 µg/l). Contrarily, a clear prevalence of dephlegmator and head cooler systems, particularly the former, in the less contaminated range (55–100 µg/l)

| Distillery | Brand[a] | EC[b] ($\mu$g/l) | Profile of copper pot still distillation[c] | | | |
|---|---|---|---|---|---|---|
| | | | Distillation scale[d] | Heating system[e] | Kettle shape[f] | Cooling system[g] (pot still column) |
| B | 02 | 55 | M | SS | O | TD |
| C | 03 | 60 | S | SS | O | TD |
| D | 04 | 75 | M | SS | O | TD |
| E | 05 | 90 | S | DF | O | HC |
| F | 06 | 100 | S | SS | O | HC |
| M | 14 | 200 | S | DF | C | HH |
| N | 15 | 220 | S | DF | C | HH |
| O | 16 | 225 | S | DF | C | HH |
| G | 18 | 235 | S | SS | O | HH |
| R | 24 | 430 | L | SS | B | HH |
| S | 25 | 700 | M | DF | O | HH |

[a]Brands of white single-distiled cachaças, according to Table 1.
[b]Ethyl carbamate levels (in increasing order) of cachaça brands.
[c]Distillation profile was obtained via data collection during visits to distilleries.
[d]Distillation scale, according to total kettle volume of distillery: small (S), 1.000–3.000 l; medium (M), 3.000–9.000 l; large (L), >9.000 l.
[e]Kettle heating system via internal steam serpentine (SS) or external fire with direct combustion of dried cane bagasse (DF).
[f]Kettle shape, according to Fig. 1: onion (O), conic (C), and boiler (B).
[g]Cooling system of pot still column (ascending parts), according to Fig. 2: tubular dephlegmator (TD), head cooler (HC), and hot head (HH).

Table 2. Ethyl carbamate (EC) levels in brands of white single-distilled *cachaças* and their corresponding distillation profile (Source: Nóbrega et al., 2009).

was observed. These observations are in line with those of Bruno et al. (2007), who also found a close connection between low levels of EC in *cachaças* and distillation in alembics using high reflux rates.

The values for the only alembic made entirely of stainless steel agree with those reported by Boscolo, (2001), emphasizing the importance of copper for the EC formation. As expected, the double distilled *cachaça* sample analyzed, produced in a large alembic imported from Scotland, also showed a low EC level. Copper levels in those two systems were 0.01 mg/L and 0.02 mg/L, respectively.

Striking differences in the EC levels measured were also observed for the different continuous distillation columns. Poorly conducted rectification, usually due to the excess of vapor and of wine feeding into the columns (Belincanta et al., 2006; Kalid, 2002) could explain the high EC levels found in most of these systems. In those cases, flooding (in the trays) and entrainment can occur at the same time (Kalid, 2002, Tham, 2006; APV, 1998).

Although there is also a significant variation on the EC levels from different distillation columns, few changes occurred during the distillation processes within these systems (continuous processes) when the temperature did not vary significantly.

## 4.1 Influence of copper contents on the formation of ethyl carbamate

The EC levels measured as a function of an increase in copper content in the experiment are shown in Figure 1. The initial value of 150 $\mu$g/L (in the y axis) was determined from other experiments with various sugar cane freshly distilled samples from other producers, for

Fig. 14. Alembic with stainless steel pan and serpentine coil (Resende-RJ).

Fig. 15. Distillation Column VMC (Bom Jesus do Itabapoana/RJ)

which an average reduction of 65% was obtained and copper levels were reduced to less than 0.01 mg/L (Bruno et al., 2007). In the previous experiment, using another PCA freshly distillate, a maximum reduction of 67% in the EC levels was observed, when the concentration of copper was reduced to 0.16 mg/L. Therefore, in order to achieve a more complete reduction of EC, the copper concentration in the distillates should be around 0.15 mg/L.

About 0.7- 0.8 mg/L was enough to promote a complete EC formation and higher concentrations of copper did not promote any additional catalytic effect.

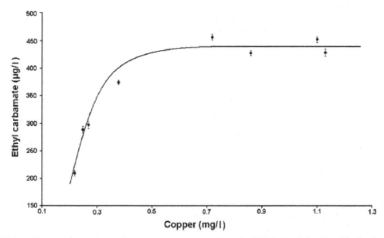

Fig. 16. EC levels as a function of copper concentration in PCA freshly distilled, eluted on Dowex Marathon C resin beds with increasing depths. The error bars show the standard deviations of EC measurements (n=3). ( Source: Bruno et al., 2007)

## 5. Conclusion

Some studies on the distillation of cachaças and how it affects the quality of these sugar cane spirits have been published and have contributed to the development of this area. However, there is still a great need for more research on the distillation processes of cachaça in order to continuously improve the quality of these products.

## 6. Acknowledgment

The author would like to thank his wife, Jovina, and sons, Saulo and Sergio Vitor for their patience, love and support. Also, thanks to Dr. Isabel Vasconcelos and Prof. Dr. Delmo Vaitsman for their continuous support during the time devoted to the studies and projects related to cachaça.

## 7. References

Almeida Lima, U. (2001) Aguardentes. In: Aquarone et al., *Biotecnologia na Produção de Alimentos*, Série Biotecnologia Industrial, 1. ed., Vol. IV, São Paulo: Edgar Blucher, pp.145-177.

Ampaq. Etapas para produção da cachaça de Minas Gerais, 21/02 2006, Available from <http://www.ampaq.com.br/arquivos/etapas_para_producao.pdf>.

Andrade-Sobrinho L. G. : Boscolo, M. ; Lima-Neto, B. S. & Franco, D. W. (2002). Carbamato de etila em bebidas alcoólicas (cachaça, tiquira, úisque e grapa). *Química Nova*, Vol. 25, No 6B, pp. 1074-1077. ISSN 0100-4042.

APV. *Distillation handbook '98.*, 4th ed., APV Crepaco Inc. 12.07.2011, Available from http://userpages.umbc.edu/~dfrey1/ench445/apv_distill.pdf

Aresta, M., Boscolo, M. & Franco, D. W. (2001). Copper(II) catalysis incyanide conversion into ethyl carbamate in spirits and relevant reactions. *Journal of Agricultural and Food Chemistry*, Vol. 49, pp. 2819-2824. ISSN 0021-8561.

Baffa Júnior, J. C.; Soares, N. F. F.; Pereira, J. M. A. T. K. & MELO, N. R. (2007). Ocorrência de Carbamato de Etila em Cachaças Comerciais da Região da Zona da Mata Mineira - MG, *Alim. Nutr.*, Araraquara, Vol. 18, No . 4, pp. 371-373. ISSN 0103-4235.

Beattie, J. K. & Polyblank, G.A. (1995). Copper-catalyzed oxidation of cyanide by peroxide in alkaline aqueous solution. *Australian Journal of Chemistry*, v. 48 n. 4, pp. 861 – 868. ISSN 0004-9425.

Belincanta, J., Kakuta Ravagnani, T. M. & Pereira, J. A. F. (2006). Hydrodynamic and tray efficiency behavior in parastillation column. *Brazilian Journal of Chemical Engineering*, Vol. 23, No 1, pp.135–146. ISSN 0104-6632

Bizelli, L. C.; Ribeiro; C. A. F. & Novaes F. V. (2000). Dupla destilação da aguardente de cana: teores de acidez total e decobre. *Scientia Agricola*, São Paulo, Vol. 57, No. 4, pp. 623-627.

Boscolo, M. Ph.D. Thesis. Caramelo e carbamato de etila em aguardente de cana. Ocorrência e quantificação. 2001. USP, Instituto de Química de São Carlos, São Paulo.

Boza, Y. & Horii, J. (1988). Influência da destilação sobre composição e a qualidade sensorial da aguardente de cana-de-açúcar. *Ciênc. Tecnol. Alim.*, Vol. 18, No 4, pp. 391-396

Brasil. Ministry of Agriculture - MAPA. *Normative Instruction N° 13* of June 29, 2005. Regulamento técnico para fixação dos padrões de identidade e qualidade para aguardente de cana e para cachaça. *Diário Oficial da República Federativa do Brasil N° 124*, of June 30, 2005, section 1, f. 3-4 (Brazilian Official Gazette)

Bruno, S. N. F. - Doctorate thesis in Food Sciences - Adequação dos Processos de Destilação e de Troca Iônica na Redução dos Teores de Carbamato de Etila em Cachaças Produzidas no Estado do Rio de Janeiro. –2006238 f. : il. – Universidade Federal do Rio de Janeiro, Chemistry Institute, Graduate Program in Food Science, Rio de Janeiro, 2006. References : f. 202 -219

Bruno, S.N.F., D.S. Vaitsman, C.N. Kunigami & Brasil, M.G. (2007). Influence of the distillation processes from Rio de Janeiro in the ethyl carbamate formation in Brazilian sugar cane spirits, *Food Chemistry*. Vol. 104, Issue 4, pp. 1345-1352. ISSN 03088146.

Bujake, J. E. (1992). Beverage spirits, distilled, In: *Kirk-Othmer Encyclopedia Of Chemical Technology*, Vol. 4, 4. ed., USA : John Wiley e Sons, Inc.

Cagnon, J.R., Cereda, M.P. & Pantarroto, S. (2002). Glicosídeos cianogênicos da mandioca: biossíntese, distribuição, destoxificação e métodos de dosagem. In: Cereda, M.P. Coord.). *Agricultura: Culturas de Tuberosas Amiláceas Latino Americanas*, SérieTuberosas Amiláceas Latino Americanas. Vol. 2, Cap. 5, pp. 83-89.

Claus, M. J.& Berglund, K.A. (2005). Fruit brandy production by batch column distillation with reflux. *Journal of Food Process Engineering*, Vol. 28, pp. 53-67. ISSN 0145-8876

Evaporation & Life Science. Rectification of a two component mixture of solvents using a rotary evaporator. Informative Bulletin, 2002. 15.10 2005. Available from <http://www.suntex.com.tw/images/e_learning/7.pdf >

Faria, J. B., Loyola, E., Lópe Faria, J. B., Loyola, E., Lopes, M. G., & Dufour, J. P. (2003). Cachaça, Pisco and Tequila. In A.G.H. Lea, J.R. Piggott (Org.). *Fermented Beverage Production* (2nd. ed.), New York: Kluwer Academic/PleniumPublishers, pp. 335-363.

Janzantti, N.S. Compostos voláteis e qualidade de sabor da cachaça. 2004. 179 f. Thesis (Doctor Degree in Food Sciences). Faculdade de Engenharia de Alimentos,Universidade Estadual de Campinas, São Paulo.

Kalid,R.A. (2002). Controle de coluna de destilação. Laboratório de Controle e Otimização Industrial – Chemical Engineering Dept., MAEQ, UFBA. Available from <www.lacoi.ufba.br/imagens_Lacoi/docs_pdf/Controle%20de%20Coluna%20de% 20Destilacao.pdf>.

Labanca, R.A. (2004). Teores de carbamato de etila, cobre e grau alcoólico em aguardentes produzidas em Minas Gerais. 2004. 62 f. Thesis (Masters in Food Sciences) Universidade Federal de Minas Gerais, MG, Brasil.

Lachenmeier, D. W. , B. ; Kuballa, T; Frank, W & Senn, T.(2005). Retrospective trends andcurrent status of ethyl carbamate in German stone-fruit spirits. *Food Additives and Contaminants*. Taylor & Francis , Vol. 22, No. 5, p.p. 397-405.

Léauté, R. (1990). Distillation in alembic. *Am. J. Enol. Viticult.*, Vol. 41, pp. 90-103,

Mackenzie, W. M.; Clyne A. H. & Mcdonald L. S. (1990). Ethyl carbamate formation in grain based spirits. Part II. The identification and determination of cyaniderelated species involved in ethyl carbamate formation in Scotch whisky. *J. Inst Brew.* , Vol. 96, pp. 223-232.

Maia, A. B. (1994). Componentes secundários da aguardente. *STAB*, Belo Horizonte, v. 12, n. 6, pp. 29-33.

Maia, A. B. (2000). Tópicos especiais em destilação e envelhecimento *Curso de Tecnologia daCachaça. Módulo III.* Amarantina, MG: [s.n.], 22 p.,. In: Doctorate Thesis by Bruno, S.N.F. Adequação dos Processos Destilação e de Troca Iônica na Redução dos Teores de Carbamato de Etila em Cachaças Produzidas no Estado do Rio de Janeiro. –2006238 f. : il. Universidade Federal do Rio de Janeiro, Rio de Janeiro, 2006. References: 202 -219.

Mutton, M. J. R. & Mutton, M. A. (2005). Aguardente. In: *Tecnologia De Bebidas. Matéria-Prima, Processamento. BPF/APPCC, Legislação e Mercado.* São Paulo: Edgar Blücher, 1st. ed., pp. 485-524.

Mutton, M. J. R. & Mutton, M. A. (2002). Cachaça: orientações técnicas para produção (CD). ABRABE -APEX –PBDAC Workshop. São Paulo, SP, BRASIL, 2002. p. 113-115. In: Bruno, S. N. F.    - Doctorate thesis in Food Sciences - Adequação dos Processos de Destilação e deTroca Iônica na Redução dos Teores de Carbamato de Etila em Cachaças Produzidas no Estado do Rio de Janeiro. – Universidade Federal do Rio de Janeiro, Chemistry Institute, Graduate Program in Food Science, Rio de Janeiro, 2006. references : f. 202 -219

Nascimento, R.F.; Cerroni, J. L.; Cardoso, D. R., Lima-Neto, B.S. & Franco, D. W. (1998). Comparação dos métodos oficiais de análise e cromatográficos para a determinação dos teores de aldeídos e ácidos em bebidas alcoólicas. *Ciência e Tecnologia e Alimentos*, Campinas, Aug/Oct., Vol. 18, No. 3, ISSN 0101-2061, pp. 350 - 355,

Nascimento, R.F. ; Cardoso, D. R., Lima-Neto, B.S. & Franco, D. W. (1998). Determination of acids in brazilian sugar cane spirits and other alcoholic beverages by HRGC-SPE. *Chromatographia*, Vol. 48, No. 11/12, pp. 751-757, DOI: 10.1007/BF02467643,

Nicol , D. A. (2003). Batch distillation. Chapter 5. In: *Whisky Technology, Production and Marketing*. Elsevier Ltd. I. Russell, G. Stewart,., Bamforth, C., & I Russell, (Eds.), pp 163- 167, ISBN 978-0-12-669202-0, Canada.

Nóbrega, I.C.C., Pereira, J., Paiva J. & Lachenmeier, D.W. (2009). Ethyl carbamate in pot still cachaças (Brazilian sugar cane spirits): Influence of distillation and storage conditions *Food Chemistry* Vol. 117, No. 4 , pp. 693-697. ISSN 0308-8146.

Nogueira , M.P & Venturini Filho W.G.(2005). Aguardente de cana. UNESP, Botucatu, SP, 12.12.2005, Available from http://www.fca.unesp.br/intranet/arquivos/waldemar/Aguardente%20de%20C ana%20-Completo.pdf>.

Novaes, F. V. (1993) Produção de aguardente de cana. *Curso de Tecnologia de Aguardente de Cana*. Delegacia Federal de Agricultura de Minas Gerais. . In: Bruno, S. N. F. - Doctorate thesis in Food Sciences - Adequação dos Processos de Destilação e de Troca Iônica na Redução dos Teores de Carbamato de Etila em Cachaças Produzidas no Estado do Rio de Janeiro. – Universidade Federal do Rio de Janeiro,Chemistry Institute, Graduate Program in Food Science, Rio de Janeiro, 2006.

Novaes, F. V. (1994). *Noções básicas sobre a teoria da destilação*. Piracicaba: ESALQ/Departamento de Ciência e Tecnologia Agroindustrial, 1994. 22p.

Novaes, F. V. (1999). Produção de aguardente de cana de alembic. In: *Programa de Fortalecimento do Setorde Aguardente de Cana de Açúcar e seus Derivados do Estado do Rio de Janeiro*. pub. SEBRAE- FAERJ-APACERJ, 45 p.

Nykänen, L. & Nykänen, (1991). Distilled beverages. In: *Volatile compounds in foods and beverages*. Marcel Dekker, Inc., Maarse, H. (Ed.), New York, pp. 548-576,

Peters, D. G., Hayes, J. M. & Hieftje, G. M. (1974). *Chemical separations and measurements* (1st ed.). Theory and practice of analytical chemistry. USA: WB Saunders Company, pp. 471–482.

Pinto, G. L. *Fabricação de aguardente*. Technical Report. Year 7, No 57, Universidade Federal de Viçosa, Nov.1986.

Reche, R.V.; Neto, AF; Silva, A.A. Galinaro C.A., Osti R.Z. & Franco, D.W. (2007). Influence of Type of Distillation Apparatus on Chemical Profiles of Brazilian Cachaças. *J. Agric. Food Chem.*, 55, p.p. 6603-6608. ISSN 00218561.

Riffkin, H. L. , Bringhurst, T.A., McDonald, A.M.L. & Howlett, S.P. (1989) . Ethyl carbamate formation in the production of pot still whisky. *J. Inst. Brew.*, Edinburgh, v. 95, pp.115-119.

Riffkin, H. L.; Wilson, R. & Bringhurst, T.A. (1999). The possible involvement of Cu II peptide/ protein complexes in the formation of ethyl carbamate. *J. Inst. Brew.*, Vol. 95, pp. 121-122. ISSN 0046-9750

Schehl, B. (2005). *Development of a genetically defined diploid yeast strainfor the application in spirit production*. Doctor Thesis, University of Hohenheim, Stuttgart. Universidade de Hohenheim, Stuttgart, Alemanha. Retrieved from <http//:www.unihohenheim.de/ub/opus/volltexte/2005/119/pdf/Dissertation_ Schehl_ B.pdf>.

Stupiello, J. P. (1992). Destilação do vinho. In: Mutton, M. J. R.; Mutton, M. A. *Aguardente De Cana: Produção E Qualidade*. Jaboticabal, São Paulo: FUNESP/ UNESP (Ed.), p. 67-78.,

Taki, N. et al. (1992) Cyanate as a precursor of ethyl carbamate in alcoholic beverages. *Jpn. J.Toxicol. Environ. Health*, Vol. 38, No 6, pp. 498-505. ISSN 0013-273X.

Tham, M.T. (nd) *Distillation, an introduction*. 22.10. 2006. Available from <http://lorien.ncl.ac.uk/ ming/distil/distil0.htm>

Tonéis & Cia (nd) . *Cachaça: aguardentes industriais e artesanais*. 11.10.2005. Available from http://toneis.com.br/modules.php?name=News&file=article&sid=105

# 8

# Spirits and Liqueurs from Melon Fruits (*Cucumis melo* L.)

Ana Briones, Juan Ubeda-Iranzo and Luis Hernández-Gómez
*Laboratory of Yeast Biotechnology,*
*University of Castilla La Mancha, Ciudad Real*
*Spain*

## 1. Introduction

Melon fruits, *Cucumis melo* L, are perishable and must be brought to market very quickly, which leads to saturation of the market and a surplus of melons. Increasing melon production has brought a serious marketing problem. Its fermentation and distillation could represent a potential solution to the problem. A review of the literature and other information sources failed to turn up prior experience with transformation of melon by means of alcoholic fermentation, making it necessary to develop a research protocol for investigating the suitability of melon juice, paste without or with skins and seeds as a fermentation substrate and behaviour of the resulting melon wine during distillation, with a view to obtaining a melon spirit or liqueur with appropriate flavour and aroma attributes.

Distillation of fermented fruit wines has been used in some countries for many years to obtain palatable beverages with high alcohol contents. The most important distilled spirits are elaborated from diverse raw material as grapes (Brandy, Grappa, Orujo), malt (Whiskey), cane sugar (Rum), etc. Other distilled beverages come from the distillation of fermented fruits, named in some zones "wine fruits" , such as cherries, apples or pears. Several countries produce spirits obtained by steam distillation of the anaerobically fermented grape pomace, left over after grapes have been crushed during wine-making. These spirits contain 30-45 % (v/v) alcohol and are highly appreciated, especially after gourmetmeals.

In Spain, the distillation of the grape wine and its by-products is frequent, but is unusual the distillation of "wine fruits". Actually, the maceration of agricultural products, generally fruits, or the addition of artificial essences and aromas to wine spirits is made to elaborate different kinds of liquors, though natural cherry or pear spirits have been developed years ago.

Most commercially available spirits (grappa, orujo, bagaceira, kirsch, plum, brandy) are filtered, bottled, and sold without maceration. However, on occasion fruits, seeds, and leaves are used as substrates to modify the product and improve its sensory qualities diversifying their range and attributes. This process is used in Spain to make a number of drinks, such as "pacharán" (maceration of sloe, *Prunus spinosa*, berries) or "anisette" (maceration of anise, *Pimpinella anisum*, seeds). During maceration, aromatic substances are leached out of the fruit into the spirit, which may then be redistilled or bottled as the finished product. Maceration time is a key factor both for component extraction and for achieving the right sugar content and colour of the spirit. The amount and the parts of the fruit used in maceration are two other aspects to be considered. For fleshy fruits like pears, apples, raspberries, strawberries, and cherries, pieces of fresh fruit may be used, or the seeds or nuts (blackthorn, hazelnuts, almonds) may be employed. Substrate conditioning is also an important factor, and may include pieces of fruit of different sizes, seeds, placenta, or skins depending on the type of product being manufactured.

Melons are a major crop in the La Mancha region (Spain) and the large crop size results in high levels of surplus production which must be commercialised in a very short period of time. The fermentation of melons and its distillation to produce genuine spirits could be a solution to the problem of the saturation of the market that would prevent wastage. The process of developing a new product has to be undertaken step by step, and for that reason trials to examine fruit processing, clarification, fermentation, column and alembic distillations were performed at laboratory and pilot scale (Briones, et al., 2002; Hernández Gómez, et al., 2003). Chemical and sensory analyses were carried out to assess the quality of the spirits and liqueurs produced comparing the results with those in other commercially available spirits.

## 2. Fruit processing

Healthy melons, *Cucumis melo* L, of the varieties *"Piel de sapo"*, *"Ruidera"* and *"Sancho"* coming from cultivars of La Mancha region in Spain were processed. Melons were washed and divided into three sets and processed in different ways to obtain three substrate types ready for their fermentation:

* Juice: The melons were hand-peeled, cut up into pieces manually, after that crushed and peeled in a horizontal des-temmer with rollers to produce a paste that was then pressed in a pneumatic vertical press

* Paste without skin ("pws"): The melons were processed as described above, but not pressed.

* Paste: The unpeeled melons were cut up into pieces directly and then crushed to form a paste that included the skins.

The yield of an industrial process must be calculated carefully since it is of great economic importance as well as conventional fermentation parameters such as °Brix, reducing sugars, volatile acidity, pH and alcohol degree. The yield (w/w) of the processed fruits depended on the substrate assayed. Therefore in the case of the paste it can reach 100%, 70% (30% of skin) for the pws and only 50% for the juice. This was mainly caused by the percentage of melon removed as skin in the case of the "pws" and by difficulties in filtering the melon paste to produce the "juice". Juice extraction yields of up to 75 % have been attained for other fermented products, such as grape because of the small weight of the skins. These yields were improved and an increase of 10 % were recorded for the juice and pws substrates, perhaps thanks to optimization of procedures. By-product yields were 21 % (w/w) of skins for the pws substrate and 40 % (w/w) skins and pressed pulp for the juice substrate (Hernández-Gómez et al, 2005a).

The initial pH of the different substrates varied between 4.4 and 5.2. and this supposes a problem for the easy growing of lactic and acetic acid bacteria; for that, pH was adjusted or not before the fermentation by adding citric acid to reach values around 4 in order to inhibit these bacteria.

## 3. Characterisation of fermented from melon fruits

Fermentation of the juice, pws and paste was carried out at 20 °C. The substrates were inoculated with a commercial yeast (*Saccharomyces cerevisiae* UCLM 325) up to a concentration of approximately $10^6$ cells/mL. The process was monitored daily by measuring residual sugars, and the end of fermentation was determined on the basis of the sugar consumption (OIV, 1969). The initial assimilable nitrogen was measured using the NitroGenius® kit.

Judging from the initial °Brix, between 10,0-10,2 an alcohol degree of 5% (v/v) could be expected. Nonetheless, experimental data showed that the ethanol yield was acceptable only in juice and pws (4.2 % v/v), being very low (3.4 %) in the case of the paste, possibly due to the complexity of the structure of the fermentation media.

The fermented paste showed the highest values of acetic acid possibly as a consequence of a contamination by acetic bacteria. Adjusting the pH, successfully bacterial growth in all three melon wines produced was diminished. Under unadjusted pH conditions, the bacterial populations increased in the pws and paste substrates but decreased in the juice. This may be attributable to a higher level of contamination from the melon skins or to sluggish fermentation in a complex media like paste substratum.

In this kind of alcoholic beverages, concentrations of the volatiles has to be refereed to the ethanol content. Otherwise, the volatile composition is closely related to the type of substratum, the conditions of the fermentation and the yeast strain used. Respecting the major volatiles, acetaldehyde ranged widely from 243 to 1196 mg/L of ethanol, being highest in the pH-unadjusted substrates, possibly due to the action of spoilage microorganism (Silva et al., 2000). Methanol is not a direct product of fermentation (Ribéreau-Gayon et al., 2000). Two types of fruit enzymes are able to act upon pectins to release methanol: polygalacturonases, by cleavage of the glycosidic bonds on the chains; and pectin-Methylesterases, by catalyzing hydrolysis of the esterified chemical function (Hernández Gómez et al., 2003). The presence of high amounts of methanol in the wine fruit produced from the paste at both pH levels may be the result of the action of these enzymes in the skin.

In general the higher alcohols (HAs) quantified [1-Propanol, isoAmyl alcohols , 1-Butanol, and 2-Methyl-1-Propanol] were higher with pH adjusted, especially in the case of the wine made from the pws. 1-Butanol and 2-Butanol were not detected, a highly positive finding, because these two substances adversely affect the final aroma of the distillate. Total esters were higher in the pH-unadjusted wine made from paste than in the rest of the wines. Ethyl lactate was the main contributor to this high value and probably depends on the initial count of lactic acid bacteria, present in this kind of substratum (Briones et al. 2002).

When ANOVA statistical analysis was applied it was noticed that except for Ethyl acetate, there were differences in the volatiles for all the melon wine types, especially between the wines made from the paste and the rest.

## 4. Distillation procedure and analysis of spirits

Upon completion of alcoholic fermentation, the fermented were immediately distilled with yeast lees in two ways in previous assays: in rectifying glass column and in French type copper pot. In the first case, a glass column of 50 cm of length and filled up to 50% with Raschig rings and a round bottomed 10 L flask were used. The flask was filled with 5 L of every type of fermented fruit. To ensure a homogenous heat distribution during the distillation process, boiling stones were added and the flow rate was adjusted at 10mL min-1. In the second case, a 30 L French type copper pot filled with 15 L of fermented fruit was used and the flow rate was adjusted at 25 mL min-1. In both cases the fermented were double distilled. The first distillation was stopped when the alcohol degree was lower than the fermented fruit, obtaining a distillate around 17-20% (v/v). In the second distillation, the first phase was the collection of 0.8% of distillate (heads) which was discarded. This distillation was stopped at around 30% (v/v), so the final distillate (heart fraction) reached an alcohol concentration around 55% (v/v). The tails were formed adding the fractions ranging from 30 % (v/v) to 5 % (v/v). The distillate was collected in fractions of different volums depending of the equipment used. In these kind of processes and in order to avoid the loss of aromas all the fractions were collected on ice and kept at 4°C until their analysis. The percentage alcohol content in all of them was determined by electronic densimetry in all the fractions (European Union, 2000). The heart fraction alcohol degree values are shown in Table 1. In both distillations (copper pot and column) the alcohol degree of the paste was lower because the initial degree of the fermented was also inferior.

|            |       | Alcohol degree (% v/v) | | | |
|------------|-------|--------------|--------------|--------------|----------------|
|            |       | 1st fraction | 2nd fraction | 3th fraction | Average value  |
|            | Juice | 68.6         | 61.5         | 44.2         | 58.1           |
| Copper pot | Pws   | 65.6         | 52.1         | 27.9         | 48.5           |
|            | Paste | 69.8         | 62.8         | 44.6         | 59.1           |
|            | Juice | 79.5         | 60.5         | 12.3         | 50.8           |
| Column     | Pws   | 80.0         | 61.3         | 11.2         | 50.8           |
|            | Paste | 75.0         | 54.0         | 10.6         | 46.5           |

Table 1. Alcohol degree of second distillation for different fractions collected in the copper pot and in the glass column

Respecting the chemical analysis of major volatiles, the glass column samples possessed higher concentrations of Propanol, 2-Methyl 1-Propanol and 3-Methyl 1-Butanol. Moreover, the sensory analysis of the first batch of distillates offered conclusive results so as to the distillation type. 100% tasters preferred the samples distilled in the copper pot due to its aroma intensity while the other ones coming from glass column were rejected on basis of their pungent and /or "not sufficiently intensive" aroma.

With regard the type of substrate, even though the "paste" type offered a better yield in the process, it does not seem to be the ideal substrate due to the sluggish of fermentation, a high methanol content and the negative sensory characteristics. Juice distillate was more appreciated due to its aroma intensity (Hernández-Gómez et al., 2008).

To compare the results with those from previous years using a glass column, and thus narrow down the production processes, a pilot-plant copper pot, *alquitara*, was used. At the same time, the volatile compounds in the distillates obtained were compared with those in other commercially available spirits.

The fermented juice was immediately distilled in a traditional 130-L "*alquitara*", (reflux still) (Silva, Macedo & Malcata, 2000) filled to 70-80 % of capacity, equipped with a series of temperature sensors. Distillation flow rate was set at 170 mL/min, and the condenser was kept at below 21 °C throughout. The distillate was collected in volumes of 1 L each, except for the head fraction. The first distillation was stopped when the alcohol content in the volume collected had reached 8-10 % (v/v), which yielded a distillate with an alcohol content of 18.5-25 % (v/v), depending on the source substrate from which it had been made. The head-fraction (200 mL), usually discarded, was not rejected.

The second distillation was carried out in a traditional 30-L alembic copper still filled with 15 L of the first distillate. Distillation flow rate was set at 35-40 mL/min. The heads, 0.8 % of the distillate, were discarded, and distillation was stopped at 40 % (v/v), thus yielding a final distillate (hearts) of 58-69 % (v/v), again depending on the source substrate from which it had been made. The tails comprised the fractions from 40 % (v/v) to 5 % (v/v). For all the distillates the alcohol content of the last volume collected was between 9.2 and 11.8 % (v/v), and the final alcohol content was between 18.5 and 25 % (v/v). In the second distillation, the alcohol content decreased from 74.5 % (v/v) to 40.0 % (v/v) in the last volume collected. The total distillation time for all the fractions was around 4 h. The highest value was for the juice distillate (pH-adjusted) and the lowest for the paste (pH-unadjusted) distillate.

After the second distillation the major volatiles present in the heads, hearts, and tails fractions of the different spirits are depicted in Figure 1.

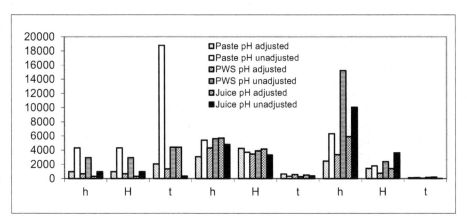

Fig. 1. Evolution in major volatile compounds during the second distillation. Methanol. Higher alcohols: 2M1P, 2M1B, 3M1B, and 1-Propanol. Esters: Ethyl lactate, Ethyl acetate, and Ethyl butyrate (mg/L of EtOH). h, heads; H, hearts; t, tails. pws = paste without skins

Methanol was collected in nearly the same proportion in all the fractions, most likely due to the formation of azeotropic mixtures (Orriols, 1994). Nevertheless, methanol concentrations were higher in the distillates from the pH-unadjusted wines except for "juice" tails. High levels of methanol in the paste distillate were observed (pH-unadjusted). The higher alcohols (HA) content was higher in the heads and hearts than in the tails. All the distillates displayed the same behaviour, with no considerable differences among them. In addition, the HA content was not related to the fermentation pH and substrate types. The ester content (Ethyl acetate, Ethyl lactate, and Ethyl butyrate) was higher in the heads, decreasing in the hearts and the tails. Fermentation pH had a pronounced influence on the ester content.

Respecting the concentrations of the volatiles in the final spirits, high amounts of methanol and 2-Butanol were noticed. They can make spirits hazardous to consumers health. Moreover, methanol imparts a cooked cabbage odour, with a threshold of 1200 mg/L (Ribéreau-Gayon, et al., 2000). The methanol content was higher in the distillates made from the pH-unadjusted in all substrates. The paste distillates exhibited the highest levels, probably owing to the action of certain pectinases on the substantial amount of melon skins present (Cortés Diéguez, et al, 2000). However, in no case did the levels exceed the limits for fruit spirits set by the legislation currently in force (European Union, 1989).

Higher alcohols are responsible for imparting complex sensory attributes to spirits (Silva et al., 2000). The Amyl alcohols and 2-Methyl 1-Propanol contribute positive to the sensory characteristics (Bertrand, 1975; Orriols, 1992, 1994). They are detectable organoleptically at concentrations below 15 mg/L of ethanol (Tourliere, 1977). 1-Propanol has a pleasant, sweetish odour, but excessive concentrations will introduce solvent notes that mask all the positive notes in distillates (Fundira, Blom, Pretorius, & van Rensburg, 2002). The highest 1-Propanol contents were recorded in the distillates made from the pH-adjusted fermentation substrates, in contrast, 2M1P was lower in the pH-adjusted distillates at levels similar to those reported in previous years. Amyl alcohol contents were very similar in all cases. 1-Butanol has a heavy, penetrating odour, and 2-Butanol is associated with low-quality raw materials (Orriols & Bertrand, 1990; Cortes Dieguez et al., 2000). 1-Butanol concentrations were higher in the pH-adjusted distillates. Conversely, 2-Butanol was not detected in any of the distillates produced. 2-Phenylethanol imparts a very clinging, rose-like aroma (Nykänen & Suomalainen, 1983), and was not influenced by pH, except in the juice distillates. 1-Hexanol, cis-3-Hexen-1-ol, and 3-Methyl-3-buten-1-ol impart strong herbaceous aromas. Hexanol and cis-3-Hexen-1-ol perception thresholds in spirits are 20 mg/L and 3.5 mg/L, respectively (Jouret & Cantagrel, 2000), and the concentrations did not exceed those values. Smoked or burnt wood aroma is conferred by 4-Methyl-guaiacol (Dubois & Dekimpe, 1982), and it was present in all distillates except the pH-adjusted juice. Benzyl alcohol is related to the quantity of benzaldehyde, the latter being important because it imparts a bitter almond aroma to wines at levels above 2-3 mg/L (Blaise, 1986). In the melon spirits, concentrations were nowhere near that perception threshold reported for wines.

Otherwise, Acetaldehyde is 90 % of the total aldehyde content in distillates (Versini, Monetti, dalla Serra & Inama, 1990, Orriols, 1991; Silva, Malcata, & Hogg, 1995). More than 1200 mg/L of ethanol is evidence of oxidation of the ethanol during alcoholic fermentation or a enzymatic pyruvic acid decarboxylation (Baro & Quiros-Carrasco, 1977; Cantagrel, Lablanquie, Snakker, & Vidal, 1993). Its importance derives from its pungent odour and its chemical reactivity (Silva et al., 2000). pH had no effect in juice and pws substrates, in contrast, it was doubled in pH-adjusted paste distillate. Furfural may be formed as a result of oxidation of ascorbic acid (Bayonove, Baumes, Crouzet, & Günata, 2000). A slightly higher furfural content in the distillates from the pH-unadjusted substrates was observed.

Esters are associated with pleasant odours. This is particularly true of Ethyl acetate, which contributes to aroma complexity and has a positive impact at very low levels (50-80 mg/L) (Steger & Lambrechts, 2000). Ethyl acetate was higher in the distillates pH-unadjusted. Ethyl lactate contributes intense, long-lasting aromas (Tourliere, 1977). The content in distillates is linked to lactic fermentation, in general it was lower in pH-adjusted distillates (Briones et al., 2002). Ethyl butyrate adversely influence the organoleptic quality of distillates (Soufleros, 1978,1987). Paste distillates at both pH had the highest concentrations. Concentrations of the minor esters, other than Ethyl lactate and Ethyl acetate, were higher in the pH-adjusted distillates, due principally to Ethyl caprylate, Ethyl palmitate, Ethyl caproate, Ethyl laurate and Ethyl decanoate.

Respecting carboxylic acids, short-chain (C4-C12) fatty acids produce unpleasant odours, and high concentrations are an indicator of poor quality (Orriols, 1992,1994). Decanoic and octanoic acids were the most abundant fatty acids and were present at higher concentrations in the pH-adjusted distillates.

### 4.1 Comparison with commercial spirits

The major volatile compounds of melon spirit were compared with other commercial spirits such grappa, orujo, and other fruit distillates from cherry, raspberry and pear (Hernández Gómez et al, 2005a) . In general, 1-Propanol was higher in melon distillates, being similar to those in grappa and cherry spirit. In like fashion, Ethyl acetate was much higher in the juice distillate with a concentration nearly three times more than in the other spirits. Acetaldehyde and Ethyl lactate were present in similar amounts in most of the spirits considered, with the exception of much higher concentration of Ethyl lactate found out in the cherry spirit, most probably as a result of the lengthy maceration time. 2-Methyl 1-Propanol and Amyl alcohols were somewhat higher in the melon distillates, pear spirit, and grappa. The melon distillates had appreciably lower methanol contents than the rest of the spirits considered, particularly as compared to the pear spirit (11216 mg / L of ethanol), most likely because of the high pectin content of pears. 1-Butanol was detectable in the pear spirit and, to a lesser extent, in the melon distillates. 2-Butanol and Ethyl butyrate were not present in detectable amounts in any of the spirits analysed.

Sensory analysis was performed in a standard tasting room [Spanish standard 87004:1979 (Aenor, 1977). Distillates were diluted with distilled water to alcohol strength of 28 % (v/v) and served at a temperature of 15 °C. Evaluations were carried out at daily sessions between 10:00 a.m. and 12:00 noon to prevent taster fatigue.

There were significant differences between juice distillates and pws and conversely, there were no significant differences between the paste distillates. The tasters preferred pws or juice distillates, but there was no preference between them. The expert tasters from a distilling company likewise expressed no preferences, possibly because the spirits being tested were new to them and had an unfamiliar flavour.

### 5. Maceration for improving melon spirit

The effect of maceration in increasing and improving aroma intensity in a melon (*Cucumis melo* L.) spirit was studied.

Double-distilled melon spirit (alcohol content 56 % v/v) was macerated using different melon parts (pieces of melon flesh, skinned and then sliced (MF) (72.5 % w/v), and seeds+placenta (MSP) (23.5 % w/v). Contact times were 8 days, 18 days, and one year. The solid percentage was selected based on literature data (Xandri, 1958) and personal communications. After steeping, the distillate was chilled and filtered and the solids pressed and 0.3 g/L bentonite was added for clarification.

Sugar levels in the double-distilled spirit macerated with sliced melon flesh (MF) were four to five times higher than in the spirit macerated with the seeds+placenta (MSP), possibly due to the higher solids content (72.5 % in the MF batches compared with 23.5 % in the MSP batches). Total sugar levels stayed practically constant throughout maceration, though after maceration for one year saccharose levels fell while glucose and fructose levels rose by 80 and 60 %, respectively, perhaps because of hydrolysis of the disaccharide.

Dry matter (DM) was much lower (0.10 g/L) in the unmacerated control batch (C) than in the macerated spirit. The spirit macerated with sliced melon flesh for one year (MF1Y) had the highest values (26.95 g/L), compared with the spirit macerated with the melon seeds+placenta also for one year (MSP1Y), (21.53 g/L), because of the larger amount of macerated solids.

The alcohol content (% v/v) decreased in the macerated double-distilled spirit, the decrease being more pronounced in the macerates made with the sliced melon flesh, possibly attributable to absorption by the melon tissue and the release of larger amounts of water from the pieces of melon flesh.

Respect to the major volatiles, Methanol levels decreased with maceration but in all cases were within the limits set by the legislation currently in force (European Union, 1989) The higher alcohols (HAs), namely, 1-Propanol, 2-Methyl-1-Propanol, 2-Methyl-1-Butanol, and 3-Methyl-1-Butanol, and the esters Ethyl lactate and Ethyl acetate fell appreciably during maceration. These decreases were much more pronounced in the spirit macerated with the sliced melon flesh (MF8 and MF18) than in the spirit macerated with the seeds+placenta (MSP8 and MSP18). Ethyl butyrate was not detected. 1-Butanol, which has a very pungent and heavy aroma (Nykänen & Suomalainen, 1983), was present in lower concentrations in the macerated spirits, and theirs decrease could have a positive influence on the aroma of the final spirit.

Levels of Benzyl alcohol, 1-Hexanol, t-3-nonenol, c-3-Hexen-1-ol, 2-Phenylethanol, and geraniol were higher in the batches macerated with the sliced melon flesh and the seeds+placenta. 2-Phenylethanol imparts an aroma of roses (Nykänen & Suomalainen, 1983), and the concentration of this compound was three to four times higher than in the control spirit. At high levels 1-Hexanol and c-Hexen-1-ol may contribute herbaceous aromas, while at low levels they may exert a positive influence (Bertrand, 1975, Orriols, 1992, 1994).

On the whole, the ester content in the macerated spirits were lower than in the control spirit, with substantial decreases in Ethyl caproate, Ethyl caprylate, Ethyl decanoate, and Ethyl laurate. At the same time, Ethyl palmitate decreased in the spirit macerated with the sliced melon flesh but remained constant in the spirit macerated with the seeds+placenta because of the quantities that leached out of the melon seed (Al-Khalifa, 1996) . Ethyl linoleate and Ethyl linoleoate were also related to the seeds and increased considerably in the spirit macerated with the seeds+placenta.

The concentration of total acids was lower in the macerated spirit, especially because of pronounced decreases in hexanoic acid, octanoic acid, and decanoic acid. Lauric acid levels increased considerably following maceration with both the sliced melon flesh and the seeds+placenta, possibly as a result of hydrolysis of Ethyl laurate.

Furfural is produced by acid hydrolysis or during the heating of polysaccharides containing hexose or pentose fragments, the highest concentrations being found in alcoholic beverages (1-33 mg/kg). This compound is currently allowed, since it is naturally present in fruits and other foodstuffs (European Union, 2002). Furfural and hydroxyMethyl furfural increased substantially in the spirit macerated with the sliced melon flesh as compared to the control spirit, but values were still within allowable limits.

The concentration of phenolic compounds was higher in the macerated spirit. This was particularly true for 4-Methyl-guaiacol, which increased appreciably in the spirit macerated with the sliced melon flesh. In wine this compound has a negative impact on aroma at levels higher than 4 mg/L but a positive impact at concentrations between 1.2 and 2.4 mg/L (Etiévant, et al., 1989).

Benzaldehyde, a carbonyl compound, decreased in the macerated spirit. Concentrations of this substance higher than 2-3 g/L are related to a bitter almond flavour in wines (Blaise & Bruns, 1986) .

Increased levels of acetoin (3-hydroxy-2-butanone) act as an indicator of oxidation of 2-3 butanediol during ageing (Jouret & Cantagrel, 2000). The increase was higher in the spirit macerated with the sliced melon flesh.

In the colour attribute determinations, the tristimulus values, to derive the rectangular (L*, a*, b*), cylindrical (L*, C*, h*) and the chromaticity (x, y, z) coordinates were used (CIE, 1986). The coordinates a*, b*, h*, and C* yielded the two-dimensional (CIELAB) colour space, where h* is the angle formed with the a* axis and C* is the distance to the origin. Sample MSP1Y had the highest colour intensity (C*) values, and in terms of chromaticity values, and MSP18 was similar to sample MF1Y. Variation in the angle (h*) was minimal, with angles in the range of 90.14 to 94.46. All the samples fell in the region pale, with macerated sample MSP1Y exhibiting the highest colour intensity and thus falling closer to the region of lightness. Thus, maceration affected both lightness (L*) and colour intensity (C*), with the unmacerated spirit and the spirit macerated with the sliced melon flesh being paler than the spirit macerated with the seeds+placenta.

Figure 2 represents the differences in colour (ΔE*) between the macerated spirit and the control spirit and also lists the gradations in visual perception according to Schmidhofer, et al., (1994). The longest maceration time (1 year) exhibited higher ΔE* values than the shorter maceration times, and the three macerations carried out with the seeds+placenta exhibited higher ΔE* values than the macerations carried out with the sliced melon flesh.

The results of sensorial analysis thus indicate that maceration did have an influence on the final product and that the panelists perceived distinct differences between the samples.The distillate macerated for 11 days (MF11) and the unmacerated control batch were significant at the 99.9-% level.

The preference tests failed to yield any preference for either the macerate spirit or the control sample. This result is ascribable; on the one hand, to the diverse make-up of the taste

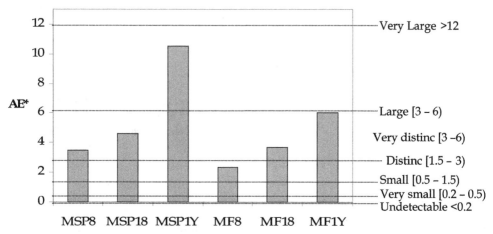

MSP8: seed+placente 8 days macerate; MSP18: seed+placente 18 days macerate; MSP1Y: seed+placente 1 year macerate; MF8: flesh 8 days macerate; MF18: flesh 18 days macerate; MF1Y: flesh 1 year macerate

Fig. 2. Colour differences ($\Delta E^*$) between the macerated double-distilled spirit samples and the unmacerated control spirit (C) and perception levels as per Schmidhofer (1994).

panel and, on the other, to the fact that some panelists preferred a "flatter" spirit with traditional distillation aromas, while others preferred a spirit with an intensely "fruity" melon aroma.

## 5.1 Optimization of maceration process

A trial run at industrial scale is the final step in new product development. Maceration time and substrate are important factors, and the latter may include pieces of fruit of different sizes, seeds, placenta, or skins depending on the type of product being manufactured. For that preliminary trial macerations using melon (Hernandez-Gomez et al., 2005b) yielded positive results in terms of extraction of colour and typical melon aromas, suggesting that macerated spirit could be used to produce an authentic liqueur reminiscent of the fruit employed in production.

Different proportions of fruit were tested based on the results of previous maceration trials, taking unmacerated distillate and adding:

-   Melon seeds + placenta in a proportion of 0.18 kg/L [MSPl1], 0.23 kg/L [MSPl2], and 0.28 kg/L [MSPl3]
-   Melon pieces + seeds + placenta in a proportion of 0.40 kg/L melon pieces + 0.10 kg/L seeds + placenta (MPSPl1) and 0.30 kg/L melon pieces + 0.20 kg/L seeds + placenta (MPSPl2)

Maceration time was 16 d, and the sugar content was measured at 0, 4, 8, 12, and 16 days of maceration.

Respect to total sugars (fructose, glucose, and saccharose) at the end of the maceration period, the total sugar content in the macerated with melon pieces and seeds+placenta was double that of the batches macerated with seeds+placenta only basically as a result of diffusion out of the melon pieces and the larger proportion of macerated substrate. Analysis of each of the sugars separately revealed that due to hydrolysis of the disaccharide by the fruit enzymes, the saccharose content gradually decreased from day 8 on, while the glucose and fructose contents rose.

According to main volatile components concentrations of esters and higher alcohols, of great organoleptic importance (Baro & Quiros-Carrasco , 1977), were higher in the batches macerated with seeds and placenta and with melon pieces + seeds and placenta. Neither n-Butanol nor Ethyl butyrate were detected in any of the batches, a highly positive finding in that these compounds are deemed to produce off flavours when they are present in distillates.

For the colour measurements, the value of L* ranges from 0 for black to 100 for white and was extremely high for the spirits considered here, between 91.9 and 95.5, the highest value being recorded for the unmacerated distillate, which also had the lowest Chromaticity (C*). C* values increased with substrate content and maceration time, while lightness (L*) decreased with maceration time and was unrelated to the amount of macerated substrate.

All the batches can be grouped close together in the greenish yellow quadrant, though in the region closer to yellow. The macerated with melon pieces were greenish in colour and those with seeds+placenta were yellow-orange in colour because of the higher proportion of carotenoids·.

Finally, the results of the preference test carried out on all the macerated batches as part of the sensory analysis indicated a slight though non-significant preference by panelists for the kirsch over the raspberry and melon spirits.

## 6. Melon liqueur

At the same time, production of an authentic melon liqueur was addressed. Three melon liqueurs were prepared (Hernández-Gómez et al., 2009). The alcohol content was adjusted to 280 ml/L (28% v/v), and the saccharose content was 100 g/L (CAE, 1997).

The resulting liqueurs were thus:

- "Melon juice liqueur": distillate + 200ml/L melon juice
- "Melon spirit liqueur": distillate + saccharose and distilled water to the corresponding alcohol content
- "Macerated liqueur": macerated spirit + saccharose

Sensory analysis of the three liqueurs was based on preference tests in the same conditions as the sensory analyses described above. Nevertheless, none of the liqueurs was statistically preferred at the 95 % significance level, though the macerated melon liqueur received the highest scores. The product was novel and unrelated to the panelists' prior experience, and as a result while some of the panelists preferred the unmacerated distillate on account of its "clean" aroma, others preferred the macerated melon liqueur on account of its fruity aroma and melon flavour.

## 7. Conclusion

Melon in the form of juice or paste without skin, constitutes an appropriate fermentation substrate, the sugars being consumed during fermentation to produce alcohol yields in accordance with expectations. The paste substrate offers a better yield in the process, but it does not seem to be an ideal substrate.

Differences were observed in the contents of the major volatiles in the different melon wines. The fermentation pH brought about perceptible differences based on both the chemical and the sensory analysis. Adjusting the pH brought about substantial decreases in the acetaldehyde and methanol contents, a facet that will have to be borne in mind in the case of methanol in view of the maximum limits set by regulations. Nevertheless, the tasters expressed no preference between paste without skin and juice according to fermentation pH.

From an industrial standpoint, the paste without skin substrate can be regarded as preferable, because it produces less waste with a lower environmental impact and it is no necessary press the paste to obtain it.

Since no preferences were observed for any of the samples of first distillate macerated in between the two distillations, this procedure would not seem to be warranted to improve distillate quality, while conversely it would increase production costs by requiring higher melon consumption during maceration, greater investment in equipment (maceration tanks, presses, and screens), longer production times, and more processing steps.

In contrast, maceration of the final double-distilled spirit did enhance the colour and aroma attributes of the final product, hence further research will be needed to focus on achieving the right sugar levels, aroma intensity, and colour.

On the basis of the results obtained in the present study, maceration is considered unsuitable in that it is too costly for industrial purposes, would yield too much production waste requiring alternative uses, and failed to attain the golden colour of the spirits with melon pieces +seeds+placenta and with melon seeds+placenta.

From a sensory standpoint the taste panelists did not evince any preference between the two spirits, and consequently we would recommend melon seeds+placenta spirit for production on account of its deeper colour, its suitability for filtration, and its typical melon aroma, as well as its lower production costs.

Lastly, comparing the melon liqueur to other commercially available fruit spirits indicated that the acceptability of this product was similar to that of the other liqueurs tasted.

Consequently, it may be concluded that using fruits for spirit production may offer a viable industrial alternative that will help keep fruit output from going to waste.

## 8. References

AENOR [Asociación Española de Normalización] (1977). *Análisis Sensorial*, vol. 1, Recopilación de Normas UNE. Ed. AENOR

Al-Khalifa, A.S. (1996). Physicochemical characteristics, fatty acid composition, and lipoxygenase activity of crude pumpkin and melon seed oils. *Journal of Agricultural and Food Chemistry*, 44, 964-966.

Bayonove, C., Baumes, R., Crouzet, J., & Günata, Z. (2000). Aromas. In: *Enología Fundamentos Científicos y Tecnológicos* (pp. 137-176) Claude Flanzy, AMV, Mundi Prensa: Madrid

Baro A.L. & Quiros-Carrasco, J.A. (1977). Les conditions de formation des aldéhydes dans les vins. *Bulletin OIV* 554, 253-264.

Bertrand, A. (1975). *Recherches sur l'analyse des vins par chromatographie en phase gaseuze.* Sc.D. Thesis, University of Bordeaux II, Bordeaux, France.

Blaise, A. & Bruns, S. (1986). Un phénomene enzymatique à L'origine du gout d'amande amére dans le vins. *CR Acad. Agric* Fr, 72 :2, 73-77

Briones, A.I., Hernández, L.F. & Úbeda, J.F. (2002). Elaboración de aguardiente de melón. *Alimentación Equipos y Tecnología*, 171, 47-52

C.A.E. (Código Alimentario Español) [Spanish Food Code]. *Manufacture of fruit juice liqueur* 330.28. 3rd Ed. Tecnos S.A. (1997).

Cantagrel, R., Lablanquie, O., Snakker, G., & Vidal, J.P. (1993). Caracteristiques analytiques d'eaux de vie nouvelles de Cognac de différentes récoltes. Relation analyse chimique-analyse sensorielle. Conaissance aromatique des cépages et qualité des vins. *Revue Française d'Oenologie* 5, 383-391.

CIE (Commission Internationale de l'Éclairage) (1986). *Technical report. Colorimetry.* 2nd Edition. CIE 15.2. Vienna.

Cortés Diéguez, S.M., Gil de la Peña, M.L. & Fernández Gómez, E. (2000). Influencia del nivel de prensado y del estado de conservación del bagazo en el contenido en metanol, acetato de etilo, 2-butanol y alcohol alílico de aguardiente de orujo. *Alimentaria* Octubre, 133-137.

Dubois, P. & Dekimpe, J. (1982). Constituans volatils odorants des vins de Bourgogne éleves en fûts de chêne neufs. *Rev Fr Oenol*, 88, 51-53.

Etiévant, P.X., Issanchou, S., Marie, S., Ducruet, V. & Flancy, C. (1989). Sensory impact of volatile phenols on red wine aroma: influence of carbonic maceration and time of storage. *Science Alimentation* 9, 19-33

European Union (1989), Council Regulation (EEC) No 1576/89 of 29 May 1989 laying down general rules on the definition, description and presentation of spirit drinks.

European Union (2002). Scientific Committee on Food. Working Group on Flavouring. SCF/CS/FLAV/FLAVOUR/11ADD1. Opinion of the Scientific Committee on Food on furfural and furfural diethylacetal.

Fundira, M., Blom, M., Pretorius, I.S., & van Rensburg. (2002). The selection of yeast Starter Culture Strains for the Production of Marula Fruit wines and Distillates. *Journal of Agricultural and Food Chemistry* 50, 1535-1542

Hernández Gómez, L.F., Úbeda, J.F. & Briones A.I. (2003). Melon fruit distillates: comparison of different distillation methods. *Food Chemistry*, 82, 539-543.

Hernández-Gómez L.F., Úbeda-Iranzo J, García-Romero E., Briones-Pérez A.I. (2005a). Comparative production of different melon distillates: chemical and sensory analyses. *Food Chemistry* 90, 115-125 .

Hernández Gómez, L.F., Úbeda, J.F., & Briones, A.I. (2005b). Role of maceration in improving melon spirit. *European Food Research and Technology*, 220, 55-62.

Hernández-Gómez L.F., Úbeda-Iranzo J, Briones-Pérez A.I. (2008). Characterisation of wines and distilled spirits from melon (*Cucumis melo* L.) *International Journal of Food Science and Technology*, 43, 644-650

Hernández-Gómez L.F., Úbeda-Iranzo J, Arévalo-Villena, M., Briones-Pérez A.I. (2009). Novel alcoholic beverages: production of spirits and liqueurs using maceration of melon fruits in melon distillates. *Jounal of Science Food and Agricultural*, 89, 1018-1022

Jouret, C., & Cantagrel, R. (2000). Aguardientes de origen vitivinícola. In *Enología Fundamentos Científicos y Tecnológicos* (pp. 659-676) Claude Flanzy, AMV, Mundi Prensa: Madrid.

Nykänen, L. & Suomalainen, H. (1983). *Aroma of beer and distilled alcoholic beverages*. D. Riedel Publishing Company, Boston

OIV [Office International de la Vigne et du Vin] (1969). *Recueil des Méthodes Internationales d´Analyse des Vins* [*Compendium of International Methods of Analysis of Wine and Musts*].

Orriols, I. (1991) Los aguardientes de orujo gallegos. Diferentes sistemas de elaboración; Componentes volátiles característicos. *Vitivinicultura 7*, 58-63.

Orriols, I. (1992). Importancia del control de las diferentes etapas de la destilación para la obtención de un aguardiente de calidad. In: *Proceedings I Jornada Técnica sobre a Aguardente*. Ourense, Galicia, Spain. pp. 85-95.

Orriols, I. (1994) Tecnología de la destilación en los aguardientes de orujo. In: *Proceedings I Congreso Internacional de la Vitivinicultura Atlántica*. Pontevedra, (pp. 291-324), Spain

Orriols, I., & Bertrand, A. (1990). Los aguardientes tradicionales gallegos. Estudio de sus componentes volátiles y de la incidencia del estado de conservación sobre ellos. *Vitivinicultura 3*, 52-58.

Ribéreau-Gayon, P., Glories, Y., Maujean, A. & Dubordieu, D. (2000). The chemistry of wine, stabilisation and treatments. In: *Handbook of Enology;* John Wiley and Sons, New York.; *Vol. 2*, 41-54

Schmidhofer, T. (1994). Métodos analíticos. In: Prandl, O., Fischer, A., Schmidhofer, T. & Shinell, H.J. *Tecnología e higiene de la carne*. Parte E. 723-803. Acribia, S.A. Zaragoza. Spain.

Silva, M.L., Malcata, F.X. & Hogg, T.A. (1995). How do processing conditions affect the microflora of grape pomace? *Food Science Technoloyi International 1*, 129-136

Silva, M.L., Macedo, A.C. & Malcata, F.X. (2000). Review: steam distilled spirits from fermented grape pomace. *Food Science and Technology 6*, 285-300.

Soufleros, E. (1978). *Les levures de la région viticole de Naoussa (Grèce). Identification et classification, etudes des produits volatides formés au cours de la fermentation*. Sc.D. Thesis, University of Bordeaux II, Bordeaux, France.

Soufleros, E. (1987). Étude sur le Tsipouro, eaux-de-vie de marc traditionelle de Grèce, précurseur del'Ouzo. *Conn Vigne Vin 21*, 93-111

Steger, C., & Lambrechts, M.G. (2000). The selection of yeast strains for the production of premium quality South African brandy base products. *Journal Industrial. Microbiology Biotechnoly 24*, 431-440

Tourliere, S. (1977). Commentaires sur la presence d'un certain nombre du composés accompagnat l'alcool dans les distillants spiritueux rectifiés et les eaux-de-vie. *Industries Alimentaires et Agricoles 94*: 565-574.

Versini, G., Monetti A., dalla Serra A. & Inama S. (1990). Analytical and statistical characterization of grappa of different Italian regions. In: Bertrand A. (ed.), *Proceedings I Symposium International sur les Eaux-de-vie Traditionnelles d'Origine Viticole*. Paris, France: Lavoisier. pp 137-150.

Xandri Tagüeña, J.M. (1958). *Elaboración de aguardientes simples, compuestos y licores*. Primera Edición. Salvat Editores, S.A. Barcelona.

# Part 3

# New Applications and Improvements

# 9

# Membrane Distillation: Principle, Advances, Limitations and Future Prospects in Food Industry

Pelin Onsekizoglu
*Trakya University Department of Food Engineering, Edirne*
*Turkey*

## 1. Introduction

Membrane separation processes have become one of the emerging technologies in the last few decades especially in the separation technology field. They offer a number of advantages over conventional separation methods in a wide variety of applications such as distillation and evaporation. Membrane processes can be easily scaled up due to their compact and modular design; they are able to transfer specific components selectively; they are energy efficient systems operating under moderate temperature conditions ensuring gentle product treatment.

Microfiltration (MF), ultrafiltration (UF), nanofiltration (NF), reverse osmosis (RO), pervaporation and electrodialysis are conventional membrane processes that have already gained wide acceptance in food processing (Bazinet et al., 2009; Couto et al., 2011; Gomes et al., 2011; Mello et al., 2010; Quoc et al., 2011; Santana et al., 2011). Membrane distillation (MD) is an emerging thermally driven membrane process in which a hydrophobic microporous membrane separates a heated feed solution and a cooled receiving phase. The temperature difference across the membrane results a water vapour pressure gradient, causing water vapour transfer through the pores from high vapour pressure side to the low one. Some of the key advantages of membrane distillation processes over conventional separation technologies are: relatively lower energy costs as compared to distillation, reverse osmosis, and pervaporation; a considerable rejection of dissolved, non-volatile species; much lower membrane fouling as compared with microfiltration, ultrafiltration, and reverse osmosis; reduced vapour space as compared to conventional distillation; lower operating pressure than pressure-driven membrane processes and lower operating temperature as compared with conventional evaporation (Bazinet et al., 2009; Couto et al., 2011; Gomes et al., 2011; Lawson & Lloyd, 1996b; Mello et al., 2010; Quoc et al., 2011; Santana et al., 2011).

Dewatering aqueous solutions is one of the key unit operations encountered in food processing, particularly in the processing of beverages, fruit juice, milk, whey, vegetable extracts, etc. The initial soluble solid contents are increased by concentration process, reducing the volume with consequent reduction of transport, storage and packaging costs. In addition, the concentrates are more resistant to microbial and chemical deterioration as a result of water activity reduction.

Today, multistage vacuum evaporation is the predominant method used for liquid concentration in food industry. The main drawbacks of this system are high energy consumption and heat induced deterioration of sensory (color changes, off-flavor formation) and nutritional characteristics (Ibarz et al., 2011; Kadakal et al., 2002; Simsek et al., 2007; Toribio & Lozano, 1986; Varming et al., 2004). Recently, technological advances related to the development of new membrane processes including membrane distillation have been proved to overcome this limitation (Bagger-Jorgensen et al., 2011; Cassano & Drioli, 2007; Hongvaleerat et al., 2008; Kozak et al., 2009; Onsekizoglu et al., 2010b; Valdes et al., 2009).

This chapter will cover the process features, theoretical aspects and the relevant mathematics related to water transport mechanism in membrane distillation. The most basic concepts of osmotic distillation, a membrane distillation variant operating at lower temperature will be also discussed. The suggestions for membrane selection taking into account the membrane material and module configuration together with contact angle and membrane wettability will be presented in detail. The process parameters affecting the transmembrane flux and the most promising applications for enhancement of flux will be highlighted. Applications in food industry and long term performance of membrane distillation systems will be evaluated. The possibility of integrating membrane distillation with other existing processes and suggestions for future work will be presented.

## 2. Process fundamentals

MD is a thermally driven process, in which water vapour transport occurs through a non-wetted porous hydrophobic membrane. The term MD comes from the similarity between conventional distillation process and its membrane variant as both technologies are based on the vapour-liquid equilibrium for separation and both of them require the latent heat of evaporation for the phase change from liquid to vapour which is achieved by heating the feed solution. The driving force for MD process is given by the vapour pressure gradient which is generated by a temperature difference across the membrane. As the driving force is not a pure thermal driving force, membrane distillation can be held at a much lower temperature than conventional thermal distillation. The hydrophobic nature of the membrane prevents penetration of the pores by aqueous solutions due to surface tensions, unless a transmembrane pressure higher than the membrane liquid entry pressure (LEP) is applied. Therefore, liquid/vapour interfaces are formed at the entrances of each pore. The water transport through the membrane can be summarized in three steps: (1) formation of a vapour gap at the hot feed solution–membrane interface; (2) transport of the vapour phase through the microporous system; (3) condensation of the vapour at the cold side membrane–permeate solution interface (Jiao et al., 2004; Peinemann et al., 2010).

Various MD configurations can be used to drive flux (El-Bourawi et al., 2006; Khayet, 2011; Lawson & Lloyd, 1997; Susanto, 2011; Zhigang et al., 2005). The difference among these configurations is the way in which the vapour is condensed in the permeate side. Figure 1 illustrates the four commonly used configurations of MD described as follows:

1.  In direct contact membrane distillation (DCMD), water having lower temperature than liquid in feed side is used as condensing fluid in permeate side. In this configuration, the liquid in both sides of the membrane is in direct contact with the hydrophobic microporous membrane. DCMD is the most commonly used configuration due to its

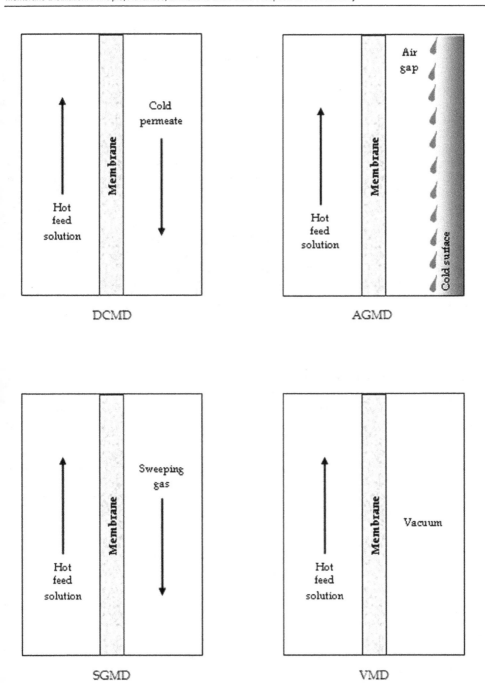

Fig. 1. Schematic representation of MD configurations

convenience to set up in laboratory. However, direct contact of the membrane with the cooling side and poor conductivity of the polymeric material results heat losses throughout the membrane. Therefore, in DCMD the thermal efficiency which is defined as the fraction of heat energy used only for evaporation, is relatively smaller than the other three configurations.

2.  In air gap membrane distillation (AGMD), water vapour is condensed on a cold surface that has been separated from the membrane via an air gap. The heat losses are reduced in this configuration by addition of a stagnant air gap between membrane and condensation surface.

3.  In sweeping gas membrane distillation (SGMD), a cold inert gas is used in permeate side for sweeping and carrying the vapour molecules to outside the membrane module where the condensation takes place. Despite the advantages of a relatively low conductive heat loss with a reduced mass transfer resistance, due to the operational costs of the external condensation system, SGMD is the least used configuration.

4.  In vacuum membrane distillation (VMD), the driving force is maintained by applying vacuum at the permeate side. The applied vacuum pressure is lower than the equilibrium vapour pressure. Therefore, condensation takes place outside of the membrane module.

Each of the MD configurations has its own advantages and disadvantages for a given application.

## 3. Osmotic distillation

Osmotic distillation (OD) is a non-thermal membrane distillation variant, in which a microporous hydrophobic membrane separates two aqueous solutions at different solute concentrations. The OD process can be operated at atmospheric pressure and ambient temperature. The driving force is the vapour pressure gradient across the membrane which is obtained by using a hypertonic salt solution on permeate side. The hydrophobic nature of the membrane prevents penetration of the pores by aqueous solutions, creating vapour/liquid interfaces at the entrance of the pores. Under these conditions, a net water flux from the high vapour pressure side to the low one occurs resulting in the concentration of feed and dilution of hypertonic salt solution. The water transport through the membrane can be summarized in three steps: (1) evaporation of water at the dilute vapour–liquid interface; (2) diffusional or convective vapour transport through the membrane pore; (3) condensation of water vapour at the membrane/brine interface (Jiao et al., 2004; Peinemann et al., 2010). In the literature the OD technique is also termed as isothermal membrane distillation, osmotic membrane distillation, osmotic evaporation and gas membrane extraction (Gryta, 2005b).

The basic requirements of osmotic agent are to be non-volatile, to have high osmotic activity in order to maintain a lower vapour pressure and to maximize the driving force and to be thermally stable to allow reconcentration of diluted stripping solution by evaporation. Other factors that should be taken into consideration are solubility, toxicity, corrosivity and cost. Although NaCl or $CaCl_2$ have chosen as osmotic agent in most of the reported studies, both of these salts have the disadvantage of being corrosive to ferrous alloys (Celere & Gostoli, 2004; Shin & Johnson, 2007). $MgCl_2$, $MgSO_4$, $K_2HPO_4$, and $KH_2PO_4$ are some other commonly used osmotic agents in OD. Potassium salts of ortho- and pyrophosphoric acid offer several advantages, including low-equivalent weight, high water solubility, steep positive temperature coefficients of solubility and safety in foods and pharmaceuticals (Jiao et al., 2004; Nagaraj et al., 2006a; Shin & Johnson, 2007)

## 4. Membrane characteristics

The selection of the membrane is the most crucial factor in MD separation performance. As stated earlier, the membrane used for MD process must be hydrophobic and porous. There are various types of membranes meeting these expectations; however the efficiency of a given MD application depends largely on additional factors such as resistance to mass transfer, thermal stability, thermal conductivity, wetting phenomena and module characterization. Membrane and module related characteristics affecting selection of the appropriate membrane are summarized in this section.

### 4.1 Membrane materials

A large variety of membranes including both polymeric and inorganic membranes of hydrophobic nature can be used in MD process; however polymeric membranes have attracted much more attention due to their possibility to modulate the intrinsic properties. Polytetrafluoroethylene (PTFE), polypropylene (PP) and polyvinylidenefluoride (PVDF) are the most commonly used polymeric membranes due to their low surface tension values (Table 1). Hydrophobic porous membranes can be prepared by different techniques like sintering, stretching, phase inversion or thermally induced phase separation depending on the properties of the materials to be used. The useful materials should be selected according to criteria that include compatibility with the liquids involved, cost, ease of fabrication and assembly, useful operating temperatures, and thermal conductivity (Li et al., 2008; Liu et al., 2011). Among them, PTFE membranes are the most hydrophobic ones showing outstanding thermal stability and chemical resistance properties (they are low soluble in practically all common solvents). The main disadvantage of PTFE membranes is the difficulty of processing. PTFE membranes are generally prepared by sintering or stretching. PP exhibits

| Polymer | Surface tension (Dynes/cm) |
|---|---|
| Polytetrafluoroethylene (PTFE) | 19 |
| Polyvinylidenefluoride (PVDF) | 25 |
| Polypropylene (PP) | 29 |
| Polyethylene (PE) | 31 |
| Polypropylene (PP) | 34 |
| Polyvinyl alcohol (PVA) | 37 |
| Polysulfone (PS) | 41 |
| Polycarbonate (PC) | 45 |
| Polyurethane (PU) | 45 |

Table 1. Critical surface tension values of some polymers (Adapted from Oliver, 2004; Pabby et al., 2009)

excellent solvent resistant properties and high crystallinity. PP membranes are generally manufactured by stretching and thermal phase inversion. PVDF membranes exhibit good thermal and chemical resistance; however this polymer easily dissolves at room temperature in a variety of solvents including dimethylformamide (DMF) and triethylphosphate (TEP). PVDF membranes are generally prepared by phase inversion (Curcio and Drioli, 2005).

There are some additional criteria that should be taken into consideration for selection of the appropriate membrane for a given MD application such as pore size, tortuosity, porosity, membrane thickness and thermal conductivity. The relationship between the transmembrane flux and the different membrane characteristic related parameters is given by (Lawson & Lloyd, 1997)

$$N \; \alpha \; \frac{<r^\alpha> \varepsilon}{\tau \delta} \qquad (1)$$

where $N$ is the molar flux, $<r^\alpha>$ is the mean pore size of the membrane pores where $a$ equals 1 for Knudsen diffusion and equals 2 for viscous flux, $\varepsilon$ is the membrane porosity, $\tau$ is the membrane tortousity and $\delta$ is the membrane thickness.

*Membrane pore size:* Membranes with pore sizes ranging from 10 nm to 1 μm can be used in MD (Pabby et al., 2009). The permeate flux increases with the increase in pore size as determined by Knudsen model. However, in order to avoid wettability, small pore size should be choosen (El-Bourawi et al., 2006; Khayet, 2011). Thus, an optimum value for pore size has to be determined for each MD application depending on the type of the feed solution.

*Membrane porosity:* Membrane porosity is determined as the ratio between the volume of the pores and the total volume of the membrane. Evaporation surface area increases with the increase in porosity level of the membrane, resulting in higher permeate fluxes (Huo et al., 2011; Susanto, 2011). Membrane porosity also affects the amount of heat loss by conduction (Lawson & Lloyd, 1996b):

$$Q_m = h_m \Delta T_m \qquad (2)$$

$$h_m = \varepsilon h_{mg} + (1-\varepsilon) h_{ms} \qquad (3)$$

where $\varepsilon$ is the membrane porosity, $h_{mg}$ is the conductive heat transfer coefficient of the gases entrapped in the membrane pores; $h_{ms}$ is the conductive heat transfer coefficient of the hydrophobic membrane material.

Conductive heat loss can be reduced by increasing porosity of the membrane, since $h_{mg}$ is generally an order of magnitude smaller than $h_{ms}$. In general, the porosity of the membranes used in MD operations lines in the range of 65%-85%.

*Pore tortuosity:* Tortuosity is the average length of the pores compared to membrane thickness. The membrane pores do not go straight across the membrane and the diffusing molecules must move along tortuous paths, leading a decrease in MD flux. Therefore, permeate flux increases with the decrease in tortuosity. It must be pointed out that this value is frequently used as a correction factor for prediction of transmembrane flux due to the difficulties in measuring its real value for the membranes used in MD. In general a value of 2 is frequently assumed for tortuosity factor. (El-Bourawi et al., 2006; Khayet et al., 2004a; Phattaranawik et al., 2003a)

*Membrane thickness:* Permeate flux is inversely proportional to the membrane thickness in MD. Therefore, membrane must be as thin as possible to achieve high permeate flux. Thickness also plays an important role in the amount of conductive heat loss though the membrane. In order to reduce heat resistances, it should be as thick as possible leading to a conflict with the requirement of higher permeate flux. Hence membrane thickness should be optimized in order to obtain optimum permeate flux and heat efficiency. The optimum thickness for MD has been estimated within the range of 30–60 μm (Lagana et al., 2000).

*Pore size distribution:* Pore size distribution affects uniformity of vapour permeation mechanism. In general, uniform pore size is preferable rather than distributed pore size (Susanto, 2011).

*Thermal conductivity:* Thermal conductivity of the membrane should be small in order to reduce the heat loss through the membrane from feed to the permeate side. Conductive heat loss is inversely proportional to the membrane thickness. However selection of a thicker membrane decreases both the flux and permeability. One promising approach may be selection of a membrane with higher porosity since thermal conductivity of polymer membrane is significantly higher than thermal conductivity of water vapour in the membrane pores (Khayet et al., 2006). The thermal conductivities of polymers used in MD generally varies in the range of 0.15–0.45 W m$^{-1}$K$^{-1}$ depending upon temperature and the degree of crystallinity (Alklaibi & Lior, 2005).

Table 2 summarizes the commercial membranes commonly used by various researchers up to date together with their principal characteristics. In fact, there is a lack of commercially available MD units and most of the MD researches use modules actually designed for other membrane operations (i.e. microfiltration) rather than MD. Design of novel membranes fabricated especially for MD purposes have been recommended by MD investigators since commercially available membranes does not meet all the requirements listed above. Novel hydrophobic membranes for MD applications can be manufactured either by hydrophobic polymers or by surface modification of hydrophilic membranes. Various surface modification applications including coating, grafting and plasma polymerization (Brodard et al., 2003; Bryjak et al., 2000; Chanachai et al., 2010; Huo et al., 2010; Kong et al., 1992; Krajewski et al., 2006; Lai et al., 2011; Li & Sirkar, 2004; Vargas-Garcia et al., 2011; Wu et al., 1992; Yang et al., 2011b) have been attempted until now. However, there is very limited number of studies on the design of MD membranes (Khayet, 2011; Khayet et al., 2010; Phattaranawik et al., 2009; Wang et al., 2009; Yang et al., 2011a). Therefore, new generation of membranes promising required features should be developed for MD applications.

## 4.2 Membrane modules

Choice and arrangement of the membrane module in a MD application is based on economic considerations with the correct engineering parameters being employed. Plate and frame, spiral wound, tubular, capillary and hollow fiber membrane modules are commonly used by MD researchers.

| Membrane module | Manufacturer | Trade name | Polymer | Membrane thickness (μm) | Nominal pore size (μm) | Porosity (%) | References |
|---|---|---|---|---|---|---|---|
| **Flat sheet** | | | | | | | |
| | 3M Corporation | 3MA | PP | 91 | 0.29 | 66 | (Kim & Lloyd, 1991; Lawson et al., 1995; Lawson & Lloyd, 1996a) |
| | | 3MB | PP | 81 | 0.40 | 76 | |
| | | 3MC | PP | 76 | 0.51 | 79 | |
| | | 3MD | PP | 86 | 0.58 | 80 | |
| | | 3ME | PP | 79 | 0.73 | 85 | |
| | Gelman | TF1000 | PTFE/PP | 60 | 0.1 | 80 | (Khayet et al., 2004b; Martinez-Diez et al., 1998; Martinez et al., 2002; Rincon et al., 1999; Rodrigues et al., 2004) |
| | | TF450 | PTFE/PP | 60 | 0.45 | 80 | |
| | | TF200 | PTFE/PP | 60 | 0.20 | 80 | |
| | | TF 200 | PTFE/PP | 178 | 0.20 | 80 | |
| | | TF 200 | PTFE/PP | 165 | 0.20 | 60 | |
| | Milipore | Durapore | PVDF | 110 | 0.45 | 75 | (Banat & Simandl, 1999; Ding et al., 2003; Khayet et al., 2004b; Phattaranawik et al., 2003b; Phattaranawik et al., 2001) |
| | | Durapore | PVDF | 100 | 0.20 | 70 | |
| | | GVHP | PVDF | 125 | 0.20 | 80 | |
| | | GVHP | PVDF | 125 | 0.22 | 75 | |
| | | HVHP | PVDF | 116 | 0.45 | 66 | |
| | Sartorious | | PTFE | 70 | 0.20 | 70 | (Phattaranawik et al., 2003b; Warczok et al., 2007) |
| | Gore | | PTFE | 64 | 0.20 | 90 | (Garcia-Payo et al., 2000; Izquierdo-Gil et al., 1999; Phattaranawik et al., 2003b) |
| | | | PTFE | 77 | 0.45 | 89 | |
| | Osmonics | | PP | 150 | 0.22 | 70 | (Cath et al., 2004) |
| | | | PTFE | 175 | 0.22 | 70 | |
| | | | PTFE | 175 | 0.45 | 70 | |
| | | | PTFE | 175 | 1.0 | 70 | |
| | Hoechst Celanese | Celgard 2400 | PP | 28 | 0.05 | 45 | (Barbe et al., 2000; Mengual et al., 1993) |
| | | Celgard 2500 | PP | 25 | 0.02 | 38 | |

| Membrane module | Manufacturer | Trade name | Polymer | Membrane thickness (µm) | Nominal pore size (µm) | Porosity (%) | References |
|---|---|---|---|---|---|---|---|
| **Flat sheet** | | | | | | | |
| | Enka | Accurel 1E-PP | PP | | 0.25 | 25 | (Mengual et al., 1993; Narayan et al., 2002) |
| | | Accurel 2E-PP | PP | | 0.48 | 90 | |
| **Capillary** | | | | | | | |
| | Membrana | Accurel S6/2 | PP | 450 | 0.20 | 73 | (Celere & Gostoli, 2004; Gryta, 2007) |
| | | Accurel Q3/2 | PP | 400 | 0.20 | 70 | |
| | Self-designed | PP | 800 | 0.40 | 73 | | (Gryta et al., 2000b) |
| | Memcor | PV 375 | PVDF | 125 | 0.20 | 75 | (Bui et al., 2004) |
| | | PV 660 | PVDF | 170 | 0.20 | 64 | |
| **Hollow fiber** | | | | | | | |
| | Hoechst-Celanese | Liqui-Cel® Extra-Flow 2.5×8 in | PP | 180 | | 40 | (Bailey et al., 2000) |
| | | | PP | 53 | 0.074 | 50 | |
| | | | PP | 50 | 0.044 | 65 | |
| | | | PP | 47 | 0.056 | 42 | |

Table 2. List of commercial membranes commonly used by various MD researchers

In plate and frame modules, the membranes which are usually prepared as discs or flat sheets are placed between two plates. The feed solution flows through flat, rectangular channels. Packing densities for flat sheet membranes may be in the range of 100–400 m³/m² (Pabby et al., 2009). Polymeric flat sheet membranes are easy to prepare, handle, and mount. The same module can be used to test many different types of MD membranes. The membrane can be supported to enhance mechanical strength. Babu et al. (2008) used a plate and frame membrane module having a membrane area of 0.01 m² for the concentration of pineapple and sweet lime juice. The module consists of a polyester mesh (0.25 mm) and a hydrophobic microporous polypropylene membrane (pore size 0.20 µm and thickness 175 µm) supported in between a viton gasket (3.0 mm) and two stainless steel frames. In spiral wound membranes, the membrane, feed and permeate channel spacers and the porous membrane support form an envelope which is rolled around a perforated central collection tube and inserted into an outer tubular pressure shell. The feed solution passes in axial direction through the feed channel across the membrane surface. The filtrate moves along the permeate channel and is collected in a perforated central collection tube. Spiral-wound modules have a packing density of 300–1000 m²/m³ depending on the channel height, which is greater than that of the plate and frame module (Pabby et al., 2009). However, the spiral-wound module is quite sensitive to fouling. Tubular, capillary or hollow fiber membrane modules are shell and tube type modules housing pressure-tight tubes. The support is not needed in this type of modules. The membranes are usually a permanent integral part of the module and are not easily replaced. Tubular membrane modules provide much higher membrane surface area to module volume ratio than plate and frame modules (Khayet, 2011). The diameter of membranes in tubular module varies within the range of 10-25 mm. The packing density is around 300 m²/m³ (Pabby et al., 2009). These modules offer higher cross-flow velocities and large pressure drop and generally used for MD of high viscous liquids. The diameters of membranes in capillary modules typically vary between 0.2-3 mm with packing densities of about 600-1200 m²/m³ (Li et al., 2008) . The production costs are very low and membrane fouling can effectively be controlled by the proper feed flow and back-flushing of permeate in certain time intervals. The main disadvantage of the capillary membrane module is the requirement of low operating pressure (up to 4 bars). The inner diameters of hollow fiber membranes is around 50-500 μm with very high packing densities of about 3000 m²/m³. Hollow fiber module has the highest packing density of all module types. Its production is very cost effective and hollow fiber membrane modules can be operated at pressures in excess of 100 bars (El-Bourawi et al., 2006). The main disadvantage of the hollow fiber membrane module is the difficult control of membrane fouling. Therefore, a proper pretreatment should be applied for separation of macromolecules. For example, in the case of fruit juice concentration by MD using a hollow fiber module, clarification is a crucial pretreatment step to enhance MD flux (Cassano & Drioli, 2007; Onsekizoglu et al., 2010b).

## 4.3 Contact angle

The contact angle is a common measurement of the hydrophobic or hydrophilic behaviour of a material. It provides information about relative wettability of membranes. The contact angle is determined as the angle between the surface of the wetted solid and a line tangent to the curved surface of the drop at the point of three-phase contact (Figure 2). The value of contact angle is greater than 90° when there is low affinity between liquid and solid; in case of water, the material is considered hydrophobic and is less than 90° in the case of high affinity. Wetting occurs at 0°, when the liquid spreads onto the surface (Curcio et al., 2010; Curcio & Drioli, 2005; Pabby et al., 2009). The wettability of a solid surface by a liquid decreases as the contact angle increases. Table 3 lists the contact angle values for few

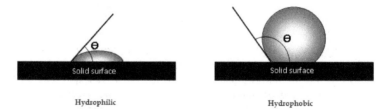

Fig. 2. Schematic representation of contact angle

| Material | Contact Angle, ° |
|---|---|
| Ordinary glass | 20 |
| Polycarbonate | 70 |
| Polyamide | 69 |
| Polyethersulphone | 54 |
| Polyethylene | 96 |
| Polypropylene | 100 |
| PTFE | 123 |
| PVDF | 111 |
| Teflon | 112 |

Table 3. Contact angle values of water on some materials at ambient temperature (Khayet & Matsuura, 2011; Sigurdsson & Shishoo, 1997)

different materials in water at ambient temperature. For example, the parameter measured on PTFE or PVDF membrane surface was 108° or 107°, respectively (Curcio et al., 2010; Hwang et al., 2011; Tomaszewska, 2000).

### 4.4 Liquid entry pressure and wetting phenomena

The hydrophobic nature of membranes used in membrane distillation prevents penetration of the aqueous solutions into the pores unless a critical penetration pressure is exceeded, as stated earlier. Liquid entry pressure (LEP) is the minimum transmembrane hydrostatic pressure that must be applied before liquid solutions penetrate into the membrane pores. LEP can be calculated using the Laplace-Young equation (Burgoyne & Vahdati, 2000; Lawson & Lloyd, 1997).

$$\Delta P = P_F - P_D - \frac{2\beta\gamma_L Cos\theta}{r_m} \tag{4}$$

where $P_F$ and $P_D$ are the hydraulic pressure of the feed and distillate side, $\beta$ is the geometric pore coefficient (equals 1 for cylindrical pores), $\gamma_L$ is the surface tension of the liquid, $\theta$ is the contact angle and $r_m$ is the maximum pore size.

LEP depends on membrane characteristics and prevents wetting of the membrane pores during MD experiments. LEP increases with a decrease in maximum pore size at the surface and an increase at the hydrophobicity (i.e., large water contact angle) of the membrane material. The presence of strong surfactants or organic solvents can greatly reduce the liquid surface tension therefore causing membrane wetting. Therefore, care must be taken to prevent contamination of process solutions with detergents or other surfacting agents.

## 5. Transport mechanisms and polarization phenomena

### 5.1 Theory of heat transfer

Heat transfer in the MD includes three main steps:

i.    Heat transfer through the feed side boundary layer

ii.    Heat transfer through the membrane
iii.   Heat transfer through the permeate side boundary layer

*Heat transfer through the feed side boundary layer* Heat transfer from the feed solution to the membrane surface across the boundary layer in the feed side of the membrane module imposes a resistance to mass transfer since a large quantity of heat must be supplied to the surface of the membrane to vaporize the liquid. The temperature at the membrane surface is lower than the corresponding value at the bulk phase. This affects negatively the driving force for mass transfer. This phenomenon is called temperature polarization (El-Bourawi et al., 2006; Pabby et al., 2009; Qtaishat et al., 2008). Temperature polarization becomes more significant at higher feed temperatures (Burgoyne & Vahdati, 2000; Lagana et al., 2000; Phattaranawik et al., 2003b).

The temperature polarization coefficient (TPC) is determined as the ratio of the transmembrane temperature to the bulk temperature difference:

$$TPC = \frac{T_{fm} - T_{pm}}{T_{fb} - T_{pb}} \tag{5}$$

where $T_{fm}$, $T_{pm}$, $T_{fb}$ and $T_{pb}$ are membrane surface temperatures and fluid bulk temperatures at the feed and permeate sides, respectively. A schematic diagram of the temperature polarization in MD is shown in Figure 3.

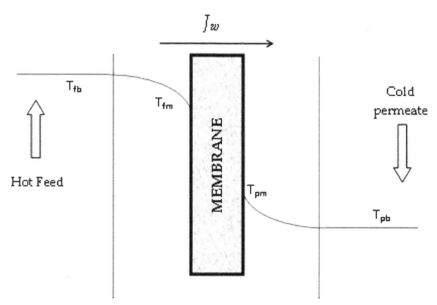

Fig. 3. Schematic diagram of temperature polarization in MD. $T_{fm}$, $T_{pm}$, $T_{fb}$ and $T_{pb}$ are membrane surface temperatures and fluid bulk temperatures at the feed and permeate sides, respectively.

Heat transfer through the feed side boundary layer can be calculated using:

$$Q_f = h_f \left( T_{fb} - T_{fm} \right) \tag{6}$$

where $h_f$ is the heat transfer coefficient of the feed side boundary layer.

*Heat transfer through the membrane:* Heat transfer through the membrane appears as a combination of latent heat of vaporization ($Q_V$) and conductive heat transfer across both the membrane matrix and the gas filled membrane pores ($Q_C$). The corresponding values can be estimated by following equations:

$$Q_V = J\Delta H_V \tag{7}$$

$$Q_C = \left( \frac{k_m}{\delta} \right) \left( T_{fm} - T_{pm} \right) \tag{8}$$

Therefore, the heat flux can be estimated by the following expression (El-Bourawi et al., 2006; Khayet & Matsuura, 2011; Lawson & Lloyd, 1997; Phattaranawik & Jiraratananon, 2001; Schofield et al., 1990a)

$$Q_m = Q_V + Q_C \tag{9}$$

$$Q_m = \frac{k_m}{\delta} \left( T_{fm} - T_{pm} \right) + J\Delta H_V \tag{10}$$

where $k_m$ is the thermal conductivity of the membrane, $\delta$ is the membrane thickness, $J$ is the permeate water vapour flux and $\Delta H_V$ is the latent heat of vaporization.

Various models have been proposed for estimation of $k_m$ in Equation [10]. Two of the most preferred ones are given below;

$$k_m = \varepsilon k_g + \left( 1 - \varepsilon \right) k_s \tag{11}$$

$$k_m = \left[ \frac{\varepsilon}{k_g} + \frac{\left( 1 - \varepsilon \right)}{k_s} \right]^{-1} \tag{12}$$

*Heat transfer through the permeate side boundary layer:* Heat transfer from the membrane surface to the bulk permeate side across the boundary layer is also related with the temperature polarization phenomenon. The temperature of membrane surface at the permeate side is higher than that of bulk permeate due to the temperature polarization effect.

Heat transfer through the permeate side boundary layer is given as:

$$Q_p = h_p \left( T_{pm} - T_{pb} \right) \tag{13}$$

where $h_p$ is the heat transfer coefficient of the permeate side boundary layer.

Both feed and permeate side boundary layers are function of fluid properties and operating conditions, as well as the hydrodynamic conditions. There are some convenient approaches in the literature to reduce the temperature polarization effects like mixing thoroughly, working at high flow rates or using turbulence promoters (Cath et al., 2004; Chernyshov et al., 2005; El-Bourawi et al., 2006; Lawson & Lloyd, 1996a; Martinez & Rodriguez-Maroto, 2006).

## 5.2 Theory of mass transfer

As mentioned above, the mass transfer in MD is driven by the vapour pressure gradient imposed between two sides of the membrane. Mass transfer in membrane distillation consists of three consecutive steps:

i.    Evaporation of water at the liquid/gas interface on the membrane surface of the feed side
ii.   Water vapour transfer through the membrane pores
iii.  Condensation of water vapour at the gas/liquid interface on the membrane surface of the permeate side

The mass flux ($J$) can be expressed as (Close & Sorensen, 2010; Zhang et al., 2010):

$$J = K\Delta P \tag{14}$$

where $K$ is the overall mass transfer coefficient which is the reciprocal of an overall mass transfer resistance. This overall resistance is the sum of three individual resistances:

$$K - \left[ \frac{1}{K_f} + \frac{1}{K_m} + \frac{1}{K_p} \right]^{-1} \tag{15}$$

where $K_f$, $K_m$ and $K_p$ are the mass transfer coefficients of feed layer, membrane and permeate layer, respectively.

*Mass transfer trough feed side boundary layer:* In membrane distillation, only water vapour transport is allowed due to the hydrophobic character of the membrane. Therefore the concentration of solute(s) in feed solution becomes higher at the liquid/gas interface than that at the bulk feed as mass transfer proceeds. This phenomenon is called concentration polarization and results in reduction of the transmembrane flux by depressing the driving force for water transport. Concentration polarization coefficient (*CPC*) is determined as the ratio of the solute concentration at the membrane surface ($C_{fm}$) to that at the bulk feed solution ($C_{fb}$):

$$CPC = \frac{C_{fm}}{C_{fb}} \tag{16}$$

The concentration gradient between the liquid/gas interface and the bulk feed results a diffusive transfer of solutes from the surface of the membrane to the bulk solution. At steady state, the rate of convective solute transfer to the membrane surface is balanced by diffusion of solute to the bulk feed.

The molar flux is expressed as follows (El-Bourawi et al., 2006; Khayet & Matsuura, 2011):

$$J = k_s \ln\left( \frac{C_{fm}}{C_{fb}} \right) \tag{17}$$

where $k_s$ is the diffusive mass transfer coefficient through the boundary layer. Several empirical correlation of dimensionless numbers, namely, Sherwood ($Sh$), Reynolds ($Re$), Schmidt ($Sc$), Nusselt ($Nu$) and Prandtl ($Pr$) numbers can be used to estimate the value of $k_s$ depending on the hydrodynamics of the system:

$$Sh = \frac{kL}{D} \qquad Re = \frac{Lu\rho}{\mu} \qquad Sc = \frac{\mu}{\rho D} \qquad Nu = \frac{hL}{k} \qquad Pr = \frac{\mu C_P}{k} \tag{18}$$

where $L$: characteristic length, $D$: diffusion coefficient, $\rho$: density, $\mu$: viscosity, $u$: feed velocity, $k$: thermal conductivity, $C_P$: specific heat, $h$: boundary layer heat transfer coefficient (Babu et al., 2008).

In other membrane separation process such as microfiltration, ultrafiltration and reverse osmosis, concentration polarization is usually considered a major cause for flux decline (Agashichev, 2006; Morao et al., 2008; Song, 2010; Wang & Tarabara, 2007; Zaamouche et al., 2009). On the other hand, it is agreed upon that concentration polarization is insignificant compared to temperature polarization in DCMD (Khayet & Matsuura, 2011; Lagana et al., 2000; Martinez & Rodriguez-Maroto, 2007).

It is worth pointing out that in osmotic distillation process, concentration polarization exists at each side of the membrane. During osmotic distillation, as mass transfer proceeds, solute concentration increases at the membrane surface due to evaporation of water vapour at the feed side. On the other hand, the solute concentration decreases due to the condensation of water vapour on the permeate side, giving rise to the difference in brine concentrations (Figure 4). The existence of concentration polarization layers at each side of the membrane

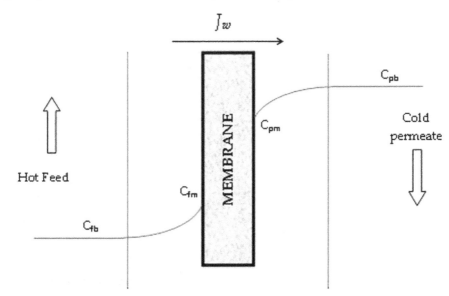

Fig. 4. Schematic diagram of concentration polarization in MD. $C_{fm}$, $C_{pm}$, $C_{fb}$ and $C_{pb}$ are membrane surface and bulk solute concentrations at the feed and permeate sides, respectively.

results in the reduction of driving force for water vapour transport leading a decrease in transmembrane flux (Babu et al., 2006; Babu et al., 2008; Nagaraj et al., 2006b).

*Mass transfer through the membrane pores:* The main mass transfer mechanisms through the membrane in MD are Knudsen diffusion and molecular diffusion (Figure 5). Knudsen diffusion model is responsible for mass transfer through the membrane pore if the mean free path of the water molecules is much greater than the pore size of the membrane and hence, the molecules tend to collide more frequently with the pore wall (Li et al., 2008; Nagaraj et al., 2006b; Pabby et al., 2009; Srisurichan et al., 2006).

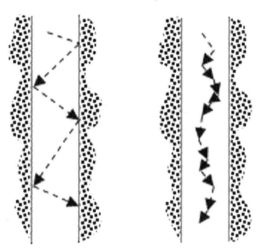

Fig. 5. Mass transfer mechanism involved in water vapour transport through membrane pores of MD module.

In this case, the membrane diffusion coefficient is calculated using equation:

$$K_m = 1.064 \frac{r\varepsilon}{\tau\delta}\left(\frac{M}{RT}\right)^{0.5} \tag{19}$$

where $\varepsilon$ is the fractional void volume, $\delta$ is the membrane thickness, $\tau$ is the tortuosity, $M$ is the molecular weight of water, $R$ is the gas constant and $T$ is the absolute temperature.

On the other hand, when the pore size is relatively large, the molecule–molecule collisions are more frequent and molecular diffusion is responsible for mass transfer through the membrane pores (Khayet & Matsuura, 2011).

$$K_m = \frac{1}{Y_{\ln}}\frac{D\varepsilon}{\tau\delta}\frac{M}{RT} \tag{20}$$

where $Y_{\ln}$ is the log mean of mole fraction of air and $D$ is the diffusion coefficient.

Both models were successfully applied for predicting the mass transfer through the membrane in DCMD systems (Babu et al., 2006; Bandini & Sarti, 1999; Chen et al., 2009; Lawson & Lloyd, 1996b; Nagaraj et al., 2006b; Srisurichan et al., 2006).

## 6. Process parameters

*Feed concentration*

Permeate flux decreases with an increase in feed concentration. This phenomenon can be attributed to the reduction of the driving force due to decrease of the vapour pressure of the feed solution and exponential increase of viscosity of the feed with increasing concentration. The contribution of concentration polarization effects is also known, nevertheless, this is very small in comparison with temperature polarization effects (Lagana et al., 2000; Pabby et al., 2009). As it is well known, MD can handle feed solutions at high concentrations without suffering the large drop in permeability observed in other pressure-driven membrane processes and can be preferentially employed whenever elevated permeate recovery factors or high retentate concentrations are requested (i.e. concentration of fruit juices) (Curcio & Drioli, 2005; Li & Sirkar, 2005; Schofield et al., 1990b).

*Feed temperature*

Various investigations have been carried out on the effect of the feed temperature on permeate flux in MD. In general, it is agreed upon that there is an exponential increase of the MD flux with the increase of the feed temperature. As the driving force for membrane distillation is the difference in vapour pressure across the membrane, the increase in temperature increases the vapour pressure of the feed solution, thus results an increase in the transmembrane vapour pressure difference.

It is worth quoting that working under high feed temperatures was offered by various MD researches, since the internal evaporation efficiency (the ratio of the heat that contributes to evaporation) and the total heat exchanged from the feed to the permeate side is high. Nevertheless, the increase in quality losses and formation of unfavorable compounds (i.e. hydroxymethyl furfural and furan) in fruit juices due to high operation temperatures restricts the temperature levels (Ciesarova & Vranova, 2009; Crews & Castle, 2007; Onsekizoglu et al., 2010b). Temperature polarization effect also increases with the increase in feed temperature (Moon et al., 2011).

*Feed flow rate & stirring*

In MD, the increase in flow and/or stirring rate of feed increases the permeate flux. The shearing forces generated at high flow rate and/or stirring reduces the hydrodynamic boundary layer thickness and thus reduce polarization effects. Therefore, the temperature and concentration at the liquid-vapour interface becomes closer to the corresponding values at the bulk feed solution (Winter et al., 2011). Onsekizoglu et al. (2010a) studied the effects of various operating parameters on permeate flux and soluble solid content of apple juice during concentration through osmotic distillation (OD) and membrane distillation (MD) processes. They observed that the effect of feed flow rate on transmembrane flux was less than half of the influence of temperature difference across the membrane.

The effect of flow rate on MD flux becomes more noticeable at higher temperatures especially associated with higher temperature drop across the membrane (Walton et al., 2004). Consequently, higher productivity can be achieved by operating under a turbulent flow regime. On the other hand, the liquid entry pressure of feed solution (LEP) must be taken into account in order to avoid membrane pore wetting when optimizing feed flow rate (Hwang et al., 2011; Khayet et al., 2006).

*Permeate temperature*

The increase in permeate temperature results in lower MD flux due to the decrease of the transmembrane vapour pressure difference as soon as the feed temperature kept constant. It is generally agreed upon that the temperature of cold water on the permeate side has smaller effect on the flux than that of the feed solution for the same temperature difference. This is because the vapour pressure increases exponentially with feed temperature (Alklaibi & Lior, 2005; El-Bourawi et al., 2006).

*Permeate flow rate*

The increase in permeate flow and/or stirring rate reduces the temperature polarization effect. Consequently, the temperature at the gas/liquid interface approaches to the bulk temperature at the permeate side. This will tend to increase driving force across the membrane; resulting an increase in MD flux (Courel et al., 2000; Hongvaleerat et al., 2008). It is important to note that as the permeate used in the MD is distilled water and in the OD is hypertonic salt solution; the extent of the effect of flow rate is more prominent in the latter configuration. This is because of the contribution of concentration polarization effects on permeate side in OD.

## 7. Flux enhancement approaches

The MD process has some significant advantages over conventional processes, however beside the lack of commercially available MD modules, one of the major technical drawbacks for the use of MD in industrial applications is the low transmembrane flux in comparison with RO. Numerous studies have been performed to reduce polarization effects and to enhance transmembrane flux including cooperation of MD with other membrane processes as well as novel MD module design approaches. Some of mentioned attempts are discussed here.

The combination of MD with other membrane systems such as RO, UF, MF, NF and OD have been well-studied by MD researchers in order to improve transmembrane flux, recovery factors and final product quality (Cabral et al., 2011; Calabro & Drioli, 1997; Cath et al., 2005; Cisse et al., 2011; Conidi et al., 2011; Gomes et al., 2011; Gryta, 2005b; Hogan et al., 1998; Mericq et al., 2009; Wang et al., 2011). Each process is unique and contributes particular advantages to the integrated system design.

UF is a powerful method for removing natural polymers (polysaccharides, proteins) that could increase the viscosity of the feed stream through the MD process. For example, pretreatment of grape juice by UF has been shown to result in an increased flux during subsequent concentration of permeate by OD. The flux increase has been attributed to the reduction in the viscosity of the concentrated juice membrane boundary layer due to protein removal (Bailey et al., 2000). Lukanin et al. (2003) have evaluated the use of an enzymatic pretreatment step before UF of apple juices. The protein level which tend to deposit on the hydrophobic surface during subsequent OD process, have been shown to decrease significantly. Such a deposition improves membrane wetting and can eventually result in a convective flow of liquid through the membrane, which is not allowable in the MD process. Onsekizoglu et al. (2010b) have proposed the use of membrane processes for the production of clarified apple juice concentrate. The efficiency of UF was improved by an additional enzymatic pretreatment and flocculation step using fining agents such as gelatine and bentonite. Hongvaleerat et al. (2008) obtained flux values of about 7-10 kg/m²h in pineapple juice concentrate production by OD which were higher than those obtained with single-strength juice.

RO or forward osmosis (FO) processes have been proposed as a pre-concentration step before OD or MD promising reduction of processing costs. High quality fruit juice concentrates can be produced economically in this manner. Therefore, an integrated process involving preconcentration of the feed by RO followed by further concentration by OD or MD should yield a high-solids product concentrate of quality comparable to that achieved by OD alone but at significant reduction in processing cost (Martinetti et al., 2009; Nayak & Rastogi, 2010; Wang et al., 2011). The combination of RO and OD processes was evaluated by Cabra et al. (2011) for concentration of Acerola juice, by Kozak et al. (2009) for concentration of Black currant juice, by Galaverna et al. (2008) for concentration of blood orange juice, by Cassano et al. (2003) for concentration of citrus and carrot juices. It is worth mentioning that in all the previously mentioned studies, a clarification pretreatment step (i.e. ultrafiltration of microfiltration) is involved in order to improve both RO and OD flux.

Criscuoli & Drioli (1999) presented a detailed energetic and exergetic analysis of both RO–MD and NF–RO–MD integrated systems. They observed an improvement in the performance of the integrated system by introducing NF as water pretreatment for the RO–MD system with almost the same energy.

The coupled operation of MD and OD processes is another promising approach to improve transmembrane flux. In this case, osmotic solution is cooled and the feed solution is slightly heated in order to provide additional driving force. Belafi-Bako & Koroknai (2006) compared MD, OD and coupled operation of OD and MD in terms of flux and final soluble solid concentration in sucrose model solutions and apple juice. Higher water flux and SSC values were achieved with coupled operation confirming an increase in driving force. More recently, Onsekizoglu (2011), have proposed the use of a coupled membrane process capable of concentrating pomegranate juice under very mild conditions. The pomegranate juice was clarified by ultrafiltration in a cross-flow membrane filtration unit (MWCO: 100 kDa). The clarified juice then concentrated by coupled operation of OD and MD, in which the feed solution is gently heated (30.0±2.0°C) and the osmotic solution (CaCl₂.H₂O) is slightly cooled (10.0±1.0°C). The final step yielded a concentration of the clarified juice (with an initial total soluble solid content of (TSS) 17°Brix up to 60-62°Brix. The experiments have proven that the driving forces were added in coupled operation, which resulted in enhanced water flux during the operation, thus the coupled process was proposed to be more effective.

Several strategies for reducing temperature polarization through membrane arrangement in MD have been proposed. Some authors have considered the use of spacer-filled channels (Chernyshov et al., 2003; Cipollina et al., 2011; Phattaranawik et al., 2001; Teoh et al., 2008; Wang, 2011). The spacers can improve the flow characteristics at the membrane surface and by promoting regions of turbulence due to the formation of eddies and wakes. Therefore, the temperature polarization can be reduced by improved boundary layer heat transfer. Various surface modification techniques including coating, grafting and plasma polymerization to reduce temperature polarization effect though improvement of membrane surface characteristics have been employed. For example, a novel hollow fiber membrane was proposed by Li & Sirkar (2005) which were commercial porous PP hollow fibres coated with a variety of ultrathin microporous silicone-fluoropolymer layer on surface by plasma polymerization. The coated fibres were arranged in a rectangular cross-flow module design, permitting the hot feed solution to flow over the outside surface of the fibres with a higher Reynolds value. Therefore, reduced temperature polarization inducing higher permeate fluxes have been reported. The reason for applying the coating layer was to provide an additional porous layer having higher hydrophobicity than PP, which itself is one of the polymeric materials with very low surface energy (Li et al., 2008). In recent years, a novel multiple-layered composite membrane have been proposed by Qtaishat et al. (2009) comprising a thin hydrophobic microporous layer and a thin hydrophilic layer. The hydrophobic side of the membrane was maintained adjacent to the hot feed, whereas the hydrophilic layer of the membrane was kept adjacent to cold water, which penetrates into the pores of the hydrophilic layer. Such membranes were found to be promising as they combine the low resistance to mass flux, achieved by the diminution of the water vapour transport path length through the hydrophobic thin top layer and a low conductive heat loss through the membrane, obtained by using a thicker hydrophilic sublayer.

## 8. Long-term performance

*Membrane fouling & Cleaning procedures*

Membrane fouling refers to the loss of membrane performance due to deposition of suspended or dissolved substances on the membrane surface and/or within its pores. There are several types of fouling in the membrane systems including inorganic fouling or scaling, particulate/colloidal fouling, organic fouling and biological fouling (biofouling) (Gryta, 2008). Inorganic fouling or scaling is caused by the accumulation of inorganic precipitates, such as calcium salts ($CaCO_3$, $CaSO_4$), and magnesium carbonates on membrane surface or within pore structure. Precipitates are formed when the concentration of these sparingly soluble salts exceeds their saturation concentrations. Particulate/colloidal fouling is mainly associated with accumulation of biologically inert particles and colloids on the membrane surface. Organic fouling is related with the deposition or adsorption of organic matters on the pores of the membrane surface. Microbial fouling however is formed due to the formation of biofilms on membrane surfaces. Such films (bacterial, algal, or fungal) grow and release biopolymers (polysaccharides, proteins, and amino sugars) as a result of microbial activity (Pabby et al., 2009).

Even though the general agreement is that the fouling phenomena is significantly lower than those encountered in other pressure-driven membrane separation processes, it is one of the major drawbacks in membrane distillation (Gryta, 2005b). The extensive research on membrane fouling has revealed that the efficiency of MD installation can be reduced by more than 50 percent after 50–100 h of process operation due to the presence of fouling effects. In fact, all of the known types of fouling have been determined to occur practically in MD operations (Gryta, 2008).

Kullab & Martin (2011) pointed out that fouling and scaling may result pore clogging in MD membranes, leading to a decrease in effective membrane area, and therefore the permeate flux. Moreover, the flow channel area may be reduced resulting higher temperature polarization due to the pressure drop across the membrane. The increased deposition of the foulant species at the membrane surface would eventually lead to an increase in the pressure drop to levels that the hydrostatic pressure may exceed the LEP of the feed or permeate solution into the membrane pores. Therefore the hydrophobic surface of membrane can be partially wetted due to very small nature of the flow channels in MD modules (especially in hollow fiber membrane modules) (El-Bourawi et al., 2006).

It should be emphasized here that although the importance of understanding the fouling phenomena in MD has been pointed out, very few studies have paid attention to long term performance. Most of the performed fouling studies so far examined fouling and scaling in seawater desalination or wastewater treatment applications.

Gryta (2005a) presented the results of the over 3 years' time research on the direct contact membrane distillation applied for production of the demineralised water using commercial capillary PP membranes. It was found that the membrane was thermally stable, maintaining its morphology and its good separation characteristics throughout the 3 years of DCMD operation. When using permeate of the RO system as DCMD feed solution, membrane pore wetting was not observed; and the DCMD flux was found to be similar to the initial permeate flux. However, precipitation of $CaCO_3$ on the membrane surface was observed when tap water was used directly as a feed. A partial wetting of the membrane was found in this case resulting in a decrease of the permeate flux from 700 to 550 L/m²day. However, the formed deposit was removed every 40–80 h by rinsing the module with a 2–5 wt% HCl solution, permitting the recovery of the initial process efficiency. On the other hand, authors reported that a multiple repetition of this operation resulted in a gradual decline of the maximum flux of permeate.

Ding et al. (2008) investigated the fouling resistance in concentrating traditional Chinese medicine (TCM) extract by DCMD. The observed permeate flux decline was attributed to membrane fouling introducing additional thermal resistance in the boundary layer. No considerable membrane wetting due to TCM deposition on the membrane surface was detected.

The membranes used in MD require regular periodic membrane cleaning to remove membrane fouling and keep the permeability loss within a given range. Durham & Nguyen (1994) evaluated the effectiveness of several cleaning agents for OD membranes fouled by tomato paste. The microporous PTFE and cross-linked acrylic-fluoroethane copolymer membranes were used in the study. The cleaning regime was determined by the membrane surface tension. The most effective cleaner for membranes with a surface tension greater than 23 mN/m was determined as 1% NaOH; however, hydrophobic integrity of these membranes was destroyed during repeated fouling/cleaning trials. On the other hand, P3 Ultrasil 56 was the most effective one for membranes with a surface tension less than 23 mN/m. Water vapour flux was maintained and no salt leakage during repeated fouling/cleaning trials were determined.

Bubbling seems to be an obvious strategy to induce flow and improve shear stress at the membrane surface to control polarization and fouling. Ding et al. (2011) successfully employed the intermittent gas bubbling method to reduce fouling layer formed in concentrating TCM extract through DCMD. To limit membrane fouling or flux decline during concentrating process, intermittent gas bubbling was introduced to the feed side of membrane module. It was confirmed by experimental results that membrane fouling was effectively controlled in the way of removing deposited foulants from membrane surface by created two phase flow.

As can be concluded from the expressed results, there is a lack of data and understanding in fouling phenomena in MD especially in the food processing field. However, the risk of fouling and wetting of membrane pores compromises the durability of the membranes limiting their applications in food industry. The long term MD performance needs to be extensively studied so as to make the MD process more challenging in food industry.

## 9. Applications in food industry

The main food-related applications of membrane distillation are the desalination and production of high purity water from brackish water and seawater. The major advantage of MD in desalination is the ability to achieve high rejection factors which cannot be accomplished by RO at high permeate fluxes. Production of high purity water is well-established with rejection factors of almost 100% of non-volatile compounds (Khayet & Matsuura, 2011). The MD process has been successfully studied for purification of waste waters of pharmaceutical (Ding et al., 2011) and textile (Criscuoli et al., 2008) industries as well as underground waters contaminated with heavy metals (Zolotarev et al., 1994) and sulfuric acid solutions (Tomaszewska, 2000). Very recently, the feasibility of applying membrane distillation process for recovering potable water from arsenic, uranium and fluoride contaminated brackish waters has been proposed (Yarlagadda et al., 2011). A high quality permeate with dissolved solids concentrations less than 20 ppm (>99% rejection of salts) along with arsenic, fluoride and uranium contaminant reductions in the range of 96.5–99.9% were reported. Ke He et al. (2011) reported a flux value of 14.36 L/m²h over one month DCMD operation of sea water at the following conditions: hot side inlet temperature of 60 °C, cold side inlet temperature of 20 °C, and hot and cold side flow rate of 0.6 L/min for PTFE pore size 0.22 µm membranes. The electrical conductivity values of permeate were determined below 8 µS/cm. Gryta (2010) evaluated the desalination of water containing up to 12 g/L of soluble salts MD for 250 h by using PP membranes. Electrical conductivity values of produced water were in the range of 3.4–4.1 µS/cm despite a ten-fold increase of salt concentration. The permeate flux during MD process lasting 250 h decreased slightly from 543 to 498 L/m²h. Thermal water pretreatment was used to prevent scaling which was formed due to decomposition of bicarbonates dissolved in water. On the other hand, the operation was found to be beneficial only for underground waters with high hardness.

One of the main advantages of MD in water purification is the lower energy consumption. Like any other distillation process MD also requires energy for evaporation of water, as stated earlier. However, MD process can effectively operate at low temperatures, which makes it possible to utilize low-grade waste and/or alternative energy sources, such as solar and geothermal energy.

MD and OD are proposed as very challenging technologies for concentration of fruit juice allowing to overcome the drawbacks of conventional thermal evaporation encountered by application of high temperatures (Ali et al., 2003; Bui & Nguyen, 2005; Cisse et al., 2005; Pabby et al., 2009; Shaw et al., 2002; Vaillant et al., 2001). The preliminary study of effective concentration of orange juice by MD was presented by Calabro et al. (1994) using a microporous PVDF membrane. Alves & Coelhoso (2006) compared MD and OD in terms of water flux and aroma retention in model orange juice. A higher retention per amount of water removal was observed with OD together with higher flux values. Very recently, Jorgensen et al. (2011) evaluated the potential of SGMD and VMD configurations for recovery of black currant and cherry juice aroma compounds. The influence of the sweeping gas flow rate (SGMD only), feed temperature and feed flow rate on the permeate flux and the concentrations factors of 12 selected aroma compounds were examined on an aroma model solution and on black currant juice in a laboratory scale set-up. At 45 °C the most volatile and hydrophobic aroma compounds was obtained with the highest concentration factors: 12.1–9.3 (black currant juice) and 17.2–12.8 (model solution). A volume reduction of 13.7% (vol.%) at 45 °C, 400 L/h, resulted in an aroma recovery of 73–84 vol.% for the most volatile compounds in black currant juice.

In concentration of fruit juices containing oily constituents (such as limonene in orange juice), membrane wetting may occur due to high affinity of hydrophobic membrane material with such compounds. Coating of membrane with hydrophilic polymers such as polyvinyl alcohol (PVA) (Mansouri & Fane, 1999) and alginate (Xu et al., 2004) has been proposed to overcome this problem. Recently, Chanachai et al. (2010) studied the coating of hydrophobic membrane PVDF with chitosan, a highly hydrophilic polymer, for protection against wetting by oils from fruit juice. The results indicated that the coated membrane well protected the membrane against wetting-out and could maintain stable flux. Coated membranes used to concentrate the oil solution (limonene 2%, v/v) for 5 h were not wetted out during flux measurement and no visual damage was observed indicating the stability on the base membrane.

It has been well-established that the combination of MD with other membrane technologies offers important benefits over stand alone use of MD in the concentration of various types of juices including grape juice (Rektor et al., 2007), pineapple juice (Hongvaleerat et al., 2008), kiwi fruit juice (Cassano & Drioli, 2007), camu-camu juice (Rodrigues et al., 2004), sugar-cane juice (Nene et al., 2004) and cactus pear juice (Cassano et al., 2007). The integration of MD with other membrane operations such as MF, UF, NF, RO and OD permits advantage of achieving high quality fruit juice concentrates with higher economic feasibility. The use of integrated membrane processes for clarification and concentration of citrus (orange and lemon) and carrot juices have been proposed by Cassano et al. (2003). A limpid phase has been produced by ultrafiltration pilot unit. The clarified permeate coming from UF has been concentrated up to 15-20 °Brix by RO with a laboratory scale unit. Finally, OD step was applied to yield 60-63°Brix concentrate with a transmembrane flux of 1kg/$m^2$h. A slight decrease in the total antioxidant activity has been reported during RO treatment, whereas no significant change was observed during OD treatment. Kozak et. al (2009) investigated an integrated approach for black currant juice concentration. The juice samples were prefiltered by MF and preconcentrated to 22°Brix by RO. A further concentration of the retentate coming from RO was obtained by MD and black currant concentrate with 58.2 °Brix was produced. Onsekizoglu et al. (2010b) have proposed the use of membrane processes for the production of clarified apple juice concentrate. The efficiency of UF was improved by an additional enzymatic pretreatment and flocculation step using fining agents such as gelatine and bentonite. The permeate coming from the UF with initial TSS contents of ca. 12 °Brix were subsequently concentrated up to TSS contents of 65 °Brix by MD, OD and coupled operation of MD & OD processes. The effect of clarification and concentration processes on formation of 5-hydroxymethylfurfural (HMF), retention of bioactive compounds (phenolic compounds, organic acids, glucose, fructose and sucrose) and their efficiency in preserving natural color and aroma (trans-2-hexenal, the most relevant compound in apple juice aroma) were evaluated in order to maintain a high quality product. The new membrane based concentration techniques have been reported to be very efficient since the concentrated juice presented nutritional and sensorial quality very similar to that of the original juice especially regarding the retention of bright natural color and pleasant aroma, which were considerably lost during thermal evaporation. Further analysis have shown that the subsequent concentration treatments by MD, OD and coupled operation of MD & OD processes did not induce any significant changes in phenolic compounds, organic acids and sugars independently on the final concentration achieved.

The MD process can be successfully applied to remove ethanol and the other volatile metabolites from the fermentation broth (Banat & Al-Shannag, 2000; Gryta, 2001; Gryta & Barancewicz, 2011; Gryta et al., 2000a; Tomaszewska & Bialonczyk, 2011). The fermentation of sugar with *Saccharomyces cerevisiae* proceeds with the formation of by-products, which tend to inhibit the yeast productivity. The removal of ethanol is usually carried out by distillation. The primary disadvantages of the conventional process of ethanol generation include high energy consumption and excessive amount of wastewater discharged from the distillation columns. The MD process provides an economical alternative to the existing distillation technique for continuous removal of fermented products. The removal of volatile metabolites from the fermentation broth by MD process enables reduction of the inhibitory effect of these compounds on microbial culture together with an increased rate of sugar conversion to ethanol and hence the cost of further concentration of alcohol can be reduced. The main advantage of MD over conventional distillation processes is that membrane distillation takes place at a temperature below the normal boiling point of broth solutions.

Other food-related applications of MD include concentration of natural food colorants (Nayak & Rastogi, 2010), dealcoholization of wine (Varavuth et al., 2009) and concentration of herbal and plant extracts (Cisse et al., 2011; Dornier et al., 2011; Johnson et al., 2002; Zhao et al., 2011).

## 10. Concluding remarks and future prospects

As a promising alternative to replace other separation processes, MD has gained much interest for its lower energy requirement in comparison with conventional distillation, lower operating pressures and higher rejection factors than in pressure driven processes such as NF, and RO. Although MD has been known for more than 40 years, a number of problems exist when MD is considered for industrial implementation. Most of the conducted MD studies are still in the laboratory scale. In recent years, some pilot plant studies have been proposed for desalination (Blanco et al., 2011; Farmani et al., 2008; Song et al., 2008; Xu et al., 2006), however long term evaluations of pilot plant applications for the concentration and recovery of aqueous solutions containing volatile solutes especially in the food industry are still scarce. Therefore, achievement of high concentration levels in certain fruit juice samples taking into account the effects on mass and heat transfer mechanisms, membrane characteristics and the quality parameters together with a detailed economical analysis should be examined on a large scale.

On the other hand, there is a lack of commercially available MD units; practically all membrane modules are designed for other membrane operations (i.e. microfiltration) rather than MD. Novel membranes specifically designed for MD applications should be fabricated in an economically feasible way. Research on transmembrane flux enhancement (i.e. acoustic field) for large scale applications is required. More attention should be paid to the possibility of integrating MD to other separation techniques in order to improve the efficiency of the overall system and to make the process economically viable for industrial applications. For fruit juice concentration, coupled operation of MD and OD seems promising to overcome high temperature related problems (i.e. aroma and colour loss) encountered in MD. However, integration of MD with other MD variants as well as conventional distillation techniques has not yet been investigated. Hence more focus on such combinations is required. In recent years, coupling MD with solar energy systems has been well studied by various researchers for desalination of sea water.

The ability to effectively operate at low temperatures makes MD process possible to utilize low-grade waste and/or alternative energy sources. In recent years, coupling MD with solar, geothermal and waste energy systems has been proposed to decrease energy consumption in desalination systems. Such an approach may be crucial for food processing systems. For example, in the case of fruit juice concentration, much lower temperatures should be applied in order to obtain stable products able to retain as much possible the uniqueness of the fresh fruit, its original color, aroma, nutritional value and structural characteristics. Thus, the possibility of operating under very mild conditions enables MD to utilize various alternative energy sources, making it more promising for industrial application. Further efforts need to be concentrated in this field, especially in utilization of waste energy and/or other renewable energy sources in the view of industrial implementation.

## 11. References

Agashichev, S. P. (2006) Modeling of the concentration polarization in a cylindrical channel of an ultrafiltration module. *Theoretical Foundations of Chemical Engineering*, 40, 215-216.

Ali, F., Dornier, M., Duquenoy, A. & Reynes, M. (2003) Evaluating transfers of aroma compounds during the concentration of sucrose solutions by osmotic distillation in a batch-type pilot plant. *Journal of Food Engineering*, 60, 1-8.

Alklaibi, A. M. & Lior, N. (2005) Membrane-distillation desalination: status and potential. *Desalination*, 171, 111-131.

Alves, V. D. & Coelhoso, I. M. (2006) Orange juice concentration by osmotic evaporation and membrane distillation: A comparative study. *Journal of Food Engineering*, 74, 125-133.

Babu, B. R., Rastogi, N. K. & Raghavarao, K. S. M. S. (2006) Mass transfer in osmotic membrane distillation of phycocyanin colorant and sweet-lime juice. *Journal of Membrane Science*, 272, 58-69.

Babu, B. R., Rastogi, N. X. & Raghavarao, K. S. M. S. (2008) Concentration and temperature polarization effects during osmotic membrane distillation. *Journal of Membrane Science*, 322, 146-153.

Bagger-Jorgensen, R., Meyer, A. S., Pinelo, M., Varming, C. & Jonsson, G. (2011) Recovery of volatile fruit juice aroma compounds by membrane technology: Sweeping gas versus vacuum membrane distillation. *Innovative Food Science & Emerging Technologies*, 12, 388-397.

Bailey, A. F. G., Barbe, A. M., Hogan, P. A., Johnson, R. A. & Sheng, J. (2000) The effect of ultrafiltration on the subsequent concentration of grape juice by osmotic distillation. *Journal of Membrane Science*, 164, 195-204.

Banat, F. A. & Al-Shannag, M. (2000) Recovery of dilute acetone-butanol-ethanol (ABE) solvents from aqueous solutions via membrane distillation. *Bioprocess Engineering,* 23, 643-649.

Banat, F. A. & Simandl, J. (1999) Membrane distillation for dilute ethanol - Separation from aqueous streams. *Journal of Membrane Science,* 163, 333-348.

Bandini, S. & Sarti, G. C. (1999) Heat and mass transport resistances in vacuum membrane distillation per drop. *Aiche Journal,* 45, 1422-1433.

Barbe, A. M., Hogan, P. A. & Johnson, R. A. (2000) Surface morphology changes during initial usage of hydrophobic, microporous polypropylene membranes. *Journal of Membrane Science,* 172, 149-156.

Bazinet, L., Cossec, C., Gaudreau, H. & Desjardins, Y. (2009) Production of a Phenolic Antioxidant Enriched Cranberry Juice by Electrodialysis with Filtration Membrane. *Journal of Agricultural and Food Chemistry,* 57, 10245-10251.

Belafi-Bako, K. & Koroknai, B. (2006) Enhanced water flux in fruit juice concentration: Coupled operation of osmotic evaporation and membrane distillation. *Journal of Membrane Science,* 269, 187-193.

Blanco, J., Guillen-Burrieza, E., Zaragoza, G., Alarcon, D. C., Palenzuela, P., Ibarra, M. & Gernjak, W. (2011) Experimental analysis of an air gap membrane distillation solar desalination pilot system. *Journal of Membrane Science,* 379, 386-396.

Brodard, F., Romero, J., Belleville, M. P., Sanchez, J., Combe-James, C., Dornier, M. & Rios, G. M. (2003) New hydrophobic membranes for osmotic evaporation process. *Separation and Purification Technology,* 32, 3-7.

Bryjak, M., Gancarz, I. & Pozniak, G. (2000) Plasma-modified porous membranes. *Chemical Papers-Chemicke Zvesti,* 54, 496-501.

Bui, A. V. & Nguyen, H. M. (2005) Scaling Up of Osmotic Distillation from Laboratory to Pilot Plant for Concentration of Fruit Juices. *International Journal of Food Engineering,* 1, -.

Bui, V. A., Nguyen, M. H. & Muller, J. (2004) A laboratory study on glucose concentration by osmotic distillation in hollow fibre module. *Journal of Food Engineering,* 63, 237-245.

Burgoyne, A. & Vahdati, M. M. (2000) Direct contact membrane distillation. *Separation Science and Technology,* 35, 1257-1284.

Cabral, L. M. C., Pagani, M. M., Rocha-Leao, M. H., Couto, A. B. B., Pinto, J. P., Ribeiro, A. O. & Gomes, F. D. (2011) Concentration of acerola (Malpighia emarginata DC.) juice by integrated membrane separation process. *Desalination and Water Treatment,* 27, 130-134.

Calabro, V. & Drioli, E. (1997) Polarization phenomena in integrated reverse osmosis and membrane distillation for seawater desalination and waste water treatment. *Desalination,* 108, 81-82.

Calabro, V., Jiao, B. L. & Drioli, E. (1994) Theoretical and Experimental-Study on Membrane Distillation in the Concentration of Orange Juice. *Industrial & Engineering Chemistry Research,* 33, 1803-1808.

Cassano, A., Conidi, C., Timpone, R., D'Avella, M. & Drioli, E. (2007) A membrane-based process for the clarification and the concentration of the cactus pear juice. *Journal of Food Engineering,* 80, 914-921.

Cassano, A. & Drioli, E. (2007) Concentration of clarified kiwifruit juice by osmotic distillation. *Journal of Food Engineering,* 79, 1397-1404.

Cassano, A., Drioli, E., Galaverna, G., Marchelli, R., Di Silvestro, G. & Cagnasso, P. (2003) Clarification and concentration of citrus and carrot juices by integrated membrane processes. *Journal of Food Engineering,* 57, 153-163.

Cath, T. Y., Adams, D. & Childress, A. E. (2005) Membrane contactor processes for wastewater reclamation in space II. Combined direct osmosis, osmotic distillation, and membrane distillation for treatment of metabolic wastewater. *Journal of Membrane Science,* 257, 111-119.

Cath, T. Y., Adams, V. D. & Childress, A. E. (2004) Experimental study of desalination using direct contact membrane distillation: a new approach to flux enhancement. *Journal of Membrane Science,* 228, 5-16.

Celere, M. & Gostoli, C. (2004) Osmotic distillation with propylene glycol, glycerol and glycerol-salt mixtures. *Journal of Membrane Science,* 229, 159-170.

Chanachai, A., Meksup, K. & Jiraratananon, R. (2010) Coating of hydrophobic hollow fiber PVDF membrane with chitosan for protection against wetting and flavor loss in osmotic distillation process. *Separation and Purification Technology,* 72, 217-224.

Chen, T. C., Ho, C. D. & Yeh, H. M. (2009) Theoretical modeling and experimental analysis of direct contact membrane distillation. *Journal of Membrane Science,* 330, 279-287.

Chernyshov, M. N., Meindersma, G. W. & de Haan, A. B. (2003) Modelling temperature and salt concentration distribution in membrane distillation feed channel. *Desalination,* 157, 315-324.

Chernyshov, M. N., Meindersma, G. W. & de Haan, A. B. (2005) Comparison of spacers for temperature polarization reduction in air gap membrane distillation. *Desalination,* 183, 363-374.

Ciesarova, Z. & Vranova, J. (2009) Furan in Food - a Review. *Czech Journal of Food Sciences,* 27, 1-10.

Cipollina, A., Micale, G. & Rizzuti, L. (2011) Membrane distillation heat transfer enhancement by CFD analysis of internal module geometry. *Desalination and Water Treatment,* 25, 195-209.

Cisse, M., Vaillant, F., Bouquet, S., Pallet, D., Lutin, F., Reynes, M. & Dornier, M. (2011) Athermal concentration by osmotic evaporation of roselle extract, apple and grape juices and impact on quality. *Innovative Food Science & Emerging Technologies,* 12, 352-360.

Cisse, M., Vaillant, F., Perez, A., Dornier, M. & Reynes, M. (2005) The quality of orange juice processed by coupling crossflow microfiltration and osmotic evaporation. *International Journal of Food Science and Technology,* 40, 105-116.

Close, E. & Sorensen, E. (2010) Modelling of Direct Contact Membrane Distillation for Desalination. *20th European Symposium on Computer Aided Process Engineering,* 28, 649-654.

Conidi, C., Cassano, A. & Drioli, E. (2011) A membrane-based study for the recovery of polyphenols from bergamot juice. *Journal of Membrane Science,* 375, 182-190.

Courel, M., Dornier, M., Herry, J. M., Rios, G. M. & Reynes, M. (2000) Effect of operating conditions on water transport during the concentration of sucrose solutions by osmotic distillation. *Journal of Membrane Science,* 170, 281-289.

Couto, D. S., Dornier, M., Pallet, D., Reynes, M., Dijoux, D., Freitas, S. P. & Cabral, L. M. C. (2011) Evaluation of nanofiltration membranes for the retention of anthocyanins of acai (Euterpe oleracea Mart.) juice. *Desalination and Water Treatment,* 27, 108-113.

Crews, C. & Castle, L. (2007) A review of the occurrence, formation and analysis of furan in heat-processed foods. *Trends in Food Science & Technology,* 18, 365-372.

Criscuoli, A. & Drioli, E. (1999) Energetic and exergetic analysis of an integrated membrane desalination system. *Desalination,* 124, 243-249.

Criscuoli, A., Zhong, J., Figoli, A., Carnevale, M. C., Huang, R. & Drioli, E. (2008) Treatment of dye solutions by vacuum membrane distillation. *Water Research*, 42, 5031-5037.

Curcio, E., Di, P. G. & Enrico, D. (Eds.). (2010). *Membrane distillation and osmotic distillation*, Elsevier.

Curcio, E. & Drioli, E. (2005) Membrane distillation and related operations - A review. *Separation and Purification Reviews*, 34, 35-86.

Ding, Z. W., Liu, L. Y., Liu, Z. & Ma, R. Y. (2011) The use of intermittent gas bubbling to control membrane fouling in concentrating TCM extract by membrane distillation. *Journal of Membrane Science*, 372, 172-181.

Ding, Z. W., Liu, L. Y., Yu, H. F., Ma, R. Y. & Yang, Z. R. (2008) Concentrating the extract of traditional Chinese medicine by direct contact membrane distillation. *Journal of Membrane Science*, 310, 539-549.

Ding, Z. W., Ma, R. Y. & Fane, A. G. (2003) A new model for mass transfer in direct contact membrane distillation. *Desalination*, 151, 217-227.

Dornier, M., Cisse, M., Vaillant, F., Bouquet, S., Pallet, D., Lutin, F. & Reynes, M. (2011) Athermal concentration by osmotic evaporation of roselle extract, apple and grape juices and impact on quality. *Innovative Food Science & Emerging Technologies*, 12, 352-360.

Durham, R. J. & Nguyen, M. H. (1994) Hydrophobic Membrane Evaluation and Cleaning for Osmotic Distillation of Tomato Puree. *Journal of Membrane Science*, 87, 181-189.

El-Bourawi, M. S., Ding, Z., Ma, R. & Khayet, M. (2006) A framework for better understanding membrane distillation separation process. *Journal of Membrane Science*, 285, 4-29.

Farmani, B., Haddadekhodaparast, M. H., Hesari, J. & Aharizad, S. (2008) Determining Optimum Conditions for Sugarcane Juice Refinement by Pilot Plant Dead-end Ceramic Micro-filtration. *Journal of Agricultural Science and Technology*, 10, 351-357.

Galaverna, G., Di Silvestro, G., Cassano, A., Sforza, S., Dossena, A., Drioli, E. & Marchelli, R. (2008) A new integrated membrane process for the production of concentrated blood orange juice: Effect on bioactive compounds and antioxidant activity. *Food Chemistry*, 106, 1021-1030.

Garcia-Payo, M. C., Izquierdo-Gil, M. A. & Fernandez-Pineda, C. (2000) Air gap membrane distillation of aqueous alcohol solutions. *Journal of Membrane Science*, 169, 61-80.

Gomes, F. D., da Costa, P. A., de Campos, M. B. D., Couri, S. & Cabral, L. M. C. (2011) Concentration of watermelon juice by reverse osmosis process. *Desalination and Water Treatment*, 27, 120-122.

Gryta, M. (2001) The fermentation process integrated with membrane distillation. *Separation and Purification Technology*, 24, 283-296.

Gryta, M. (2005a) Long-term performance of membrane distillation process. *Journal of Membrane Science*, 265, 153-159.

Gryta, M. (2005b) Osmotic MD and other membrane distillation variants. *Journal of Membrane Science*, 246, 145-156.

Gryta, M. (2007) Influence of polypropylene membrane surface porosity on the performance of membrane distillation process. *Journal of Membrane Science*, 287, 67-78.

Gryta, M. (2008) Fouling in direct contact membrane distillation process. *Journal of Membrane Science*, 325, 383-394.

Gryta, M. (2010) Desalination of thermally softened water by membrane distillation process. *Desalination*, 257, 30-35.

Gryta, M. & Barancewicz, M. (2011) Separation of volatile compounds from fermentation broth by membrane distillation. *Polish Journal of Chemical Technology*, 13, 56-60.

Gryta, M., Morawski, A. W. & Tomaszewska, M. (2000a) Ethanol production in membrane distillation bioreactor. *Catalysis Today*, 56, 159-165.

Gryta, M., Tomaszewska, M. & Morawski, A. W. (2000b) A capillary module for membrane distillation process. *Chemical Papers*, 54, 370-374.

He, K., Hwang, H. J., Woo, M. W. & Moon, I. S. (2011) Production of drinking water from saline water by direct contact membrane distillation (DCMD). *Journal of Industrial and Engineering Chemistry*, 17, 41-48.

Hogan, P. A., Canning, R. P., Peterson, P. A., Johnson, R. A. & Michaels, A. S. (1998) A new option: Osmotic distillation. *Chemical Engineering Progress*, 94, 49-61.

Hongvaleerat, C., Cabral, L. M. C., Dornier, M., Reynes, M. & Ningsanond, S. (2008) Concentration of pineapple juice by osmotic evaporation. *Journal of Food Engineering*, 88, 548-552.

Huo, R. T., Gu, Z. Y., Zuo, K. J. & Zhao, G. M. (2010) Preparation and Humic Acid Fouling Resistance of Poly(vinylidene fluoride)-Fabric Composite Membranes for Membrane Distillation. *Journal of Applied Polymer Science*, 117, 3651-3658.

Huo, R. T., Gu, Z. Y., Zuo, K. J. & Zhao, G. M. (2011) Fouling resistance of PVDF-fabric composite membrane in membrane distillation desalination. *Advances in Composites, Pts 1 and 2*, 150-151, 334-339.

Hwang, H. J., He, K., Gray, S., Zhang, J. H. & Moon, I. S. (2011) Direct contact membrane distillation (DCMD): Experimental study on the commercial PTFE membrane and modeling. *Journal of Membrane Science*, 371, 90-98.

Ibarz, A., Garza, S., Garvin, A. & Pagan, J. (2011) Degradation of mansarin juice concnetrates treated at high temperatures. *Journal of Food Process Engineering*, 34, 682-696.

Izquierdo-Gil, M. A., Garcia-Payo, M. C. & Fernandez-Pineda, C. (1999) Air gap membrane distillation of sucrose aqueous solutions. *Journal of Membrane Science*, 155, 291-307.

Jiao, B., Cassano, A. & Drioli, E. (2004) Recent advances on membrane processes for the concentration of fruit juices: a review. *Journal of Food Engineering*, 63, 303-324.

Johnson, R. A., Sun, J. C. & Sun, J. (2002) A pervaporation-microfiltration-osmotic distillation hybrid process for the concentration of ethanol-water extracts of the Echinacea plant. *Journal of Membrane Science*, 209, 221-232.

Kadakal, C., Sebahattin, N. & Poyrazoglu, E. S. (2002) Effect of commercial processing stages of apple juice on patulin, fumaric acid and hydroxymethylfurfural (HMF) levels. *Journal of Food Quality*, 25, 359-368.

Khayet, A., Matsuura, T., Mengual, J. I. & Qtaishat, M. (2006) Design of novel direct contact membrane distillation membranes. *Desalination*, 192, 105-111.

Khayet, M. (2011) Membranes and theoretical modeling of membrane distillation: A review. *Advances in Colloid and Interface Science*, 164, 56-88.

Khayet, M., Cojocaru, C. & Garcia-Payo, M. C. (2010) Experimental design and optimization of asymmetric flat-sheet membranes prepared for direct contact membrane distillation. *Journal of Membrane Science*, 351, 234-245.

Khayet, M., Khulbe, K. C. & Matsuura, T. (2004a) Characterization of membranes for membrane distillation by atomic force microscopy and estimation of their water vapor transfer coefficients in vacuum membrane distillation process. *Journal of Membrane Science*, 238, 199-211.

Khayet, M., Velazquez, A. & Mengual, J. I. (2004b) Direct contact membrane distillation of humic acid solutions. *Journal of Membrane Science*, 240, 123-128.

Khayet, M. S. & Matsuura, T. 2011. *Membrane distillation: principles and applications*. Elsevier.

Kim, S. S. & Lloyd, D. R. (1991) Microporous membrane formation via thermally-induced phase separation. 3. Effect of thermodynamic interactions on the structure of isotactic polypropylene membranes. *Journal of Membrane Science*, 64, 13-29.

Kong, Y., Lin, X., Wu, Y. L., Chen, J. & Xu, J. P. (1992) Plasma Polymerization of Octafluorocyclobutane and Hydrophobic Microporous Composite Membranes for Membrane Distillation. *Journal of Applied Polymer Science*, 46, 191-199.

Kozak, A., Bekassy-Molnar, E. & Vatai, G. (2009) Production of black-currant juice concentrate by using membrane distillation. *Desalination*, 241, 309-314.

Krajewski, S. R., Kujawski, W., Bukowska, M., Picard, C. & Larbot, A. (2006) Application of fluoroalkylsilanes (FAS) grafted ceramic membranes in membrane distillation process of NaCl solutions. *Journal of Membrane Science*, 281, 253-259.

Kullab, A. & Martin, A. (2011) Membrane distillation and applications for water purification in thermal cogeneration plants. *Separation and Purification Technology*, 76, 231-237.

Lagana, F., Barbieri, G. & Drioli, E. (2000) Direct contact membrane distillation: modelling and concentration experiments. *Journal of Membrane Science*, 166, 1-11.

Lai, C. L., Liou, R. M., Chen, S. H., Huang, G. W. & Lee, K. R. (2011) Preparation and characterization of plasma-modified PTFE membrane and its application in direct contact membrane distillation. *Desalination*, 267, 184-192.

Lawson, K. W., Hall, M. S. & Lloyd, D. R. (1995) Compaction of Microporous Membranes Used in Membrane Distillation .1. Effect on Gas-Permeability. *Journal of Membrane Science*, 101, 99-108.

Lawson, K. W. & Lloyd, D. R. (1996a) Membrane distillation .1. Module design and performance evaluation using vacuum membrane distillation. *Journal of Membrane Science*, 120, 111-121.

Lawson, K. W. & Lloyd, D. R. (1996b) Membrane distillation .2. Direct contact MD. *Journal of Membrane Science*, 120, 123-133.

Lawson, K. W. & Lloyd, D. R. (1997) Membrane distillation. *Journal of Membrane Science*, 124, 1-25.

Li, B. & Sirkar, K. K. (2004) Novel membrane and device for direct contact membrane distillation-based desalination process. *Industrial & Engineering Chemistry Research*, 43, 5300-5309.

Li, B. & Sirkar, K. K. (2005) Novel membrane and device for vacuum membrane distillation-based desalination process. *Journal of Membrane Science*, 257, 60-75.

Li, N. N., Fane, A. G., Ho, W. S. W. & Matsuura, T. (Eds.). (2008). *Advanced Membrane Technology and Applications*, John Wiley & Sons Inc., New Jersey.

Liu, F., Hashim, N. A., Liu, Y. T., Abed, M. R. M. & Li, K. (2011) Progress in the production and modification of PVDF membranes. *Journal of Membrane Science*, 375, 1-27.

Lukanin, E. S., Gunko, S. M., Bryk, M. T. & Nigmatullin, R. R. (2003) The effect of content of apple juice biopolymers on the concentration by membrane distillation. *Journal of Food Engineering*, 60, 275-280.

Mansouri, J. & Fane, A. G. (1999) Osmotic distillation of oily feeds. *Journal of Membrane Science*, 153, 103-120.

Martinetti, C. R., Childress, A. E. & Cath, T. Y. (2009) High recovery of concentrated RO brines using forward osmosis and membrane distillation. *Journal of Membrane Science*, 331, 31-39.

Martinez-Diez, L., Vazquez-Gonzalez, M. I. & Florido-Diaz, F. J. (1998) Study of membrane distillation using channel spacers. *Journal of Membrane Science*, 144, 45-56.

Martinez, L., Florido-Diaz, F. J., Hernandez, A. & Pradanos, P. (2002) Characterisation of three hydrophobic porous membranes used in membrane distillation - Modelling and evaluation of their water vapour permeabilities. *Journal of Membrane Science*, 203, 15-27.

Martinez, L. & Rodriguez-Maroto, J. M. (2006) Characterization of membrane distillation modules and analysis of mass flux enhancement by channel spacers. *Journal of Membrane Science*, 274, 123-137.

Martinez, L. & Rodriguez-Maroto, J. M. (2007) On transport resistances in direct contact membrane distillation. *Journal of Membrane Science*, 295, 28-39.

Mello, B., Petrus, J. C. C. & Hubinger, M. D. (2010) Performance of nanofiltration concentration process in propolis extracts. *Ciencia E Tecnologia De Alimentos*, 30, 166-172.

Mengual, J. I., Dezarate, J. M. O., Pena, L. & Velazquez, A. (1993) Osmotic Distillation through Porous Hydrophobic Membranes. *Journal of Membrane Science*, 82, 129-140.

Mericq, J. P., Laborie, S. & Cabassud, C. (2009) Vacuum membrane distillation for an integrated seawater desalination process. *Desalination and Water Treatment*, 9, 287-296.

Meyer, A. S., Bagger-Jorgensen, R., Pinelo, M., Varming, C. & Jonsson, G. (2011) Recovery of volatile fruit juice aroma compounds by membrane technology: Sweeping gas versus vacuum membrane distillation. *Innovative Food Science & Emerging Technologies*, 12, 388-397.

Moon, I. S., Hwang, H. J., He, K., Gray, S. & Zhang, J. H. (2011) Direct contact membrane distillation (DCMD): Experimental study on the commercial PTFE membrane and modeling. *Journal of Membrane Science*, 371, 90-98.

Morao, A. I. C., Alves, A. M. B. & Geraldes, V. (2008) Concentration polarization in a reverse osmosis/nanofiltration plate-and-frame membrane module. *Journal of Membrane Science*, 325, 580-591.

Nagaraj, N., Patil, B. S. & Biradar, P. M. (2006a) Osmotic Membrane Distillation - A Brief Review. *International Journal of Food Engineering*, 2.

Nagaraj, N., Patil, G., Babu, B. R., Hebbar, U. H., Raghavarao, K. S. M. S. & Nene, S. (2006b) Mass transfer in osmotic membrane distillation. *Journal of Membrane Science*, 268, 48-56.

Narayan, A. V., Nagaraj, N., Hebbar, H. U., Chakkaravarthi, A., Raghavarao, K. S. M. S. & Nene, S. (2002) Acoustic field-assisted osmotic membrane distillation. *Desalination*, 147, 149-156.

Nayak, C. A. & Rastogi, N. K. (2010) Comparison of osmotic membrane distillation and forward osmosis membrane processes for concentration of anthocyanin. *Desalination and Water Treatment*, 16, 134-145.

Nene, S., Kaur, S., Sumod, K., Joshi, B. & Raghavarao, K. S. M. S. (2004) Membrane distillation for the concentration of raw cane-sugar syrup and membrane clarified sugarcane juice (vol 147, pg 157, 2002). *Desalination*, 161, 305-305.

Oliver, M. R. 2004. *Chemical-mechanical planarization of semiconductor materials* Springer.

Onsekizoglu, P., Bahceci, K. S. & Acar, J. (2010a) The use of factorial design for modeling membrane distillation. *Journal of Membrane Science*, 349, 225-230.

Onsekizoglu, P., Bahceci, K. S. & Acar, M. J. (2010b) Clarification and the concentration of apple juice using membrane processes: A comparative quality assessment. *Journal of Membrane Science*, 352, 160-165.

Onsekizoglu, P. (2011) A novel integrated membrane process for pomegranate juice concentration, *Proceedings of Novel Approaches in Food Industry*, Cesme, Izmir, Turkey, May 2011

Pabby, A. K., Rizvi, S. S. H. & Sastre, A. M. (Eds.). (2009). *Handbook of Membrane Seperations*, CRC Press, New York.

Peinemann, K.-V., Nunes, S. P. & Giorno, L. (Eds.). (2010). *Membrane Technology: Volume 3: Membranes for Food Applications*, Wiley-VCH Verlag GmbH & Co. KGaA, Germany.

Phattaranawik, J., Fane, A. G., Pasquier, A. C. S., Bing, W. & Wong, F. S. (2009) Experimental Study and Design of a Submerged Membrane Distillation Bioreactor. *Chemical Engineering & Technology*, 32, 38-44.

Phattaranawik, J. & Jiraratananon, R. (2001) Direct contact membrane distillation: effect of mass transfer on heat transfer. *Journal of Membrane Science*, 188, 137-143.

Phattaranawik, J., Jiraratananon, R. & Fane, A. G. (2003a) Effect of pore size distribution and air flux on mass transport in direct contact membrane distillation. *Journal of Membrane Science*, 215, 75-85.

Phattaranawik, J., Jiraratananon, R. & Fane, A. G. (2003b) Heat transport and membrane distillation coefficients in direct contact membrane distillation. *Journal of Membrane Science*, 212, 177-193.

Phattaranawik, J., Jiraratananon, R., Fane, A. G. & Halim, C. (2001) Mass flux enhancement using spacer filled channels in direct contact membrane distillation. *Journal of Membrane Science*, 187, 193-201.

Qtaishat, M., Khayet, M. & Matsuura, T. (2009) Guidelines for preparation of higher flux hydrophobic/hydrophilic composite membranes for membrane distillation. *Journal of Membrane Science*, 329, 193-200.

Qtaishat, M., Matsuura, T., Kruczek, B. & Khayet, A. (2008) Heat and mass transfer analysis in direct contact membrane distillation. *Desalination*, 219, 272-292.

Quoc, A. L., Mondor, M., Lamarche, F. & Makhlouf, J. (2011) Optimization of electrodialysis with bipolar membranes applied to cloudy apple juice: Minimization of malic acid and sugar losses. *Innovative Food Science & Emerging Technologies*, 12, 45-49.

Rektor, A., Kozak, A., Vatai, G. & Bekassy-Molnar, E. (2007) Pilot plant RO-filtration of grape juice. *Separation and Purification Technology*, 57, 473-475.

Rincon, C., de Zarate, J. M. O. & Mengual, J. I. (1999) Separation of water and glycols by direct contact membrane distillation. *Journal of Membrane Science*, 158, 155-165.

Rodrigues, R. B., Menezes, H. C., Cabral, L. M. C., Dornier, M., Rios, G. M. & Reynes, M. (2004) Evaluation of reverse osmosis and osmotic evaporation to concentrate camu-camu juice (Myrciaria dubia). *Journal of Food Engineering*, 63, 97-102.

Santana, I., Gurak, P. D., da Matta, V. M., Freitas, S. P. & Cabral, L. M. C. (2011) Concentration of grape juice (Vitis labrusca) by reverse osmosis process. *Desalination and Water Treatment*, 27, 103-107.

Schofield, R. W., Fane, A. G. & Fell, C. J. D. (1990a) Gas and Vapor Transport through Microporous Membranes .2. Membrane Distillation. *Journal of Membrane Science*, 53, 173-185.

Schofield, R. W., Fane, A. G., Fell, C. J. D. & Macoun, R. (1990b) Factors Affecting Flux in Membrane Distillation. *Desalination*, 77, 279-294.

Shaw, P. E., Lebrun, M., Ducamp, M. N., Jordan, M. J. & Goodner, K. L. (2002) Pineapple juice concentrated by osmotic evaporation. *Journal of Food Quality*, 25, 39-49.

Shin, C. H. & Johnson, R. (2007) Identification of an appropriate osmotic agent for use in osmotic distillation. *Journal of Industrial and Engineering Chemistry*, 13, 926-931.

Sigurdsson, S. & Shishoo, R. (1997) Surface properties of polymers treated with tetrafluoromethane plasma. *Journal of Applied Polymer Science*, 66, 1591-1601.

Simsek, A., Poyrazoglu, E. S., Karacan, S. & Velioglu, Y. S. (2007) Response surface methodological study on HMF and fluorescent accumulation in red and white grape juices and concentrates. *Food Chemistry*, 101, 987-994.

Song, L. F. (2010) Concentration Polarization in a Narrow Reverse Osmosis Membrane Channel. *Aiche Journal*, 56, 143-149.

Song, L. M., Ma, Z. D., Liao, X. H., Kosaraju, P. B., Irish, J. R. & Sirkar, K. K. (2008) Pilot plant studies of novel membranes and devices for direct contact membrane distillation-based desalination. *Journal of Membrane Science*, 323, 257-270.

Srisurichan, S., Jiraratananon, R. & Fane, A. G. (2006) Mass transfer mechanisms and transport resistances in direct contact membrane distillation process. *Journal of Membrane Science*, 277, 186-194.

Susanto, H. (2011) Towards practical implementations of membrane distillation. *Chemical Engineering and Processing*, 50, 139-150.

Teoh, M. M., Bonyadi, S. & Chung, T. S. (2008) Investigation of different hollow fiber module designs for flux enhancement in the membrane distillation process. *Journal of Membrane Science*, 311, 371-379.

Tomaszewska, M. (2000) Membrane distillation - Examples of applications in technology and environmental protection. *Polish Journal of Environmental Studies*, 9, 27-36.

Tomaszewska, M. & Bialonczyk, L. (2011) The investigation of ethanol separation by the membrane distillation process. *Polish Journal of Chemical Technology*, 13, 66-69.

Toribio, J. L. & Lozano, J. E. (1986) Heat-induced browning of clarified apple juice at high temperatures. *Journal of Food Science*, 51, 172-&.

Vaillant, F., Jeanton, E., Dornier, M., O'Brien, G. M., Reynes, M. & Decloux, M. (2001) Concentration of passion fruit juice on an industrial pilot scale using osmotic evaporation. *Journal of Food Engineering*, 47, 195-202.

Valdes, H., Romero, J., Saavedra, A., Plaza, A. & Bubnovich, V. (2009) Concentration of noni juice by means of osmotic distillation. *Journal of Membrane Science*, 330, 205-213.

Varavuth, S., Jiraratananon, R. & Atchariyawut, S. (2009) Experimental study on dealcoholization of wine by osmotic distillation process. *Separation and Purification Technology*, 66, 313-321.

Vargas-Garcia, A., Torrestiana-Sanchez, B., Garcia-Borquez, A. & Aguilar-Uscanga, G. (2011) Effect of grafting on microstructure, composition and surface and transport properties of ceramic membranes for osmotic evaporation. *Separation and Purification Technology*, 80, 473-481.

Varming, C., Andersen, M. L. & Poll, L. (2004) Influence of thermal treatment on black currant (Ribes nigrum L.) juice aroma. *Journal of Agricultural and Food Chemistry*, 52, 7628-7636.

Walton, J., Lu, H., Turner, C., Solis, S. & Hein, H. 2004. *Solar and waste heat desalination by membrane distillation*. El Paso: College of Engineering University of Texas.

Wang, C. C. (2011) On the heat transfer correlation for membrane distillation. *Energy Conversion and Management*, 52, 1968-1973.

Wang, F. L. & Tarabara, V. V. (2007) Coupled effects of colloidal deposition and salt concentration polarization on reverse osmosis membrane performance. *Journal of Membrane Science*, 293, 111-123.

Wang, K. Y., Foo, S. W. & Chung, T. S. (2009) Mixed Matrix PVDF Hollow Fiber Membranes with Nanoscale Pores for Desalination through Direct Contact Membrane Distillation. *Industrial & Engineering Chemistry Research*, 48, 4474-4483.

Wang, K. Y., Teoh, M. M., Nugroho, A. & Chung, T. S. (2011) Integrated forward osmosis-membrane distillation (FO-MD) hybrid system for the concentration of protein solutions. *Chemical Engineering Science*, 66, 2421-2430.

Warczok, J., Gierszewska, M., Kujawski, W. & Gueell, C. (2007) Application of osmotic membrane distillation for reconcentration of sugar solutions from osmotic dehydration. *Separation and Purification Technology*, 57, 425-429.

Winter, D., Koschikowski, J. & Wieghaus, M. (2011) Desalination using membrane distillation: Experimental studies on full scale spiral wound modules. *Journal of Membrane Science*, 375, 104-112.

Wu, Y. L., Kong, Y., Lin, X., Liu, W. H. & Xu, J. P. (1992) Surface-Modified Hydrophilic Membranes in Membrane Distillation. *Journal of Membrane Science*, 72, 189-196.

Xu, J. B., Lange, S., Bartley, J. P. & Johnson, R. A. (2004) Alginate-coated microporous PTFE membranes for use in the osmotic distillation of oily feeds. *Journal of Membrane Science*, 240, 81-89.

Xu, Y., Zhu, B. K. & Xu, Y. Y. (2006) Pilot test of vacuum membrane distillation for seawater desalination on a ship. *Desalination*, 189, 165-169.

Yang, X., Wang, R. & Fane, A. G. (2011a) Novel designs for improving the performance of hollow fiber membrane distillation modules. *Journal of Membrane Science*, 384, 52-62.

Yang, X., Wang, R., Shi, L., Fane, A. G. & Debowski, M. (2011b) Performance improvement of PVDF hollow fiber-based membrane distillation process. *Journal of Membrane Science*, 369, 437-447.

Yarlagadda, S., Gude, V. G., Camacho, L. M., Pinappu, S. & Deng, S. G. (2011) Potable water recovery from As, U, and F contaminated ground waters by direct contact membrane distillation process. *Journal of Hazardous Materials*, 192, 1388-1394.

Yasuda, T., Okuno, T. & Yasuda, H. (1994) CONTACT-ANGLE OF WATER ON POLYMER SURFACES. *Langmuir*, 10, 2435-2439.

Zaamouche, R., Beicha, A. & Sulaiman, N. M. (2009) Cross-Flow Ultrafiltration Model Based on Concentration Polarization. *Journal of Chemical Engineering of Japan*, 42, 107-110.

Zhang, J. H., Dow, N., Duke, M., Ostarcevic, E., Li, J. D. & Gray, S. (2010) Identification of material and physical features of membrane distillation membranes for high performance desalination. *Journal of Membrane Science*, 349, 295-303.

Zhao, Z. P., Zhu, C. Y., Liu, D. Z. & Liu, W. F. (2011) Concentration of ginseng extracts aqueous solution by vacuum membrane distillation 2. Theory analysis of critical operating conditions and experimental confirmation. *Desalination*, 267, 147-153.

Zhigang, L., Biaohua, C. & Zhongwei, D. 2005. *Special distillation processes*. The Netherlands: Elsevier.

Zolotarev, P. P., Ugrozov, V. V., Volkina, I. B. & Nikulin, V. N. (1994) Treatment of Waste-Water for Removing Heavy-Metals by Membrane Distillation. *Journal of Hazardous Materials*, 37, 77-82.

# 10

# The Separation of Tritium Radionuclide from Environmental Samples by Distillation Technique

Poppy Intan Tjahaja and Putu Sukmabuana
*Nuclear Technology Center for Materials and Radiometry,*
*National Nuclear Energy Agency of Indonesia, Bandung*
*Indonesia*

## 1. Introduction

Tritium is an isotope of hydrogen which emit beta particles in energy range of 5 - 18,6 keV, and has relatively long half life of 12.3 years. In the environment, tiritum is originated both from the nature and human activities. Naturally, tritium is produced in the atmosphere by interaction of cosmic rays with the nuclei of atoms present in the air, and estimated to contribute the annual effective dose equivalent about 0.01 µSv (United Nation, 1988). The other natural sources of tritium are nuclear reactions in earth's crust and environmental materials.

The sources of tritium in the environment relating to human activities, includes atmospheric nuclear testing, continuous tritium release from nuclear reactor under normal operation or accident, and incidental release from consumer products such as paint, watch materials, and other fluoresence goods (Glastone & Jordan, 1980; Puhankainen & Heikkinen, 2007). The food, drink and the use of such products can cause tritium contamination to the human body (Konig, 1990) and contribute the internal radiation doses to the tissues since tritium is distributed to the whole body (Trivedi et al., 2000).

As tritium in the environment possibly contaminates human, the measurement of tritium content in environmental materials as well as in human body samples is needed to be conducted to evaluate the tritium β radiation doses probably received by the body. Unfortunately, β radiation from tritium is not so simple to be determined without sample preparation for separating tritium from the sample bulk. Since tritium is a hydrogen isotope, it is possible to take tritium out from the samples as tritiated water molecule lead to an easy measurement using liquid scintillation counting (LSC) method.

Distillation technique has been choosen as a simplest technique to be used in tritium separation from the environmental samples. Some distillation techniques have been developed to take the tritium out from various environmental material including air, water, soil, plants, and biological materials such as human urine.

This chapter describes the developed distillation technique applied in environmental tritium analysis. The performance of the developed distillation techniques in tritium separation from the environmental samples are discussed based on the experimental test.

## 1.1 Environmental tritium

The majority of tritium in the environment, both naturally and anthropogenic, is present in the form of water molecule (named HTO), while the others are gas (named HT and $CH_3T$) and organic molecules. As an isotope of hydrogen, tritium can easily enter various environmental materials containing hydrogen. In the environmental component tritium is bonded to water molecule, called free-water tritium (FWT) and also to the organic compound as organically bound tritium (OBT).

Tritium in the environmental material is possible to contaminate human body through the food chain. The drinking water, food, inhaled air, and skin absorption are the main sources of tritium for human body contamination. Tritium is uniformly distributed in the body liquid 2 until 4 hours after its inhalation, ingestion or absorption through the skin (Taylor, 2003). At the same time, some percents of tritium atoms reversibly replaced H in OH, NH and SH bonds of organic molecules in the body, while about 1% of the activity is incorporated into stable CH bonds. The physical and chemical properties of tritium are almost the same as that of hydrogen.

The biological half times of tritiated water (HTO) are 10 days (97%) and 40 days (3%), whereas for OBT are 10 days (50%) and 40 days (50%) (ICRP, 1997 & Taylor, 2003). A major part (>90%) of an OBT intake is oxidized and excreted as HTO, this tritium is called the exchangeable tritium fraction. The nonexchangeably bound tritium fraction is normally released only as a result of enzymatic breakdown of the molecules containing this carbon-tritium bound. The HTO is excreted in urine, feces, sweat, and breath in which about a half is excreted via urine.

The tritium toxicity in the human body is contributed by β particle emitted by tritium bonded in biochemical structure of the cells that leads to genetic mutation and oncogenic effect of the long live cells. Such as it was described previously, the existence of tritium in the environment possibly lead to the internally radiation effect to human body. Therefore, the existence of tritium in the environment should be monitored. The environmental tritium monitoring generally is carried out by taking environmental materials samples, such as air, water, soil, and plants for measurement using liquid scintillation technique.

## 1.2 Tritium analysis of environmental samples

### 1.2.1 Sample preparation

The tritium concentration in the water fraction as well as in the organic fraction of the sample is measured in the form of water molecule. Hence the sample preparation is needed in order to isolate the tritium from the samples bulk and convert to liquid form to be possible for LSC measurement. Distillation is one suitable technique to be applied in that purpose.

The sample is heated and the vapor is condensed to obtain the liquid form of free water tritium. In the case of OBT the sample needs to be combusted and the vapor released is condensed at low temperature (Tjahaja et al., 2004). Some modifications in distillation technique have been developed to facilitate the tritium samples preparation such as described in the following sections.

## 1.2.2 Sample measurement for tritium

The tritium in the environmental and human body samples are measured using liquid scintillation technique after it is prepared according to its chemical form. In this technique the tritium in the water molecule form is mixed with a liquid detector called scintillator to detect the emitted β particle, and then measured using Liquid Scintillation Counter device. In this measurement, β particle from tritium interact with the molecule of the liquid detector and produce photon emission which then is detected by a photo multiplier tube in the device. In order to obtain good measurement result it is important to prepare the environmental sample in the form of clear and colorless liquid (Rusconi et al., 2006; Hariharan & Mishra, 2000). Therefore, the sample distillation is an important step in the environmental tritium analysis.

Two mL of liquid from prepared environmental sample is added to 13 mL scintillator (as liquid detector) in a 20 mL scintillation vial. The mixture is shaked vigorously and kept in the dark and cool storage for 24 hours to obtain the stable mixture. The tritium content in the environmental samples is measured using LSC for 1 hour.

# 2. Separation of tritium from the environmental samples

## 2.1 Separation of tritium from air

The measurement of tritium in the atmosphere is begun with air sampling where the solid absorbent such as silica gel and molecular sieve generally used to collect air sample. The silica gel is an oxide silicate having bee nest like structure, porous, and has very wide surface (Patton et al, 1995). Each gram of silica gel has great affinity toward water molecule, so it is very effective for water absorption. Generally, the silica gel contains blue cobalt salt as an indicator of water content that can exhibit color change from blue to pink after absorbing water or water vapor. The disadvantage of silica gel use in atmospheric tritium sampling is the bond of tritiated water to silica gel grain that is take place in the surface of the grain causing an easy evaporation if it is used in high temperature environment (Patton et al, 1995).

The other solid absorbent can be used for atmospheric tritium sampling is molecular sieve that made of aluminosilicate (zeolite). The aluminosilicate is a chelat compound that can absorb the chemical substances both organic and inorganic as well as water molecule to the cavity of its molecular structure (Iida et al., 1995). The cavity of molecular sieve structure have various sizes such as 3Å (MS3A), 4Å (MS4A), 5Å (MS5A) and 10Å (MS13X).

The sampling apparatus for tritiated water vapor collection is composed of a sucking pump, a quartz tube of 0.80 m length and 0.05 m diameter, and a water bubler bottle installed at the end of the circuit to confirm the air flow (Figure 1). The atmospheric air is drawn by the pump with low flow rate of about 15 L/h and flown to the tube packed with the silica gel or molecular sieve pelet. Sampling is conducted for about 15 h, the time when the 50 g packed absorbent has been saturated absorbing the tritiated water vapor.

The tritium water vapor absorbed in the solid absorbent is then recovered by developed distillation technique using an apparatus shown in Figure 2. The distillation apparatus is composed of nitrogen gas source, a tube furnace, and a condensation system. The condensation system consists of two condensation tubes that placed in a serial arrangement

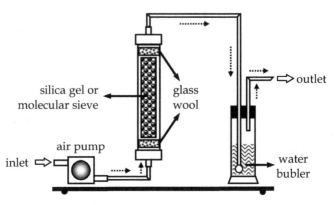

Fig. 1. The apparatus for tritiated water vapor sampling

in an ice box. The quartz tube with the tritiated water vapor saturated absorbent is inserted into the tube furnace, and connected to nitrogen gas source using teflon pipe, while the other side is connected to the condensation system. The solid absorbent is heated at 400 ºC under nitrogen gas flow until stable mass of solid absorbent is gained. For 50 g of absorbent about 24 h heating is needed. The vapor released from the absorbent is flowed by nitrogen gas to the condensation tubes in the ice box and condensed to liquid form that trapped in the bottom of the tubes. The trapped water is then ready to be measured using liquid scintillation counting method, such as described in section 1.2.2.

The whole system, both tritiated water vapor sampling and separation method have been tested by introducing the various activities of tritiated water vapor to the air sampling apparatus for 15 h. The absorbed tritium in the solid absorbent is then recovered using the distillation apparatus in Figure 2. The measurement of the recovered tritiated water indicates the linearity correlation between the tritiated water vapor radioactivities with those in the recovered water.

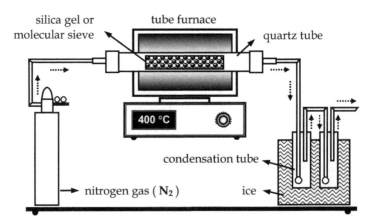

Fig. 2. Distillation apparatus to separate tritiated water vapor from the absorbent bulk.

Figure 3 and 4 show the linear relationship between the tritiated water radioactivity in the air with those recovered from the absorbent with the correlation coefficient of 0.985 and approximately 1 for silica gel and molecular sieve, respectively. The capacity of silica gel and molecular sieve in tritiated water vapor absorption has been tested gravimetrically, and the capacity of 48.6% and 43.6% were obtained for silica gel and molecular sieve, respectively.

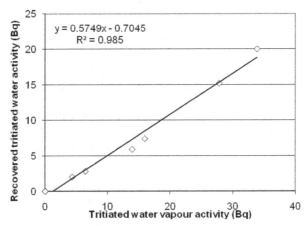

Fig. 3. The linear relationship between the tritiated water activity in the air with those recovered from silica gel

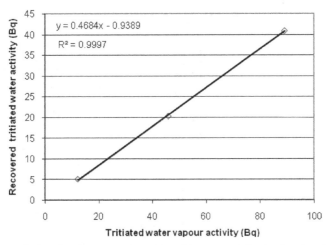

Fig. 4. The linear relationship between the tritiated water activity in the air with those recovered from molecular sieve.

## 2.2 Separation of tritium from water

As stated in previous section, the tritium measurement using LSC method needs clear and colurless liquid, hence the environmental water sample with low organic matter can directly prepared using simple water distillation. The distillation should be conducted in relatively low temperature of about 70 ℃ and the water have to be completely distilled to near dryness. The complete distillation is neccessary as tritium which has relatively higher atomic mass compared to its isotopes (hydrogen and deuterium) will be distributed in the bottom of the distillation flask and evaporated later.   For the environmental water samples with high organic matter content, the distillation should be done with the addition of $KMnO_4$ and activated carbon to decompose and absorb the organic impurities. The organic matter content in the water samples can be determined visually or using UV vis spectrophotometery method. The complete distillation is carried out at 70℃ temperature.

The apparatus for environmental water sample distillation is shown in Figure 5. The apparatus is composed of round bottom flask for water samples, a glass water circulation condensor, a glass vessel with a valve in its end opening for collecting distillate, an adapter between the sample flask and the condensor that equipped with a thermometer. In the top end of the condensor vessel a glass vessel filled with silica gel absorbent is installed to avoid the laboratory atmospheric water vapour contaminates the distillate obtained. The sample heating is conducted using a mantle heater.

In order to obtain the clear and colourless ditillate, the distillation can be done once, but if the water samples contain high organic matter the distillation is conducted more than once until the clear and colurless distillate is gained.

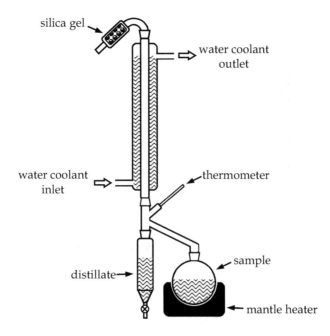

Fig. 5. The distillation apparatus for environmental water samples preparation

## 2.3 Separation of tritium from soil

Tritium in the soil is present as both free water tritium (FWT) and organically bound tritium (OBT). The soil tritium is separated from the soil bulk using modified distillation technique to obtain tritium samples as water to be possible for LSC measurement.

### 2.3.1 Free water tritium

The FWT sample from the soil represents tritium in the water molecule in soil particle surface and interstitial particles (Tjahaja et al., 1995). The FWT from soil sample is collected using distillation technique by heating the soil sample at 150 °C in a quartz tube and condensing the vapor released from the samples at low temperature of about 10 °C. The sample heating is conducted in a tube furnace. The distillation apparatus for soil FWT collection consists of oxygen and nitrogen gas flow system, tube furnace, and condensation system as shown in Figure 6. The condensation system consists of two condensation tubes in serial arrangement inside an ice box. A quartz tube packed with 250 g of soil samples is inserted into the furnace and connected to the gas source system in one side and to the condensation tube in another side. The soil sample is heated at 150 °C under the slow nitrogen gas flow and the vapor from the heated soil is transferred to the condensation tube to be condensed in liquid form. The distillation process of 250 g soil sample take about 30 h, nevertheless it depends on the soil water content. The FWT obtained from this process possibly contains organic material that is visually seen as unclear and yellow color liquid. In this condition, FWT distillation is necessary to remove the impurities of the FWT so clear and colorless water can be obtained.

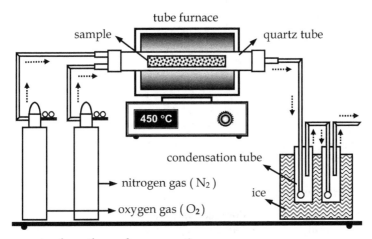

Fig. 6. The apparatus for soil samples preparation

### 2.3.2 Organically bound tritium

The dried soil sample as the residue of the FWT distillation process is combusted in the same quartz tube by increasing the temperature of the furnace to 450 °C and applying the oxygen gas to the tube to obtain OBT as liquid form. The combustion results ($H_2O$ and $CO_2$) is flowed to the condensation tubes where the $H_2O$ gas will be condensed as liquid while the

$CO_2$ will be flushed out since its condensation point is lower than the temperature in the condensation system. The combustion is held for 30 hours until all soil samples become ash. The condensed water (OBT) contains the organic matter, especially if the combustion process is not perfect. Therefore, a distillation with $KMnO_4$ and activated charcoal addition is necessary to obtain the clear and colorless water for the LSC measurement.

The performance of the distillation apparatus for soil trtium separation has been tested by preparing 250 g of soil samples with varied tritium concentration, i.e. 180, 270, 555, and 740 Bq. A good performance of this apparatus system is indicated by the recovery efficiency of 92 % for total tritium (FWT and OBT) in the soil. The measured tritium activity in soil samples both for FWT and OBT linearly increased with the amount of given tritium activity in the soil. The correlation of tritium activities in the soil with those measured after sample preparation shows linear correlation with the coefficient of 0.99, such as shown in Figure 7.

The distillation of FWT and OBT after the heating and combustion process is necessary to be conducted to obtain the suitable sample for LSC measurement. The distillation of FWT and OBT water after the sample heating or combustion can increase the tritium analysis efficiency of about 10%.

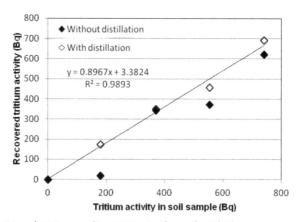

Fig. 7. The correlation of tritium radioactivity in the soil with those measured (FWT and OBT) after sample preparation.

## 2.4 Separation of tritium from plant

As the same as with the soil, the tritium radionuclide in plant occur in the form of free water tritium and organically bound tritium (Konig, 1990), hence the sample preparation for tritium measurement using LSC method is the same as with those for soil sample.

The performance of the apparatus for tritium separation from plant has been tested (Tjahaja et al., 2005) using varied tritium concentration in the grass of 225, 450, 670, and 960 Bq in 230 g of fresh grass. The tritium contaminated fresh grass was packed in a quartz tube and heated at 150 ºC for 30 h using tube furnace that was arranged as in Figure 6. The vapor released was condensed in condensation tube cooled by ice cube. The recovered vapor as FWT was completely distilled by $KMnO_4$ and activated charcoal addition until clear and colorless distillate was obtained and could be measured using LSC technique. The dried grass sample in the quartz tube was then combusted at 450 °C under oxygen gas flow using tube furnace in the same apparatus. The gas released ($H_2O$ and $CO_2$) from the combusted grass samples was flowed to the combustion tube and condensed at ice temperature. The condensed water representing the OBT of the grass as the same as the FWT was completely distilled and measured using LSC method.

The increase of tritium activity in the grass sample is followed with the increase of the recovered FWT and OBT. The correlation of tritium radioactivity in the grass with those measured after sample preparation shows linear correlation with the coefficient of 0.98, such as shown in Figure 8. A good performance of this apparatus system is indicated with the recovery efficiency of 80% for total tritium in the grass (FWT and OBT). The performance of the separation method for plant samples relatively low compared with those for soil samples relating to the complexity of organic compound in the grass leading to the incomplete combustion process in OBT recovery.

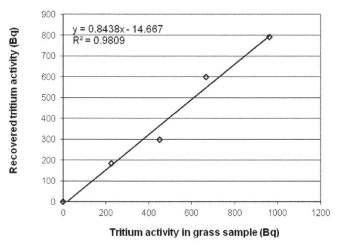

Fig. 8. The correlation of tritium radioactivity in the grass with those measured (FWT and OBT) after sample preparation.

## 2.5 Separation of tritium from human urine

The presence of tritium in the air, drinking water, foods, and other consumer products is the source of human contamination with tritium (Puhakainen et al., 2007 and Trivedi et al, 2000). The detection of tritium contamination in the human body can be carried out using urine samples measurement. Nevertheless, tritium measurement in urine has a limitation relating to the body metabolite content in the urine that signicantly affect the LSC measurement (Trivedi et al., 2000; Rusconi et al, 2006). The measurement of such urine possibly to increase counting rate of tritium in the LSC measurement because of chemical luminesence from samples impurities and from β radiation emmitted by [14]C bounded in organic matter of body metabolite (Rusconi et al., 2006; Groning, 2004). Such a counting result cause over estimation of tritium concentration in the urine especially in low level tritium measurement (Groning, 2004; Momoshima et al., 1986). Therefore, urine sample preparation is important to be applied before LSC measurement to reduce body metabolite content affecting the measurement result.

Several urine sample preparation methods have been performance tested for low level tritium (Tjahaja, 2009). Six methods, named A, B, C, D, E, and F, were applied to two urine samples already contaminated with tritium in the concentration of 0.5 Bq/mL and 1 Bq/mL. The method efficiency was analyzed based on the tritium activity recovered in the urine sample after the preparation.

In A - E method, sample preparations were applied such as described in the scheme in the Figure 9-13 (Momoshima et al, 1986; Bhatt et al., 2002), while in the F method an amount of urine sample was directly measured using LSC.

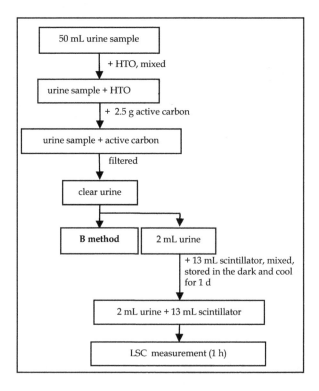

Fig. 9. The scheme of urine sample preparation by A method

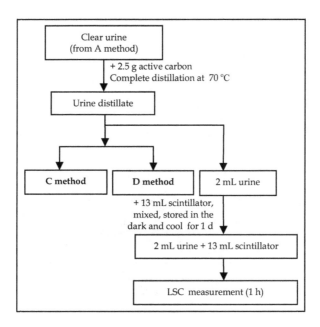

Fig. 10. The scheme of urine sample preparation by B method

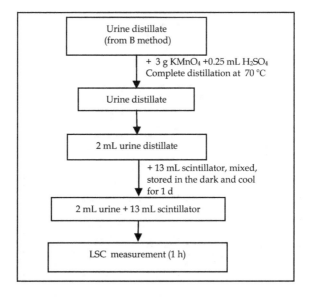

Fig. 11. The scheme of urine sample preparation by C method

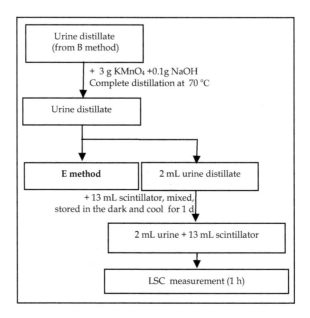

Fig. 12. The scheme of urine sample preparation by D method

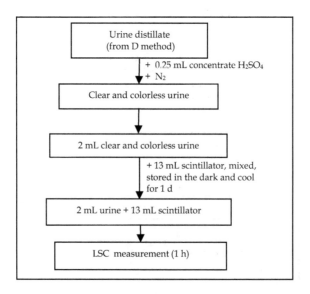

Fig. 13. The scheme of urine sample preparation by E method

The performance test result can be seen in Figure 14 and 15 for tritium concentration in the urine of 0.5 Bq/mL and 1 Bq/mL, respectively. They indicate that the E method has the highest eficiency with the average of 35% and 58% for tritium concentration of 0.5 Bq/mL and 1 Bq/ mL. In the F method that is direct LSC measurement, the tritium activity is not detected in the recovered urine as the same as with those for A and B method. The separation procedure shown in Figure 10-13 is carried out using distillation apparatus shown in Figure 5 in section 2.2.

For tritium measurement in the urine the E method is used. The 24 h urine sample is collected in a plastic bottle, and about 25 mL of the sample is added with 1.25 mg of active charcoal, shaked vigorously for about 30 minutes, then filtered using filter paper to obtain clear and colorless filtrate that free of particulate matter. The filtrate is then again added with active charccoal and distilled completely at 70 ºC to eliminate the yellow color of the first filtrate that can alter measurement using LSC device (Puhakainen et al., 2007; Momoshima et al, 1986). The distilate obtained is then added with 1.5 g $KMnO_4$ and 0.05 g NaOH to destruct the urine organic compound, and again distilled completely at 70 ºC. The distillate is then added with 0.125 mL concentrate sulfuric acid to destruct carbonate compound resulted from previous reaction. The $CO_2$ released from the reaction is then flushed out by nitogen gas flow for about 10 minutes.

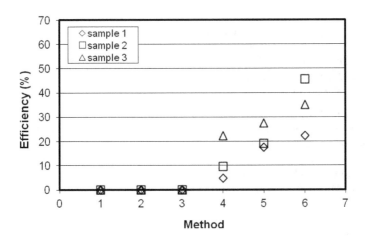

Fig. 14. The method efficiency of 0.5 Bq/mL urine sample preparation. Number 1 represents the F method, numbers 2-6 represent the A – E method.

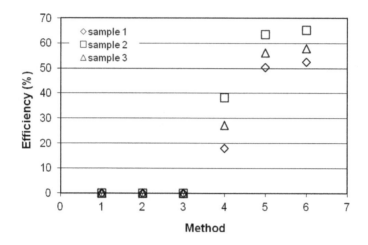

Fig. 15. The method efficiency of 1 Bq/mL urine sample preparation. Number 1 represents the F method, numbers 2-6 represent the A – E method.

## 3. Conclusion

As tritium naturally is present in environmental material, human is normally exposed internally with tritium via foods, drinking water, inhaled air, and also skin absorption. The tritium in the human body contributes radiation doses that affecting human health, hence the monitoring of tritium concentration in the environmental samples becomes interesting to avoid tritium intake by human.

This chapter has described the environmental sample preparation needed in tritium monitoring. Several developed distillation techniques in the preparation of environmental sample including air, water, soil, grass and urine have indicated a good performance result. Nevertheless, the sample preparation method for environmental tritium measurement is still interested to be developed to obtain the better performance. It is hoped that the chapter is technically worth in solving environmental and health problems arise in this industrial era.

## 4. References

Bhatt, A.R., Greenhalgh, R., Ortel, C.P. 2002. Determination of tritium in urine samples and validity of results, Idaho National Engineering and Environmental Laboratory.

Glastone, S., Jordan, W.H. 1980. *Nuclear Power Reactor and Its Environmental Effects*, American Nuclear Society, 0-89448-024-3, La Grange Park, Illinois.

Groning, K.R.M. 2004. Quantifying uncertainties of Carbon-14 assay in water samples using benzene synthesis and liquid scintillation spectrometry, www.iaea.or.at/ programmes/ rial/ pci/ isotope hydrology/ docs/intercomparison/ Vienna/ 30-04-2004.

Iida, T., Yokoyama, S., Fukuda, H., and Ikebe, Y. 1995. A simple passive method of collecting water vapour for environmental tritium monitoring, *Radiation Protection Dosimetry*, 58, 1, pp 23-27.

Harihan, C., Mishra, A.K. 2000. Quenching of liquid scintillator fluorescence by chloroalkanes and chloroalkenes, *Radiat. measurement*, 32, pp. 113-121.

Harrison, J. D., Khursheed, A., Lambert, B. E. 2002. Uncertainties in dose coefficients for intakes of tritiated water and organically bounds forms of tritium by members of th public. *Rad. Prot. Dos.*, 98, 3, pp. 299 - 311

Konig, L.A. 1990. Tritium in the food chain, Radiation Protection Dosimetry, 30 2, pp. 77-86.

Momoshima, N., Nagasato, Y., Takashima, Y. 1986. A sensitive method for the determination of tritium in urine. *Environmental Tritium Measurement*, Kyushu University, Fukuoka, pp. 78 – 84.

Patton, G.W., Cooper, A.T., and Tinker, M.R. 1995. *Evaluation of an ambient air sampling system for tritium using silica gel adsorbent columns*, Pacifics Northwest Laboratory Richland, Washington.

Puhakainen, M., Heikkinen, T. 2007. Tritium in the urine in Finnish people. *Radiat. Prot. Dos.* Advance Access, pp. 1-4.

Rusconi, R., Forete, M., Caresana, M., Bellinzona, S., Cazzaniga, M.T., Sgorbati, G. 2006. The evaluation of uncertainty in low-level LSC measurement of water samples. *Appl. Radiat. and Isot.*, 64, pp. 1124-1129.

Taylor, D.M., A biokinetic model for predicting the retention of $^3$H in the human body after intakes od tritiated water, *Rad. Prot. Dos.*, 105, 1-4, pp. 225-228.

Tjahaja, P.I. 1995. The distribution of organically bound tritium concentration in soil (in Indonesian), *Proc. of National Seminar on Science and Nuclear Technology*, ISSN 1858-3601, Research Center for Nuclear Technique of National Nuclear Energy Agency of Indonesia, Bandung, 1995

Tjahaja, P.I., Sukmabuana, P., Zulfakhri, Widanda, Zulaika, S. 2004. A $^3$H sampling method for soil sample using heating and combustion method (in Indonesian), *Proc. Seminar on Basic Research and Nuclear Science Technology* , ISSN 0216-3128, Yogyakarta Nuclear Research Center, Yogyakarta July, 2004.

Tjahaja, P.I. , Sukmabuana, P. 2005. The Method of Tritium Sampling from Grass (in Indonesian), *Proc. of National Seminar on Science and Nuclear Technology*, ISSN 1858-3601, Nuclear Technology Center for Materials and Radiometry of National Nuclear Energy Agency of Indonesia, Bandung, June, 2005.

Tjahaja, P.I, Sukmabuana, P. 2009. Method of sample preparation for low level tritium concentration in urine (in Indonesian), *Proc. of National Seminar on Science and Nuclear Technology*, ISSN 1858-3601, Research Center for Nuclear Technique of National Nuclear Energy Agency of Indonesia, Bandung, June, 2009.

Trivedi, A., Galeriu, D., Lamoth, E.S. 2000. Dose contribution from metabolized organically bound tritium after chronic tritiated water intakes in human. *Health Phys.*, 78, pp. 2-7.

United Nations. 1988. *Sources, effects and risks of ionizing radiation: United Nations Scientific Committee on the Effects of Atomic Radiation. Report to the General Assembly, with annexes.* United Nations Publication, Sales No. E.88.IX.7. New York .

US Environmental Protection Agency, Radioactivity in drinking water, EPA Report 570/9-81-002 (Office of Drinking Water (WH-550), Washington DC), 1981.

# 11

# Separation of Odor Constituents by Microscale Fractional Bulb-To-Bulb Distillation

Toshio Hasegawa
*Saitama University*
*Japan*

## 1. Introduction

An important task in natural products chemistry is finding new compounds with novel properties and structures. Moreover, in fragrance chemistry, the odors of constituents are evaluated and key compounds are identified that contribute to the scent profiles of fragrance materials. Figure 1 shows the general investigative process for achieving these goals of fragrance chemistry.

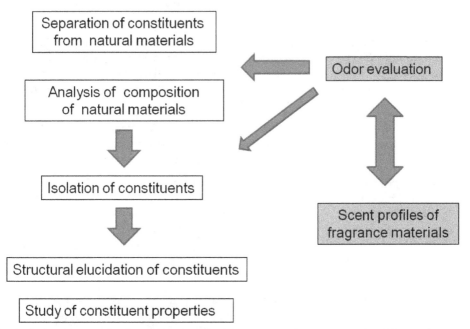

Fig. 1. General investigative process for identifying new odor compounds with novel properties and structures in fragrance organic chemistry.

Extraction of odor plants gives oils containing many compounds. A popular and useful method for analyzing the composition is gas chromatography/mass spectrometry (GC-MS), which provides important information about the scent profiles of fragrant plants. The analytical advances in the latest GC-MS technology are remarkable. In testing for the presence of a particular compound, detection of trace amounts of the substance has become possible. Recently, the olfactometry (O) functionality was incorporated into the GC-MS method. That analytical method, GC-MS-O, enables the odors of each constituent in a mixture to be precisely evaluated. These advances in analytical technology are associated with the discovery of natural products with important properties, for instance, a unique flavor. However, because the level of analysis is extremely detailed, the number of components in the characteristic scent of a plant can become huge. For this reason, indentifying the important scent components out of the many components is nontrivial. As a solution, a method for measuring the relative strength of each constituent's scent has recently been developed, and key findings have been reported. However, this method has the following problems.

1.   As the number of components increases, the effort required for analysis becomes greater.
2.   In a mixture of multiple components with similarly strong scents, the essential components are difficult to identify.

Such difficulties arise in elucidating the key fragrance constituents of essential oils obtained from fragrance materials, such as sandalwood and vetiver, which are currently used as base notes.

We found that the fragrances of these scent materials can be expressed by combining groups of scent components. These groups consist of several organic compounds that have similar structures and thus similar odor. The interaction between these constituents is vital to the odor. From an oil, it is important to separate the *minimum odorant groups* necessary to retain the characteristic fragrance of the scent material.

In this chapter, I will explain how to divide the odor oils from fragrant plants into groups with similar properties (structure, odor, etc.). I will also present the results of experiments conducted in my laboratory, which used popular and important scent materials.

Sandalwood (*Santalum album, L.*), vetiver (*Vetiveria zizanioides*), patchouli (*Pogostemon patchouli*), and frankincense (*Boswellia papyrifera*) are typical materials in traditional Japanese incense, and the essential oils obtained from these materials possess unique and valuable features such as providing the base notes for perfume. Although many studies on the constituents of these materials have been reported, the key compounds of their scent profiles remain unknown. Determining these scent profiles is a prerequisite to elucidating the key components of the characteristic odors of these materials.

In attempting such a determination, however, the following main issues must be overcome. Firstly, the fragrances of the scent materials are not formed by a mere superposition of individual scents. Secondly, the scent components of these materials have weak odors. Almost all researchers have recognized these issues, but previously developed methods have not been suitable for clarifying the odor characteristics. In our work, we reconsidered and revised previous methods and were able to successfully evaluate the scent profiles of these materials

## 2. Microscale fractional bulb-to-bulb distillation

### 2.1 General procedure of microscale fractional bulb-to-bulb distillation

Commercially available, a bulb-to-bulb distillation apparatus (see Fig. 2) is usually used to purify a small amount of organic material (10-500 mg) by distillation under vacuum. Glass bulbs of about 3-3.5 cm in diameter are connected in series. The joint between bulbs, for example, between bulb A and bulb B, is made of ground glass. The general procedure for performing a distillation on this apparatus is given below.

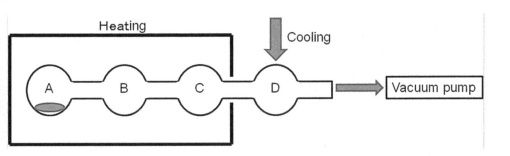

Fig. 2. Sketch of bulb-to-bulb distillation apparatus.

1.  Oily sample is placed in bulb A.
2.  The apparatus is evacuated.
3.  Bulbs A, B, and C are placed into an oven and heated slowly. While the other bulbs are being heated, bulb D is placed in a cooling bath (e.g., ice or water).
4.  The first fraction is collected from bulb D.
5.  Next, bulb C is removed from the oven and becomes a collection bulb. After cooling to room temperature, bulb C, along with bulb D, is placed in a cooling bath. Then, bulbs A and B are heated.
6.  Finally, bulb A is heated and bulbs B, C, and D are cooled.

Cooling all the bulbs outside the oven is a critical part of this procedure. Failure to do so will result in low recovery and poor separation.

Using this method, we can perform separations on small amounts of material (down to 10 mg) to obtain fractions with different boiling points.

### 2.2 Separation of odor constituents of representative incense by bulb-to-bulb distillation technique

We used fractional bulb-to-bulb distillation to evaluate the odors of oils obtained from incense materials (sandalwood, frankincense, etc.). We extracted the essential oils of the incense first with hexane and then with methanol. We compared the odors of the hexane-extracted essential oils to the original base materials. We found that the odors of the extracted oils were similar to the odors of the base materials. We performed the following fractional distillations to obtain the minimum odorant groups constituting the fragrances of the scented materials (Fig. 3). We found that the fragrances of these scent materials could be expressed by combining these groups. Each group consists of several organic compounds with similar structures and thus similar odors.

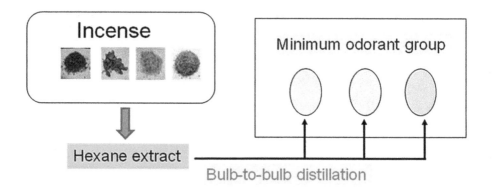

Fig. 3. General procedure for separating constituents from extract by microscale fractional bulb-to-bulb distillation.

### 2.2.1 Sandalwood

High-quality sandalwood—from which the essential oil is collected—is a valuable and expensive material because it can only be obtained from mature sandalwood trees. Sandalwood is a medium-sized evergreen parasitic tree and is found in India, Malaysia, and Australia. The highest quality of Sandalwood trees for incense and perfume are grown in India (especially East India). Many investigations on the composition of sandalwood essential oils have been carried out, and more than 300 constituents have been identified. The main constituents are α-santalol and β-santalol. These compounds have distinctive woody odors. Many studies have been done on sandalwood, and the structure–odor relationships of β-santalol and its related compounds have been investigated in detail.

Recently, we reported that the odor of sandalwood chips is formed by a combination of santalols and their aldehyde and formate derivatives (Hasegawa et al., 2011). Here, we will examine the interesting relationship between the structure and odor of α-santalol and its derivatives having modified side chains. Recently, we identified new odor constituents of sandalwood by the method shown in Fig. 4

We applied the distillation method introduced in this chapter to the evaluation of sandalwood odor.

We collected the hexane extract and the steam-distilled oil from sandalwood chips and compared their odors with the odor of sandalwood chips. The odor of the extracted oil was found to be similar to the odor of the base material.

The $^1$H NMR spectroscopy revealed that the main constituents were α-santalol and β-santalol with an extremely small amount of compounds with a formyl group (Fig. 5).

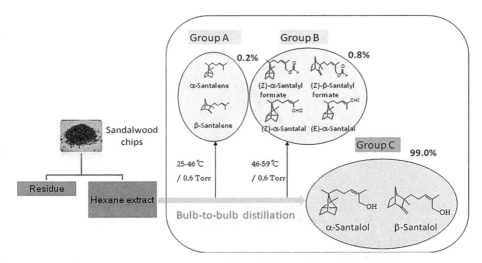

Fig. 4. Separation of odor constituents from hexane extract of sandalwood chips.

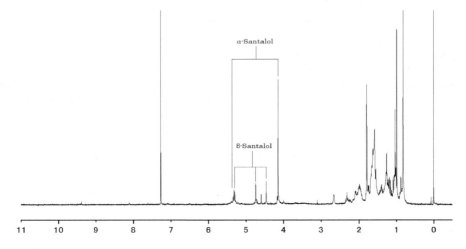

Fig. 5. $^1$H NMR Spectrum of the hexane extract of sandalwood chips (200 MHz, CDCl$_3$).

Generally, compounds with a formyl group (aldehyde and formate) are important odor constituents. These compounds, especially aldehydes, are common decomposition products of the corresponding carboxylic acids. Aldehydes, because they are prone to decomposition, are difficult to collect from an extract by chromatography or distillation. The bulb-to-bulb distillation method, however, is suitable for handling these compounds, because the heating time is shorter than that in a typical distillation.

We performed bulb-to-bulb distillation of the hexane extract. Two fractions were obtained, and the residue was composed of α-santalol and β-santalol. The first fraction was a mixture of santalol hydrocarbon derivatives and the second fraction was santalyl aldehydes and formates, as determined by NMR spectroscopy (Fig. 6).

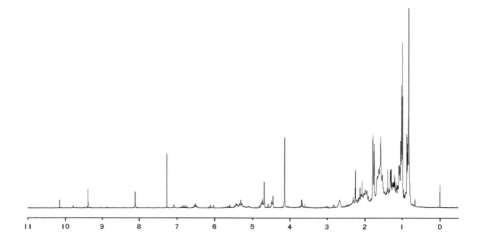

Fig. 6. $^1$H NMR spectrum of group B obtained from hexane extract by bulb-to-bulb distillation (200 MHz, CDCl$_3$).

The obtained fractions were analyzed, and santalol derivatives with a formyl group (Group B) were found to play an important role in the odor of sandalwood chips. The diagram (Fig. 4) indicates the scent profile of sandalwood obtained by this method. β-Santaol has been reported to be the principal constituent of sandalwood odor. In contrast, α-santalol has been reported to be only a supporting component of sandalwood odor because α-santalol has a weaker odor than does β-santalol. However, we found that both α-santalol and β-santalol derivatives with a formyl group were important constituents of sandalwood odor.

This result demonstrates that bulb-to-bulb distillation is useful for collecting very small fractions from a mixture.

### 2.2.2 Patchouli

The unique woody aroma of patchouli is one of the four major woody notes derived from essential oils, and serves as an indispensable scent in modern fragrance. Although many studies have been performed, the key components that constitute this odor have not been successfully identified (Nabeta et al., 1993; Singh et al., 2002). Suitably appraising patchouli fragrance is crucial in order to produce potentially useful synthetic compounds.

Despite being a topic of investigation for many years, the complete odor profile of patchouli remains elusive for three reasons: first, the scents of individual compounds are weak; second, the compounds are structurally diverse and complex; and third, the aroma changes over time. To overcome these obstacles, we performed bulb-to-bulb distillation of the patchouli hexane extract, which had a similar odor as the base material.

The composition of the hexane extract of patchouli leaves was analyzed by $^1$H and $^{13}$C NMR spectroscopy. One constituent of the extract was found to be patchoulol, but its content was low and the other constituents were unidentified. We presumed that the extract contained a large amount of odorless constituents and thus attempted to collect only odor constituents from the hexane extract. Bulb-to-bulb distillation of the hexane extract produced two groups (group A and B) with characteristic odors; the residue did not have a significant odor.

Group A consisted of several sesquiterpenes and anethole (Fig. 7). Group B consisted of almost entirely patchoulol. These two groups were found to contain the key compounds that contribute to patchouli odor.

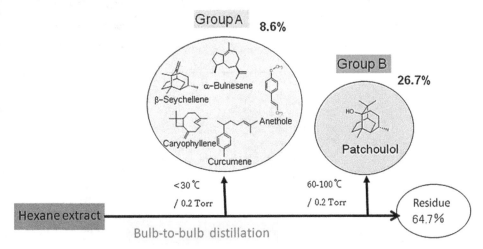

Fig. 7. Separation of scent components from hexane extract from patchouli leaves.

### 2.2.3 Frankincense

The resin of frankincense is obtained from many species of trees in the genus Boswellia. Frankincense has been used as a valuable fragrance source since ancient times, and has been reported to possess a wide range of bioactivity. Many compounds have been identified in frankincense (Hamm et al., 2005; Mertens et al., 2009). To our knowledge, however, the effects of particular odor components have not been clarified in detail. There are two representative species of frankincense. The main components of frankincense are markedly different between these two species. One has many monoterpenes (e.g., α-pinene) as key compounds that contribute to frankincense odor. The other contains diterpenes as the main constituents, along with octyl acetate and octanol. This latter species is used in traditional Japanese incense.

The hexane extract of frankincense is a highly viscous oil, suggesting that it contains a large amount of compounds that contribute relatively little to the characteristic odor of frankincense. The NMR spectrum of the extract (Fig. 8) supports this assessment. We did a bulb-to-bulb distillation to evaluate the key compounds of frankincense odor. First, fraction 1 was obtained from distillation below 124 °C at 0.09 Torr. The constituents were octanol and octyl acetate (Fig. 9). Then, the temperature and pressure were maintained at 124 °C (0.09 Torr), and highly similar constituents were collected in the three different bulbs according to the slightly different boiling point of each constituent (Fig. 9).

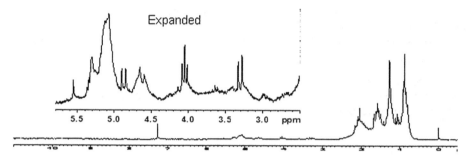

Fig. 8. ¹H NMR spectrum of hexane extract from frankincense resin (200 MHz, CDCl₃)

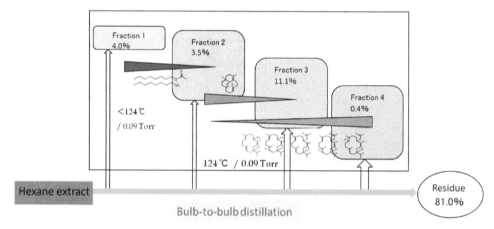

Fig. 9. Separation of scent components from the hexane extract of frankincense resin.

We evaluated the odor of this species by focusing on the difference in odor between the hexane extract and the steam-distilled oil and separation of odor components by bulb-to-bulb distillation.

NMR revealed that each fraction contained different components (Fig. 10). The odor of fraction 4 was similar to that of the hexane extract. The main components of fraction 4 were found to be diterpenes — in particular, incensole derivatives. This result shows that these incensoles make a key contribution to the odor of frankincense (Hasegawa et al., 2011).

### 2.2.4 Vetiver

Vetiver essential oil is a spice that provides base notes, similarly to materials such as sandalwood and patchouli. Vetiver is used as an essential material for providing fragrance, for instance in perfume, that emerges comparatively late and lasts a long time. Vetiver is said to provide the heart of the perfume. Although vetiver is important in terms of fragrance, there is still much research to do into its odor properties.

Conventionally GC-MS has been used to analyze scent components (Anonis, 2004; Weyerstahl et al., 2000), but we thought that this method would be insufficient for

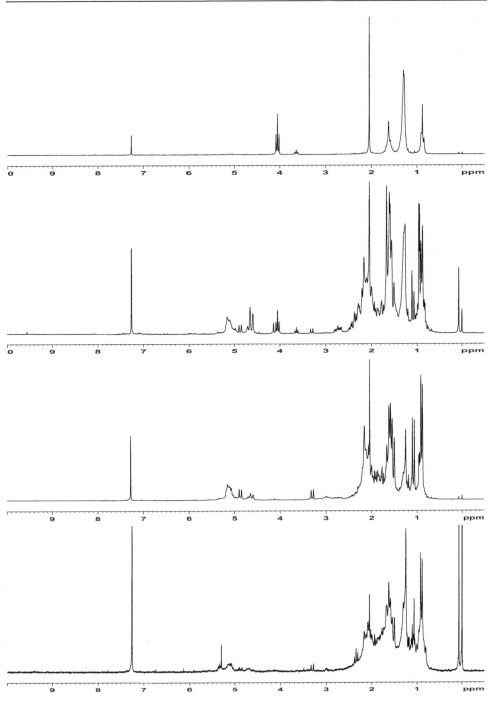

Fig. 10. 1H NMR spectra of fractions 1, 2, 3, and 4 from top to bottom (200 MHz, CDCl₃)

identifying the complex combinations of odorants that form the overall vetiver scent. Then, we used bulb-to-bulb diffraction to divide the hexane extract into groups with different characteristic odors; each group contained structurally similar compounds (Fig. 11).

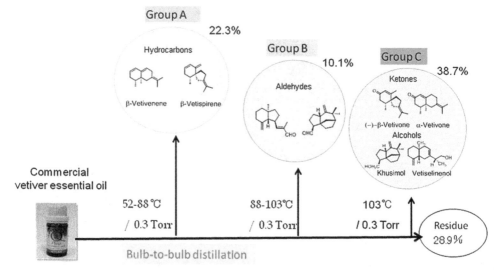

Fig. 11. Separation of scent components from commercial vetiver essential oil.

The $^1$H NMR spectrum of a commercial vetiver essential oil (Quinessence Aromatherapy Ltd.) indicates that this oil was constituted by many compounds (Fig. 12). The separation of groups with different odors and different structures could not be accomplished by the aforementioned distillation procedure. In this case, the lower boiling group, group A, was first separated from the oil; then the distillation was stopped and bulbs B and C were replaced with fresh ones. We succeeded in the separation of the oil (Fig. 13). If this change was not done, the separation was poor.

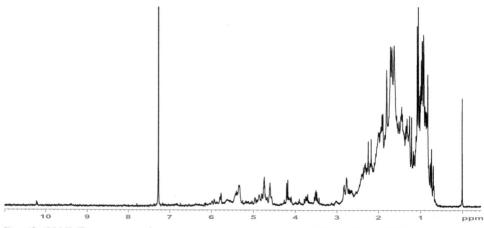

Fig. 12. $^1$H NMR spectrum of commercial vetiver essential oil (200 MHz, CDCl₃).

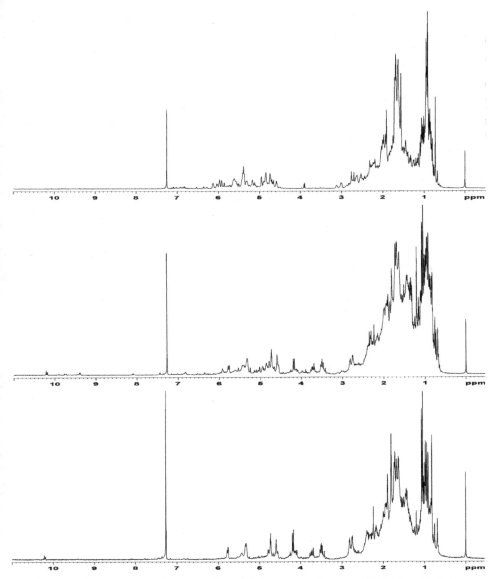

Fig. 13. $^1$H NMR spectrum of Group A, B, and C from top to bottom (200 MHz, CDCl$_3$)

## 3. Conclusion

In the field of fragrance chemistry, the determination of scent profiles is a fundamental and vital endeavor. However, almost all essential oils that are used as fragrances are composed of many types of odorants, and the fragrances of the scent materials are not formed by mere superposition of individual odorants. This problem has hindered the precise evaluation of the base constituents of fragrance. We have demonstrated that the complex scent profiles of materials can be clarified by microscale fractional bulb-to-bulb distillation. To determine important components in natural materials with complex compositions, the method presented in this chapter is useful not only for perfumery chemicals but for many substances in natural products chemistry.

## 4. Acknowledgments

I would like to thank Hideo Yamada at Yamada-matsu Co., Ltd., for providing plant specimens and contributing to sensory evaluations and fruitful discussions. I also would like to thank Shohei Tamada for work in early experiments on vetiver.

## 5. References

Anonis, D. P. (2004) *Perfumer & Flavorist*, Vol. *29*, pp. 30-36, ISSN 0272-2666.

Hamm, S; Bleton, J; Connan, J.; Tchapla A. (2005) *Phytochemistry*, pp. 1499-1514, ISSN 0031-9422.

Hasegawa, T.; Kikuchi, A.; Yamada, H. (2011) *The Journal of Essential Oil Research, 2011 ISEO Abstracts*, p. 8, ISSN 1041-2905.

Hasegawa, T.; Toriyama, T.; Ohshima, N.; Tajima, Y.; Mimura, I.; Hirota, K.; Nagasaki, Y.; Yamada, H. (2011) *Flavour and Fragrance Journal*, Vol.26, pp. 98-100, ISSN 0882-5734.

Mertens, M.; Buettner, A.; Kirchhoff, E. (2009) *Flavour and Fragrance Journal* Vol. 24, pp. 279-300, ISSN 0882-5734.

Nabeta, K.; Kawauchi, J.; Sakurai, M. (1993) *Development in Food Science*, Vol. *32*, pp. 577-589, ISSN 0167-4501.

Singh, M.; Sharma, S.; Ramesh, S. (2002) *Industrial Crops and Products*, Vol. *16*, pp. 101-107, ISSN 0926-6690.

Weyerstahl, P.; Marschall, H.; Splittgerber, U.; Wolf, D.; Surburg, H. (2000) *Flavour and Fragrance Journal* Vol. 15, pp. 395-412, ISSN 0882-5734.

# 12

# Mass Transport Improvement by PEF – Applications in the Area of Extraction and Distillation

Claudia Siemer, Stefan Toepfl and Volker Heinz
*German Institute of Food Technologies*
*Germany*

## 1. Introduction

Mass transport processes in the food industry are mostly based on the diffusion of soluble products out of food tissue. The main barrier for the diffusion is the biological membrane separating the inner cellular material from the outside. A rupture of the membrane results in an enhanced diffusion rate resulting in a higher yield of the product located in the cell. Most methods used for disintegration of cellular material are mechanical, chemical or thermal based treatments. A new promising technique for cell rupture is the application of pulsed electric fields (PEF). The product is treated with pulses of microseconds at a high electric field strength. The electric field affects the cell membrane of the biological tissue in order to increase the permeability resulting in pore formation. Pore formation facilitates the diffusion process. Moderate PEF settings are used to achieve a disintegration of the cellular material. Some researchers define a moderate PEF treatment by applying a field strength of 0,5 to 1,0 kV/cm and treatment times in a range of 100 and 10.000 µs. The same effects were obtained by other researchers using electric field strengths of 1 to 10 kV/cm and shorter treatment times in the range of 5 to 100 µs (Schilling et al., 2007; Corrales et al., 2008; López et al., 2009).

It has been reported that PEF induces an increase of the mass transfer process resulting in a higher extraction of different intracellular materials, such as sucrose from sugar beet, betalains from red beetroot or polyphenols from grapes during the wine production. An increase in the extraction yield of juices from different fruits and vegetables has also been noted. The drying process can also be improved by PEF application. The reduction in the drying time yields a better end product quality. For example an accelerated osmotic dehydration and drying was reported for different fruits and vegetables, such as potatoes and pepper. An improvement of the distillation processes by PEF, for example distillation of rose oil was also reported.

The following chapter describes in detail the effect of PEF on the mass transfer and the related application fields.

## 2. Principle of action of cell disintegration using PEF technology

The application of PEF using short pulses of a high voltage affects the membrane of a cell resulting in a permeabilisation of the biological membranes.

The membrane acts as a semi permeable barrier for the intra- and extracellular transport of ions and macromolecules. The membrane can be considered as a capacitor filled with dielectric material of low electrical conductance and a dielectric constant in the range of 2 (Castro et al., 1993; Barbosa-Cánovas et al., 1999). The accumulation of the free charges in- and outside the cell leads to the formation of a concentration gradient, which is called trans membrane potential. The resulting voltage varied in a range from 20 to 50 mV (Wilhelm et al., 1993).

The exposure of the cell to an electric field with a specific intensity causes a movement of the charges along the electric field lines, which results in an induction of an additional potential. The trans membrane potential increased due to the electric field up to a value of 1 V (Zimmermann et al., 1976; Weaver 2000). The action of polarizing the cell is illustrated in Fig. 1.

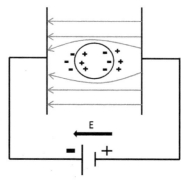

Fig. 1. Induction of cell polarization caused by PEF application (adapted from (Dimitrov 1995))

The polarization of the membrane is related to an increase of conductivity of the membrane in the range of more than 1 S/cm² (Dimitrov 1984; Tsong 1991; Wilhelm et al., 1993). Due to this increase the resistance of the membrane decreases and results in dramatic change in permeability for potassium and sodium (Pliquett et al., 2007). The higher permeability leads to a dielectric breakdown of the cell, which occurs at different electric potentials depending on the electrical pulse duration (Dimitrov 1984).

Due to the polarization of the membrane, which is related to the movement of free charges, forms electro compressive forces causing the local dielectric rupture of the membrane. Dielectric rupture is the formation of pores and a drastic increase of permeability, which can be termed as dielectric breakdown (Zimmermann et al., 1976). This electro mechanical process is still the most accepted explanation for describing the effect of applying an external electric field on biological cells. Other theories describe the membrane as a viscoelastic model with fluctuating surfaces. The external applied electric field induces reorientation and deterioration of the membrane molecules resulting in pore formation and also the expansion of the pores leading to mechanical breakdown of the membrane (Dimitrov 1984).

The electric breakdown occurs if the applied electrical potential exceeds a critical value, which is termed the critical potential. Temperature, cell size and shape as well as medium and process conditions are influencing factors for the critical potential. The pulse duration as a process condition leads to a lower critical potential if long pulse durations are applied (Dimitrov 1984).

The induced pore formation by the external electric field leads to an increase of the permeability of the membrane. Due to the intensity of the pulses reversible and irreversible pores can be formed. Reversible pores are formed, applying less energy than the critical potential. Using that intensity of the electric pulses stress reactions in plant cells can be induced and the production of secondary metabolites can be stimulated (Schilling et al., 2007). If the intensity of the electric pulses exceeds the critical potential irreversible pores are formed, which causes lethal cell damages. This potential can be applied to inactivate pathogen and non-pathogenic microorganisms or to facilitate the mass transfer and to improve the diffusion of intra- and extracellular liquids. The result of a facilitated mass transfer and improved diffusion is a facilitated distillation process due to the disintegration of cellular material by PEF (Dobreva et al., 2010).

Consequently applying PEF using intensities higher than the critical value two different effects can be observed. On the one hand inactivation of microorganisms (Heinz et al., 2001; Toepfl et al., 2007) and on the other hand the cell disintegration (Angersbach&Heinz 1997; Angersbach et al., 2000). The difference between these two effects is the applied intensity of the electric field and the specific energy, which represents two of the main process parameters of a PEF treatment. The electric field strength is defined as the electric potential difference (U) for two given electrodes in space divided by the distance (d) between them, separated by a non-conductive material (Zhang et al., 1995). The specific energy describes the intensity of the treatment with high intensity pulses and is expressed in kJ/kg. Because of the small size of the microorganisms and the composition of their membranes, a high energy input is required for a successful inactivation (Glaser et al., 1988).

For a microbiological inactivation of different vegetative microorganisms, which has been widely demonstrated by various researchers (Álvarez et al., 2000; Toepfl et al., 2007) an electric field strength of 15 to 20 kV/cm and a specific energy of 40 to 1000 kJ/kg is required. In contrast for a disintegration of cellular material lower values for electric field strength (range of 0,7 to 3 kV/cm) and a specific energy (range of 1 to 20 kJ/kg) is required (Corrales et al., 2008). The comparison of the values shows less energy is required for cell disintegration.

## 3. Determination of cell disintegration index

The application of PEF induces a polarization of the cellular membrane resulting in an increased trans membrane potential. The increase leads to a rapid electrical breakdown and to local structural changes in the membrane. The result of the applied external electric field is the formation of pores related to an increase of permeability of the membrane. A permeabilisation influences the diffusion processes, for example increasing the extraction yield and shortening the distillation time. To make the best use of the permeabilisation, a maximum of cell disintegration is required. But it has to be noted that cell permeabilisation by PEF leads to a softer structure. An optimum for the PEF settings (process parameters) has to be defined as for some liquid solid separation techniques a too soft texture may be limiting. Up to now different methods are available to measure the permeability of membranes. Microscopic methods can be used when the analysed tissue was stained. It is also possible to measure the content of the extracted material (sugar, ions) (Angersbach&Heinz 1997). Mostly these two methods are not exact enough to determine the disintegration of biological membranes. A better method for determining the cell disintegration is the electrical method based on measuring the frequency dependent conductivity.

An external electric field induces a polarization of the membranes. A result of the polarization is an increase of the membrane current and a simultaneously decrease of the resistance of the membrane, which corresponds to an increase of the conductivity (Angersbach et al., 2000). The increase of the conductivity up to 5 mS in less than 5 µs (Angersbach et al., 2000) can be measured and be set in relation to the conductivity of intact membrane. Consequently analyzing the frequency dependent conductivity offers the possibility to estimate the amount of intact cellular material. The relation of the conductivity of the intact and the disintegrated cells is termed as the cell disintegration index $Z_p$. The cell disintegration index allows an exact definition of the amount of ruptured cellular material. If there is no difference between the conductivity of the untreated and PEF treated cell membrane, the $Z_p$ value is 0 and the cell can be defined as an intact cell. In contrast to that a $Z_p$ value of 1 indicates the maximum of cell disintegration due to the PEF treatment (Angersbach&Heinz 1997).

The frequency dependent conductivity allows a detection of product properties. Fig. 2. illustrates an example of conductivity of various treated and untreated tissues at different frequencies.

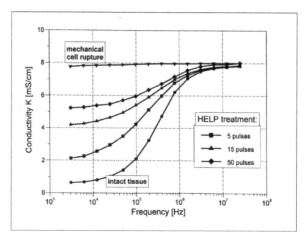

Fig. 2. Measured conductivity at different frequencies (Knorr&Angersbach 1998)

As it can be seen in Fig. 2. the conductivity changes with increasing frequency. At low frequency ranges the conductivity of PEF treated cell material and intact tissue is different, because of the capacitive properties of intact tissue. The highest value for conductivity relates to mechanically ruptured cells. From the graph it can also be seen that as the frequency increases, the conductivity of all cellular material reaches the same conductivity value. The reason being the relationship between conductivity and the frequency. At high frequency ranges (5-10 MHz) the cell membranes, whether damaged or not, are not resistant to the electric current (Angersbach&Heinz 1997; Knorr&Angersbach 1998).

Besides the relation between the conductivity and the frequency, the Fig. 2. also displays the impact of pulse number on the effect of the permeabilisation of cellular material. A higher number of applied pulses leads to a higher conductivity, which shows a higher permeabilisation of the membrane (Knorr&Angersbach 1998). Not only does the number of pulses affect the efficiency of the PEF treatment, the applied electric field strength amplitude also plays a part. Higher electric field strength causes a more rapid dynamics of pore formation and a faster initiation of conductive channels (Angersbach et al., 2000). To compare the relation between number of pulses and electric field strength; typically a lower number of pulses is necessary to reach a high disintegration of cellular material when high electric field strength is applied (Knorr&Angersbach 1998; Angersbach et al., 2000). The effect of the electric field strength in relation to the permeability of the membrane is illustrated in Fig. 3.

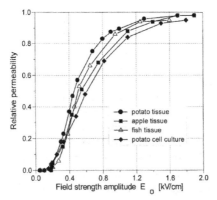

Fig. 3. Relation between the electric field strength amplitude and the relative permeability (Angersbach et al.,2000)

Using low field strength amplitudes no significant differences between the samples were observed, because the electric field strength did not reach the critical value and no electrical breakdown is induced. This phenomenon occurs if the applied field strength is higher than the critical field strength. With increasing electric field strength the relative permeability of the membrane increases up to the maximum, where all cells are disintegrated.

Fig. 3. also displays the dependence of the applied electric field amplitude to the type of product. Four different tissue types (potato-, apple- and fish tissue and potato cell culture) were tested and different values of electric field strength were necessary for a disintegration of these tissues. The reason for this is based on dependence of the membrane to rupture on the trans membrane voltage and the cell size distribution (Angersbach et al., 2000).

By measuring the conductivity the optimum electric field strength can be determined. The aim is to reach a maximum cell disintegration. Lebovka et al. 2002 (Lebovka et al., 2002) detected an electric field strength of 400 V/cm for maximum of cell disintegration of apple-, carrot and potato tissue. Comparing these tissues treated with 400 V/cm different treatment times are necessary to reach a high cell disintegration. Apple tissue required the highest treatment time and consequently the highest energy consumption in contrast to potato, which has the lowest energy consumption (Lebovka et al., 2002).

In conclusion, by measuring the conductivity changes induced by PEF treatment it is possible to determine the amount of disintegrated cells.

## 4. Influence of PEF on mass transfer

The application of pulsed electric fields allows a disintegration of cellular material. The result of disintegrated cells is an improved mass transfer, which affects the following processes:

- Dehydration/Freezing
- Extraction
- Distillation

The drying process can be described as a preservation method removing water from food. Water removal, lowers the water activity and increases the shelf life of the products. It can be used as a pre-treatment to improve nutritional, organoleptic and functional properties of the food products (Torreggiani 1993). An important parameter for the drying process is the mass transfer of water. A high mass transfer rate results in a better quality of dried product due to the shortening of the drying time and a reduction in the drying temperature. The drying process can be regarded as a complex process step including momentum, heat and mass transfer. The moisture in food can be classified in two main groups. First the "immobilized" water, which is retained in fine capillaries and is adsorbed at the surface, and second the "free" water, which retains in voids in foods. The drying process is based on the equilibrium of moisture content. The food looses or gains moisture over a period of time to attain a new equilibrium status (Sharma et al., 2000). For drying of heat sensitive products the osmotic dehydration (OD) can be used as a pre-treatment in order to reduce the drying time. Other pre-treatments, like microwave or conventional thermal heating, lead to a thermal destruction of the nutrients and organoleptic quality of the food (Ade-Omowaye et al., 2001).

For this process, food is placed in an osmotic solution resulting in the formation of a water and a solubility gradient across the cell membrane. The cell membrane separates cell content, mostly water, and the osmotic solution. Consequently two fluxes are formed; the water out of the cell and the osmotic solution into the cell. This flux is a mass transfer process and is a function of the difference in chemical potential. Because of the composition of the cell membrane and the better permeability to water, more water is removed from the cell than less solute goes into the cell. The following Figure (Fig. 4.) represents the flux of water and solute in and out the cell (Torreggiani 1993; Sharma et al., 2000).

The OD process depends on different parameters, like concentration of osmotic solution, contact time and process temperature as well as the exposed surface area. The most widespread problem is the simultaneous solute transfer in the food countercurrent to the water flow. This influences the product quality in a negative way resulting in a candy or salty taste of the product, as well as an altered sugar-to-acid ratio. Different kinds of osmotic solutions can be used. The choice depends on parameters, like organoleptic quality, preservative effect and the product taste. Mostly a sucrose solution is used, because it is very effective and convenient. But because of the sweetness the application is limited to vegetables and fruits. The second most used substance for osmotic solutions is sodium chloride.

In comparison to convection drying the OD process uses less energy, because a lower temperature can be used. Due to the reduced temperature, less heat damage is found on the

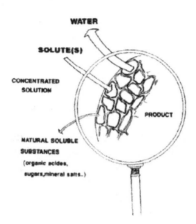

WATER

SOLUTE(S)

CONCENTRATED
SOLUTION

PRODUCT

NATURAL SOLUBLE
SUBSTANCES
(organic acides,
sugars,mineral salts..)

Fig. 4. Mass transfer during OD process (Torreggiani 1993)

product and the high concentration of the osmotic solution prevents discoloration (Torreggiani 1993).

As the OD process is directly dependent on the mass transfer, a PEF treatment can enhance OD. The first time facilitated mass transfer during OD using PEF was reported by Rastogi et al. (Rastogi et al., 1999). Because of the structural changes of the cell membrane due to PEF application, the mass transfer is facilitated (Ade-Omowaye et al., 2001). Consequently a facilitated, fast exit of the water in the osmotic solution is possible, but no solid uptake, because of the selective permeabilisation of the cell membrane (Taiwo et al., 2003). Using PEF before the OD process the drying time can be significantly reduced and a better structural food quality can be obtained.

Different product types were treated with PEF before OD. For example, the pre-treatment of red pepper with PEF shows a facilitated moisture removal and an improvement of the quality of the dried products. Using OD in combination with PEF a preserved color quality of the red pepper could be obtained (Ade-Omowaye 2003). Carrots can also be pre treated with PEF resulting in an increased diffusion coefficient and a reduction of the OD time from 4 h to 2 h (Rastogi et al., 1999). The PEF induced enhanced mass transfer depends on the electric field strength and the applied number of pulses (Rastogi et al., 1999). The same reduction in OD was reported (Rastogi et al., 1999; Amami et al., 2007a; Amami et al., 2007b) for the treatment of apple tissue (Amami et al., 2006).

In addition to drying process the freezing process can be regarded as a mass transport dependent preservation method. Suitable products for this type of treatment are fruits (strawberries and raspberries), vegetables (peas, green beans) as well as fish and meat products (Sharma et al., 2000). The process is based on a decrease of temperature under the freezing point and combines the effect of low temperature with the conversion of water into ice (Delgado&Sun 2001). The advantage of the freezing process is the reduced chemical reactions and the delay of cellular metabolic reactions (Delgado&Sun 2001).

In general the freezing process can be separated into three main phases. The first phase is called the pre-cooling or chilling phase, where the product is cooled down to the freezing point. During the second phase (phase change period) most of the water in the product crystalizes. The final end temperature is reached during the last phase termed as the tempering phase (Delgado&Sun 2001).

The time, which is required to lower the temperature, is called the freezing time. This time can be predicted with mathematic modeling. Therefore the heat and mass transfer phenomena has to be considered. Two possible models to describe the freezing process are the heat transfer model and coupled heat and mass transfer model. The application of PEF improves the mass transfer. The pre-treatment of potato (Jalté et al., 2009) for example induces a higher freezing rate with a shorter freezing time. Due to PEF treatment the cell membrane gets porous with the result of an enhanced diffusion mass exchange of extracellular and intracellular water as well as a reduced freezing time. Using PEF as a pretreatment can achieve a better quality of the frozen food as smaller ice molecules are also formed. The improved quality can be seen in Fig. 5. The structure and form of the PEF pre-treated potato was much better in comparison to the untreated ones.

Fig. 5. Pre-treatment of potato discs in comparison to untreated discs after freezing (Jalté et al., 2009)

Besides drying and freezing, PEF treatment influences the extraction process. In general, extraction is a separation process separating a substance from a matrix. Two main extraction processes are liquid-liquid extraction and solid-liquid extraction. In the following the soli-liquid extraction is described in detail as well as the impact of PEF on the solid-liquid expression.

The aim of a solid-liquid extraction is the separation of the desired extract located in the solid by using a solvent in which the extract is soluble. Mostly, the extract is located in the pore or cell structure of the solid. This technique has been used a long time for the extraction of plant oils, for example sucrose from sugar beet or tanning- or color extracts. Today the solid-liquid extraction is mostly used for (Bouzrara&Vorobiev 2003)

- Extraction of fruit juices and vegetable oils
- Production of wine
- Dewatering of fibrous materials (sugar beet)
- Dehydration of organic wastes

The principle of an extraction process is based on the diffusion characteristics of the solid and the solubility of the extract. The solvent is added to the solid and during the extraction the extract diffuses from the solid in the solvent. This mass transport from the solid in the solvent follows the diffusion law. After this process a mechanical separation of the solid and liquid is required as well as a separation of the extract and the extraction agent. A facilitated extraction can be reached by chopping or mashing the solid with the result of an increased surface area and shorter capillary ways. Another possibility to increase the extraction velocity is an increase of the concentration gradient. This can be achieved by using a high volume of solvent (Vauck&Mueller 2000).

As described before the principle of the extraction process is based on the mass transfer of the extract from the solid in solvent. The extract is mostly located in cells, which are surrounded by membranes. Consequently, the amount of compounds released to the solvent depends on the degree of the damaged cellular material. The effect of PEF can be defined as electroporation resulting in a disintegration of cellular material and an improved mass transfer. Regarding the extraction process a facilitated extraction is possible using PEF as a pre-treatment, because the extract can easily diffuse into the solvent. There is no force required to open the cellular material. Using PEF as a pre-treatment, the extraction time can be reduced. The extraction yield can be increased and important quality ingredients can be easily extracted.

PEF can be used for an improved extraction process in different application fields. One example is the extraction of calcium from bones (Yin&He 2008). There are different methods used for the extraction of calcium from bones, for example boiling or microwave treatment, often leads to negative effects on the product and the extracted concentration is low. Using PEF, the temperature can be reduced and the calcium concentration increased (Yin&He 2008).

Another example for a solid-liquid expression assisted by PEF treatment is sugar beet. Using PEF as a pre-treatment the extraction time and temperature can be lowered (López et al., 2009) as well as a higher sucrose yield can be obtained (Eshtiaghi&Knorr 2002; El-Belghiti&Vorobiev 2004) resulting in more efficient sugar production with lower energy requirements (Lebovka et al., 2007; Loginova et al., 2011). Due to the lower extraction temperature the PEF process offers the possibility to improve the extraction of colorants from red beet root in order to increase the extraction yield and a better stability of the colorants (Fincan et al., 2004; Loginova et al., 2011). An increased juice yield plays an important role in juice industry. In general mechanical, enzymatic or temperature based methods were used to rupture the apple cells for a facilitated extracting, but mostly they lead to a degradation of important juice components as well as a high energy consumption (Toepfl 2006). A PEF treatment of the apples increases the yield from 1,7 to 7,7 % in comparison to an enzymatic treatment (4,2 %) (Schilling et al., 2007) and reduces the energy consumption. Additional the pomace can be used after the treatment for pectin extraction.

Another process limited by mass transfer is the distillation process. In this process a high temperature is applied to the product for a defined time in order to separate two components with different boiling points. The high temperature denatures the cellular material and facilitates the extraction of the product. The high heat load degrades the heat sensitive ingredients minimizing the quality of the product. The membrane of the cells represents a semipermeable barrier through which the desired extract cannot diffuse because of the size. After the cells are ruptured the extract can easily diffuse along the concentration gradient. To rupture the cells, PEF can be used. Using PEF a disintegration of the cells could be observed at moderate temperatures. Consequently, the heat load could be minimized and no change of quality of the extract could be detected. In addition to that, the extraction yield could also be increased. In summary, the result of using pulsed electric fields instead of high temperature treatment is a reduced temperature and distillation time. A possible application being the treatment of roses to gain a higher yield of rose oil while reducing the distillation time (Dobreva et al., 2010).

## 5. Application

### 5.1 Extraction

A facilitated extraction of valuable components, like sucrose from sugar beet, red colorants from red beet roots and polyphenols from wine as well as juice from fruits, can be achieved by using PEF as a pre treatment.

The traditional sucrose extraction process is a thermal one treating the cossettes at 70 to 75 °C for 1 to 1,5 h. The heating process permits a denaturation of the sugar beet cells and facilitates the sugar extraction. This process uses a significant amount of energy and water and assists the growth of spoilage microorganisms. Lopez et al. (López et al., 2009) studied the influence of PEF on the extraction yield dependent on the electric field strength and pulse number as well as temperature. As a result applying 20 pulses with a field strength of 7 kV/cm allows a temperature reduction from 70 °C to 40 °C in a 60 min extraction process with yield of 80 %. The related thermal energy consumption is lowered by more than 50 %. In addition the extraction is dependant on the field strength, specific energy as well as temperature and independent from pulse shape and pulse duration (López et al., 2009). Agitation during extraction leads to an increase of the sucrose yield (El-Belghiti&Vorobiev 2004). In this study the sugar beets were treated with 0,9 kV/cm and 250 pulses in a cylindrical batch treatment chamber. For an extraction yield of 30 %, an extraction time of 120 min with 100 rpm agitation is required in comparison to 500 min extraction time without agitation. The sugar extraction process assisted by PEF can be improved by additional agitation, as well as by applying pressure. Using a pressure of 30 MPa during the extraction process the extraction yield can be increased by 20 % (Eshtiaghi&Knorr 2002).

Besides the yield of the PEF assisted extraction process, the quality of the sucrose was analysed. The treatment of sugar beets with an electric field strength of 0,6 kV/cm and 500 pulses with a pulse duration of 100 μs at 30 °C leads to comparable results of thermal process at 70 °C (Loginova et al., 2011).

The thermal treatment at 70 °C leads to a sucrose content of 14 % and a purity of more than 90 %. Equal values have been reached using a PEF treatment at 30 °C. The energy consumption required by PEF treatment is 5,4 kWh/t versus 46,7 kWh/t for the thermal treatment. In conclusion, the application of PEF lowers the energy consumption without influencing the sucrose content or the purity of the extracted sucrose solution (Loginova et al., 2011). Increasing the temperature of the PEF treatment leads to a reduction of the required electric field strength without decreasing the important quality parameters. Another study of Loginova et al. (Loginova et al., 2011) shows an increased sucrose content and °Brix values of extracted product assisted by PEF in a parallel stainless steel electrode with a gap of 7 cm applying 0,6 kV/cm at 50 °C. The end result is a higher purity and

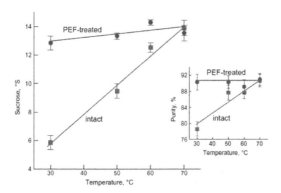

Fig. 6. Influence of temperature and PEF on the sucrose concentration and the purity of the sucrose solution (Loginova et al., 2011)

improved filter capability of the product in comparison to the thermally treated product at 70 °C.

The thermal treatment is dependent on the holding time (equal with treatment time for the extraction process) and aiming on an almost complete thermo-break. The application of PEF reduces the treatment time resulting in a high cell disintegration index of approximately 0,95 in several milliseconds in contrast to a cell disintegration index of 0,85 in more than 100 s for a thermal treatment. A higher cell disintegration index can be reached by using a higher electric field strength (Lebovka et al., 2007). The researchers Eshtiaghi et al. (Eshtiaghi&Knorr 2002) indicated a cell disintegration index of approximately 0,7 by treating the slices with an electric field strength of 2,4 kV/cm and 20 pulses at 20 °C. The same cell disintegration index can be achieved by a heat treatment at 72 °C for 15 min.

In addition to the extraction of sucrose from sugar beet the extraction of industry relevant colorants from red beet roots can be improved. Red beet concentrate can be used as a food additive to color food. Betalains are the major colorants, and are composed of the red –violet substance betacyanin and the yellow-orange betaxanthin. The main function of betalains beside the color, is antiviral and antimicrobial activity (Azeredo 2009). Most extraction processes contains a thermal treatment leading to a lower extraction yield and high color degradation, because of the sensitivity of the betalains towards high temperature. For example a thermal treatment at 30 °C for 5 h leads to a degradation of 20 % of the betalains. In comparison an extraction at 80 °C for 1 h results in a complete degradation of the betalains (Loginova et al., 2011). The extraction assisted by PEF lowers the temperature without decreasing the extraction yield and destruction of the color (Fincan et al., 2004; Loginova et al., 2011). The same extraction rate can be achieved using a pre-treatment with PEF at 30 °C of the red beet using a field strength of 1,5 kV/cm and 20 pulses with a duration of 100 µs in comparison to a thermal treatment at 80 °C for 40 min and in addition with less color degradation (Loginova et al., 2011).

Another main product function is the solid-liquid juice extraction process assisted by PEF. The effects of PEF on the extraction of apple juice will be described. The left side of the Fig. 7. below shows a current process for apple juice extraction.

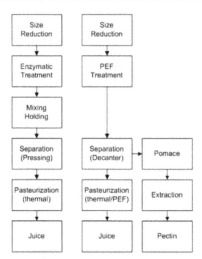

Fig. 7. Left side the traditional apple juice process, right side the process using PEF (adapted from (Toepfl 2006))

The current extraction process contains an enzymatic treatment to rupture the apple cells. This step can be replaced by a PEF treatment (see right side of Fig. 7.) in order to increase the extraction yield without a destruction of the juice quality (Bazhal&Vorobiev 2000; Bazhal et al., 2001; Toepfl&Heinz 2011). The extraction yield can be increased to 1,7-7,7 % using field strengths of 1 and 5 kV/cm and 1 to 30 kJ/kg using a decanter centrifuge. In comparison, an increased extraction yield of 4,2 % can be achieved using enzyme treatment (Schilling et al., 2007). For the enzymatic treatment the apples were treated with a pectolytic enzyme for 1 h at 30 °C , the pectin is degraded using an enzymatic treatment for increasing the extraction yield. After the PEF treatment, native pectin can be produced from the pomace. An increase of the nutrient content of the juice was not observed (Schilling et al., 2007). Grimi et al. (Grimi et al., 2011) studied the extraction yield, clarity, polyphenol content and the antioxidant activity of the pre-treated apples. Additionally they differ between the treatment of whole apples and sliced apples. Treatment of sliced apples with a field strength of 0,4 kV/cm and a treatment time of 0,1 s, resulted in a higher extraction yield than the treatment of whole apples using the same conditions. The cell disintegration index, clarity of the juice, polyphenol content and the antioxidant activity showed no difference between the two different apple type treatments.

The studies of Grimi et al. (Grimi et al., 2011) show an increase of the polyphenol content in apple juice in comparison to the non PEF treated sample. Similar results were reported by Puertolas et al. and Corrales et al. (Corrales et al., 2008; Puertolas et al., 2010) studying the extraction of polyphenols from red wine grapes. The polyphenols are the most important components for the color and the quality of the red wine. A rupture of the grape skin, where the polyphenols are located, leads to a release of the polyphenols resulting in a facilitated extraction. A PEF induced cell disintegration using an electric field strength of 3 kV/cm and 10 kJ/kg results in an increase of the phenolic compounds as well as a higher antioxidant content (7841 $\mu$molTE g$^{-1}$DM in comparison to 187 $\mu$molTE g$^{-1}$DM thermal treatment at 70 °C for 1 h) (Corrales et al., 2008).

The oil extraction process also contains a solid-liquid extraction step, which can be improved by PEF application. Guderjan et al. (Guderjan et al., 2005) studied the effect of PEF on the oil yield of maize and olives. Pulsing the maize with an electric field strength of 0,6 kV/cm and a specific energy of 0,62 kJ/kg at 20 °C yields an extraction of 43,7 % in comparison to a yield of 23,2 % of the untreated maize. The same effect of increased extraction yield is observed when treating olives with 1,3 kV/cm and 100 pulses. An increase of 7,4% with PEF in contrast to an increase of 5,3 % for a heat treatment (50 °C for 30 min) was determined (Guderjan et al., 2005). Beside olive and maize, rape seed can be treated to enhance the extraction yield and the quality of the oil. In that case, an improved oil yield was observed as well as an increase of the polyphenol-, phytosterol and tocopherol content. The rape seeds were pulsed with field strength of 7 kV/cm and 120 pulses (Guderjan et al., 2007).

## 5.2 Distillation

Distillation is a separation process, where the product is exposed to a defined temperature for a specific time. Because of the different boiling point of the components, a separation of these substances is possible.

The traditional rose oil distillation process is characterized by exposing a product to a high temperature for a defined time. The temperature causes a denaturation of the cell membrane and substances diffuse out of the cells. Due to the temperature sensitivity of the rose oil during the distillation process, the critical threshold should not be exceeded as the quality of the rose oil will be reduced. A lower temperature results in reduced disintegration of the rose cells and in a low extraction yield. To improve the rupture of the cell membrane, the cells can be exposed to pulsed electric fields. The electric field induces a polarization with a resulting breakdown of the cell membrane and an increased permeability. The process can be used for the rose oil production. Dobreva et al. (Dobreva et al., 2010) studied different PEF treatment conditions varying the specific energy and different distillation times. The aim was to detect the influence of PEF on the extraction yield and distillation time.

The PEF treatment was performed in parallel treatment chamber with a gap of 5 cm. The white oil-bearing roses (*Rosa alba* L.) were pulsed with an electric field strength of 4 kV/cm and a specific energy of 10 and 20 kJ/kg at room temperature. The distillation time using high temperature for cell rupture was 2,5 h. For the analysis the oil yield was determined after 0,5, 1, 1,5 and 2,5 h (Fig. 8.).

After a distillation time of 1 h equivalent oil yields for the PEF and control samples were found. An increase of energy input from 10 to 20 kJ/kg causes an increase of oil yield of 20 %. Treating the roses with a specific energy of 10 kJ/kg, the yield was increased by 35 % compared to the control. In conclusion, using PEF as a pre-treatment the distillation time can be reduced resulting in lower energy consumption and a higher productivity of the distillation process. A higher yield can be achieved by treating the roses with PEF and a specific energy of 10 kJ/kg and a distillation time of 2,5 h.

Besides the improvement of yield and the distillation time, the quality of the oil analysis also showed no undesirable changes of any of the essential components in comparison to the reference extraction method.

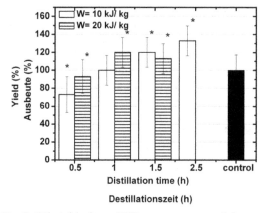

Fig. 8. Oil yield after a PEF pre-treatment of the roses in comparison to untreated after different distillation times (Dobreva et al., 2010)

### 5.3 Drying/Freezing

Drying is regarded to be an important preservation step. Temperature, relative humidity and time are the most important processing paramters. The drying process can be separated into three main parts. The first describes the evaporation from free surface, which is mostly influenced by heat and mass transfer. The second part includes the liquid flow from the capillaries and the third the diffusion of liquid or vapour (Sharma et al., 2000). The mass transfer, which is important for the first step of drying, is determined by the structure of the cellular material. By influencing the structure for example a disintegration of the cells, a facilitated mass transfer can be obtained. A cell disintegration of biological cell membranes can be induced by exposing the material to an electric field. The application of pulsed electric fields induces structural changes that leads to a rapid cellular breakdown and an increased permeability resulting in an enhanced mass transfer. The idea to use the PEF process in order to facilitate the drying process is described by several researchers and food products.

One product example using PEF as a pre-treatment before drying is red pepper. Ade-Omowaye et al. (Ade-Omowaye et al., 2000) investigated an increased water loss using the PEF process. For drying, a fluidized bed was used with a temperature of 60 °C for 6 h. The researchers examined the influence of the electric field strength and pulse number on the efficiency of the drying process. After 1,5 h drying the moisture content of the untreated red pepper sample was 2,06 kg/kg, of the sample treated with 1 kV/cm it was 1,45 kg/kg and the moisture content of the red pepper sample treated with 2 kV/cm it was 1,09 kg/kg. That corresponds to a decrease of moisture of 47 % at the same drying time. In summary an increasing electric field strength leads to a higher reduction of moisture content. In comparison to that shows the pulse number not that high influence so that a high pulse number has only a minimal effect on the electro permeabilisation (Ade-Omowaye et al., 2003). The same effect is reported by Knorr 1998 (Knorr&Angersbach 1998). Regarding the distillation time, after 3 h drying the drying rate was not higher as compared to the rate at 1,5 h. The reason for that is based on the fact, that nearly all water was evaporated at that time and the driving force was reduced.

Besides the use of PEF as a pre-treatment for red pepper, reports have shown that it can be used for treating carrots in order to increase the drying efficiency. Amami et al. (Amami et al., 2007a; Amami et al., 2007b) treated the sliced carrots with an electric field strength of 0,6 kV/cm and a specific energy of 19 kJ/kg as well as a pulse duration of 100 µs. After the PEF treatment different osmotic dehydration trials were performed to evaluate the influence of salt and sucrose concentration as well as the additional effect of a centrifugal instead of a static osmotic dehydration process. The water loss and °Brix of the solution was determined after defined time periods. For the OD process the ratio of carrots to solution was 1:3 and the duration was 240 min at 20 °C. The water loss of the untreated carrots after 2 h static OD process was 42 %. A pre-treatment of the carrots with PEF induces a water loss of 38 % and a sugar concentration of 45 °Brix. Consequently, the PEF treatment positively affects the mass transfer process. Besides the analysis of the effect of PEF on the water loss, the effect of the addition of sucrose and additional salt concentration was determined. As the sugar concentration is increased, the water loss was also increased. The addition of salt (NaCl) to the sucrose solution leads to different water loss values. An overview of the results is shown in Table. 1.

| Composition of the OD solution [%salt/%sucrose] | water loss [%] untreated | water loss [%] PEF treated |
|---|---|---|
| 0%/65% | 48,5 | 50 |
| 5%/60% | 50,9 | 54,6 |
| 15%/50% | 54,5 | 58,8 |

Table 1. Water loss of untreated and PEF treated (0,6 kV/cm, 19 kJ/kg) carrots slices using different salt concentrations for the osmotic solution in a OD process after 4 h (summarized from (Amami et al., 2007b))

By increasing the salt concentration a higher water loss can be obtained and the water loss can be improved even more by using PEF as a pre-treatment. A higher salt concentration leads to a higher diffusion gradient and combining a higher salt concentration with a PEF process, which induces an improved mass transfer, leads to a better diffusion with regard to the dehydration process. The addition of NaCl lowers the water activity, which increases the driving force for the drying process. Using a centrifugal OD process the water loss of the untreated carrot slices can be increased from 48,5 %, 50,9 % and 54,5 % in 0%/65%, 5%/60% and 15%/50% to 56,5 %, 58,1 % and 61,9 %. As a result, centrifugation leads to an improved OD. The best result for an OD can be reached using PEF, a salt/sucrose solution as osmotic solution and centrifugation. The temperature influence was also evaluated with the result of an improved OD using 40 °C in comparison to 20 °C (Amami et al., 2007a; Amami et al., 2007b). Similar results could be obtained using the PEF pre-treatment of apples. The mass transfer of apples is increased by the PEF treatment (Chalermchat et al., 2010) and the kinetics of water and solute transfer were accelerated during the convective and diffusion stages of OD process (Amami et al., 2006). Also the treatment of strawberries with PEF leads to an improvement of the OD process. Taiwo et al. (Taiwo et al., 2003) treated strawberries placed in tap water in a parallel stainless steel electrode with a gap of 3 cm, 5 pulses with a duration of 350 µs were applied to the product using an electric field strength of 1,2 kV/cm and 100 J/pulse. For the OD different osmotic solutions were tested to find out the best solution with the highest efficiency (Fig. 9.).

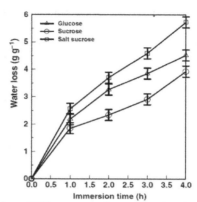

Fig. 9. Extent of water loss from PEF pre-treated strawberries during OD process in different osmotic solutions (Taiwo et al., 2003)

From the above examples it can be said that pre-treatment with PEF increases mass transfer during the OD. When considering different osmotic solutions containing glucose, sucrose or salt/sucrose results show that the salt/sucrose solution shows the best result regarding the water loss.

In addition to PEF treatment of carrots in order to improve the OD, potatoes can also be regarded as suitable products for a PEF pre-treatment. The application of PEF on potatoes with an electric field strength of 1,5 kV/cm induces an enhanced drying process, because of the disintegration of the potato cells induced by PEF. The comparison of a freeze dried potato without PEF and a PEF freeze dried potato shows less structure damage (Lebovka et al., 2007). Besides the structure also the diffusion characteristics in potatoes can be improved using PEF. For example a PEF treatment using 1,5 kV/cm and 20 applied pulses leads to a release of glucose and fructose, which are precursors for Maillard reaction, and a facilitated uptake of sodium chloride (Janositz et al., 2010). A reduced browning and acrylamide formation has been reported (Lindgren et al., 2002). A comparison of the treatment of potatoes and apples shows different diffusion coefficients during OD for these two different food products. The treatment of apples showed an increase of diffusion coefficient. The treatment of potatoes by PEF showed that the diffusion coefficient could be improved. The reason for this difference is based on the morphological structure of potato and apple. The tissue of potato is more tightly packed in contrast to the apple tissue (Arevalo et al., 2004).

Using PEF as a pre-treatment the drying and freezing rate can be improved. With increasing cell disintegration the freezing rates increased. Regarding the freezing process cellular water flows easily out of the cell and ice nucleation outside the cell starts. SEM pictures show the impact of PEF on the freezing process (Jalté et al., 2009).

The untreated potato cells show perturbation of the original polyhedral arrangement of the cells and also disrupted cell walls. The structure of PEF treated potato indicates a structural damage in order to a partial destruction of the polyhedral shape. Fig. 10. (right) shows voids, which can be explained by the formation of ice crystals outside the cell (Jalté et al., 2009).

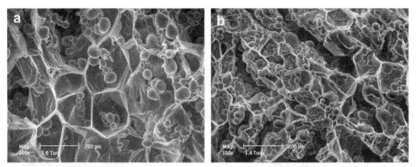

Fig. 10. Cellular structure of a untreated (a) and a PEF treated (b) potato after freeze and thaw process (Jalté et al., 2009)

## 5.4 Process and equipment design

Most of the PEF applications described as a pre-treatment in order to increase the extraction yield, to decrease the distillation time and/or process temperature, were performed with discontinuous PEF systems using a parallel or cylindrical electrode configurations for the treatment chambers. A parallel electrode configuration of the treatment chamber allows a homogenous electric field distribution. All tests with the different food types and systems were performed on a lab-scale. Toepfl et al. (Toepfl&Heinz 2011) studied a scale up from pilot plant to production scale for the apple juice production. They showed nearly the same increase of juice yield compared to the lab scale trials without influencing the taste or nutritional content. An increase of total acids, glucose, fructose and saccharose as well as total polyphenols was observed. Because of the high quality of the juice and the accordance of all quality parameters within the Europrean Code of Practice for Fruit Juices (AIJN 1996), the production including the PEF process step is equal to the common used process without PEF (Schilling et al., 2007; Toepfl&Heinz 2011).

A continuous treatment up to an industrial scale has been realized by DIL. At the moment three different systems are available for laboratory scale, semi-industrial scale and industrial scale. The 5 kW system has a maximum voltage of 30 kV and maximum current of 200 A. The pipe diameter ranges from 10 to 30 mm. Due to the small pipe diameter it is not possible to treat tubers or whole fruits. An increased pipe diameter can be used in 30 kW systems with a maximum voltage of 30 kV and a maximum current of 700 A. The pipe diameter can be increased up to 100 mm and for the application cell disintegration the capacity is 10.000 kg/h. A further increase of capacity up to 50.000 kg/h can be realized using the 80 kW systems with a maximum voltage of 60 kV and a maximum current of 5.000 A. The pipe diameter of these systems is in range of 200 mm, as an alternative belt type chambers can be used. At a belt width of 1 m the processing capacity can be up to 50.000 kg/h.

The 30 kW system and the belt system and a PEF system with a pipe diameter of 50 mm are shown in Fig. 11.

The PEF treatment consumes less energy compared to for example thermal treatments. For the treatment of sliced apples, the energy consumption is 12,5 ±0,5 kJ/kg and for whole apples 7,9±0,4 kJ/kg (Grimi et al., 2011). The typical energy requirement is in a range of

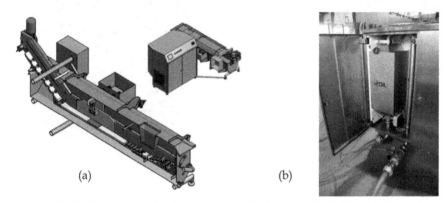

(a)                                                        (b)

Fig. 11. (a) 30 kW belt system for the treatment of tubers or whole fruits, (b) PEF system with a pipe diameter of 50 mm

3 kWh/ton for a complete tissue disintegration. Compared to other cell disruption methods (mechanical: 20-40 kJ/kg; enzymatic: 60-100 kJ/kg; heating, freezing/thawing: >100 kJ/kg) the energy consumption is very low (Toepfl 2006). For tissue softening, where a lower extent of cell disintegration is required, the typical specific energy input for a PEF treatment is in a range of 1 kJ/kg. The same effect can be observed during the sugar extraction process. More than 50 % of the thermal energy can be saved using a PEF treatment (temperature: 40 °C) instead of a thermal treatment (70 °C) in 60 min extraction process with a yield of 80 % (López et al., 2009).

In conclusion the application of PEF offers the possibility to decrease the energy consumption and a continuous scale up is possible.

## 6. Conclusion

Food processing is a wide field containing many different process steps and techniques based on the principles of process engineering. The industry is searching for new and innovative techniques to improve the quality of the food and to introduce new products with a simultaneous cost reduction. One of the most promising new novel food processing techniques is the application of pulsed electric fields (PEF). This non-thermal treatment is based on the application of pulses with a certain voltage and short duration times (μs) to the product located between two electrodes. The product is located between the electrodes and is exposed to the electric field. The cells and the microorganisms in the product are affected by PEF. Membranes of the cells are destroyed.

For some food processing steps, especially extraction, dehydration and distillation, a rupture of the tissue is required. Many different cell rupturing techniques are available based on mechanical, chemical or thermal treatments, but they often induce a quality loss of the product or the rupture is not sufficient resulting in a low product yield. Using PEF a more efficient rupture of the cells can be achieved. PEF leads to a poration of the cell membrane resulting in a facilitated diffusion out of the cell. The application field has a very wide range such as the extraction yield of juices, oils, sugar and the reduction of drying time and distillation time. Industrial scale PEF systems are available and can be implemented in the industry.

Other products as described in the chapter can be analysed as well. Many different studies showed an increased extraction yield, cost reduction and improvement of product quality. Consequently, the PEF process can be used for other product fields, like the extraction of essential oils from plants. Besides products mentioned in that chapter, many other raw materials are eligible for a PEF treatment.

# 7. References

Ade-Omowaye, B. I. O., A. Angersbach, N. M. Eshtiaghi and D. Knorr (2000). Impact of high intensity electric field pulses on cell permeabilisation and as pre-processing step in coconut processing. *Innovative Food Science & Emerging Technologies* Vol.1, No.3, pp. 203-209.

Ade-Omowaye, B. I. O., A. Angersbach, K. A. Taiwo and D. Knorr (2001). Use of pulsed electric field pre-treatment to improve dehydration characteristics of plant based foods. *Trends in Food Science & Technology* Vol.12, No.8, pp. 285-295.

Ade-Omowaye, B. I. O., N. K. Rastogi, A. Angersbach and D. Knorr (2003). Combined effects of pulsed electric field pre-treatment and partial osmotic dehydration on air drying behaviour of red bell pepper. *Journal of Food Engineering* Vol.60, No.1, pp. 89-98.

Álvarez, I., J. Raso, A. Palop and F. J. Sala (2000). Influence of different factors on the inactivation of *Salmonella senftenberg* by pulsed electric fields. *International Journal of Food Microbiology* Vol.55, No.1-3, pp. 143-146.

Amami, E., A. Fersi, L. Khezami, E. Vorobiev and N. Kechaou (2007a). Centrifugal osmotic dehydration and rehydration of carrot tissue pre-treated by pulsed electric field. *LWT - Food Science and Technology* Vol.40, No.7, pp. 1156-1166.

Amami, E., A. Fersi, E. Vorobiev and N. Kechaou (2007b). Osmotic dehydration of carrot tissue enhanced by pulsed electric field, salt and centrifugal force. *Journal of Food Engineering* Vol.83, No.4, pp. 605-613.

Amami, E., E. Vorobiev and N. Kechaou (2006). Modelling of mass transfer during osmotic dehydration of apple tissue pre-treated by pulsed electric field. *LWT - Food Science and Technology* Vol.39, No.9, pp. 1014-1021.

Angersbach, A. and V. Heinz (1997). Elektrische Leitfähigkeit als Maß des Zellaufschlussgrades von zellulären Materialien durch Verarbeitungsprozesse. *Lebensmittelverfahrenstechnik* Vol.42, pp. 195-200.

Angersbach, A., V. Heinz and D. Knorr (2000). Effects of pulsed electric fields on cell membranes in real food systems. *Innovative Food Science & Emerging Technologies* Vol.1, No.2, pp. 135-149.

Arevalo, P, Ngadi, O. M, Bazhal, I. M, Raghavan and V. G. S (2004). *Impact of pulsed electric fields on the dehydration and physical properties of apple and potato slices.* Philadelphia, PA, ETATS-UNIS, Taylor & Francis.

Azeredo, H. M. C. (2009). Betalains: properties, sources, applications, and stability - a review. *International Journal of Food Science & Technology* Vol.44, pp. 2365-2376.

Barbosa-Cánovas, G. V., M. M. Góngora-Nieto, U. R. Pothakamury and B. G. Swanson (1999). Biological Principles for Microbial Inactivation in Electric Fields. *Preservation of Foods with Pulsed Electric Fields.* San Diego, Academic Press: pp. 47-75.

Bazhal, M. and E. Vorobiev (2000). Electrical treatment of apple cossettes for intensifying juice pressing. *Journal of the Science of Food and Agriculture* Vol.80, No.11, pp. 1668-1674.

Bazhal, M. I., N. I. Lebovka and E. Vorobiev (2001). Pulsed electric field treatment of apple tissue during compression for juice extraction. *Journal of Food Engineering* Vol.50, No.3, pp. 129-139.

Bouzrara, H. and E. Vorobiev (2003). Solid/liquid expression of cellular materials enhanced by pulsed electric field. *Chemical Engineering and Processing: Process Intensification* Vol.42, No.4, pp. 249-257.

Castro, A. J., G. V. Barbosa-Canovas and B. G. Swanson (1993). Microbial Inactivation of Foods by Pulsed Electric Fields. *Journal of Food Processing and Preservation* Vol.17, pp. 47-73.

Chalermchat, Y., L. Malangone and P. Dejmek (2010). Electropermeabilization of apple tissue: Effect of cell size, cell size distribution and cell orientation. *Biosystems Engineering* Vol.1 0 5 pp. 357-366.

Corrales, M., S. Toepfl, P. Butz, D. Knorr and B. Tauscher (2008). Extraction of anthocyanins from grape by-products assisted by ultrasonics, high hydrostatic pressure or pulsed electric fields: A comparison. *Innovative Food Science & Emerging Technologies* Vol.9, No.1, pp. 85-91.

Delgado, A. E. and D.-W. Sun (2001). Heat and mass transfer models for predicting freezing processes - a review. *Journal of Food Engineering* Vol.47, No.3, pp. 157-174.

Dimitrov, D. S. (1984). Electric Field-Induced Breakdown of Lipid Bilayers and Cell Membranes: A Thin Viscoelastic Film Model. No.78, pp. 53-60.

Dimitrov, D. S. (1995). Electroporation and Electrofusion of Membranes. *Handbook of Biological Physics*, Elsevier Science B. V. Vol.1: pp.851-900

Dobreva, A., F. Tintchev, V. Heinz, H. Schulz and S. Toepfl (2010). Effect of pulsed electric fields (PEF) on oil yield and quality during destillation of white oil-bearing rose (Rosa alba L.). *Zeitschrift für Arznei- und Gewürzpflanzen* Vol.15, No.3, pp. 127-131.

El-Belghiti, K. and E. Vorobiev (2004). Mass transfer of sugar from beets enhanced by Pulsed electric field. *Food and Bioprocess Technology* Vol.82, No.C2, pp. 226-230.

Eshtiaghi, M. N. and D. Knorr (2002). High electric field pulse pretreatment: potential for sugar beet processing. *Journal of Food Engineering* Vol.52, No.3, pp. 265-272.

Fincan, M., F. DeVito and P. Dejmek (2004). Pulsed electric field treatment for solid-liquid extraction of red beetroot pigment. *Journal of Food Engineering* Vol.64, No.3, pp. 381-388.

Glaser, R. W., S. L. Leikin, L. V. Chernomordik, V. F. Pastushenko and A. I. Sokirko (1988). Reversible electrical breakdown of lipid bilayers: formation and evolution of pores. *Biochimica et Biophysica Acta (BBA) - Biomembranes* Vol.940, No.2, pp. 275-287.

Grimi, N., F. Mamouni, N. Lebovka, E. Vorobiev and J. Vaxelaire (2011). Impact of apple processing modes on extracted juice quality: Pressing assisted by pulsed electric fields. *Journal of Food Engineering* Vol.103, No.1, pp. 52-61.

Guderjan, M., P. Elez-Martinez and D. Knorr (2007). Application of pulsed electric fields at oil yield and content of functional food ingredients at the production of rapeseed oil. *Innovative Food Science & Emerging Technologies* Vol.8, No.1, pp. 55-62.

Guderjan, M., S. Toepfl, A. Angersbach and D. Knorr (2005). Impact of pulsed electric field treatment on the recovery and quality of plant oils. *Journal of Food Engineering* Vol.67, No.3, pp. 281-287.

Heinz, V., I. Alvarez, A. Angersbach and D. Knorr (2001). Preservation of liquid foods by high intensity pulsed electric fields--basic concepts for process design. *Trends in Food Science & Technology* Vol.12, No.3-4, pp. 103-111.

Jalté, M., J.-L. Lanoisellé, N. I. Lebovka and E. Vorobiev (2009). Freezing of potato tissue pre-treated by pulsed electric fields. *LWT - Food Science and Technology* Vol.42, No.2, pp. 576-580.

Janositz, A., A. K. Noack and D. Knorr (2010). Pulsed Electric Fields and their impact on the diffusion characteristics of potato slices. *LWT - Food Science and Technology* Vol.In Press, Accepted Manuscript, pp.

Knorr, D. and A. Angersbach (1998). Impact of high-intensity electric field pulses on plant membrane permeabilization. *Trends in Food Science & Technology* Vol.9, No.5, pp. 185-191.

Lebovka, N. I., M. I. Bazhal and E. Vorobiev (2002). Estimation of characteristic damage time of food materials in pulsed-electric fields. *Journal of Food Engineering*, No.54, pp. 337?346.

Lebovka, N. I., M. V. Shynkaryk, K. El-Belghiti, H. Benjelloun and E. Vorobiev (2007). Plasmolysis of sugarbeet: Pulsed electric fields and thermal treatment. *Journal of Food Engineering* Vol.80, No.2, pp. 639-644.

Lebovka, N. I., N. V. Shynkaryk and E. Vorobiev (2007). Pulsed electric field enhanced drying of potato tissue. *Journal of Food Engineering* Vol.78, No.2, pp. 606-613.

Lindgren, M., K. Aronsson, S. Galt and T. Ohlsson (2002). Simulation of the temperature increase in pulsed electric field (PEF) continuous flow treatment chambers. *Innovative Food Science and Emerging Technologies* Vol.3, No.3, pp. 233-245.

Loginova, K. V., N. I. Lebovka and E. Vorobiev (2011). Pulsed electric field assisted aqueous extraction of colorants from red beet. *Journal of Food Engineering* Vol.106, No.2, pp. 127-133.

Loginova, K. V., E. Vorobiev, O. Bals and N. I. Lebovka (2011). Pilot study of countercurrent cold and mild heat extraction of sugar from sugar beets, assisted by pulsed electric fields. *Journal of Food Engineering* Vol.102, No.4, pp. 340-347.

López, N., E. Puértolas, S. Condón, J. Raso and I. Alvarez (2009). Enhancement of the extraction of betanine from red beetroot by pulsed electric fields. *Journal of Food Engineering* Vol.90, No.1, pp. 60-66.

Pliquett, U., R. P. Joshi, V. Sridhara and K. H. Schoenbach (2007). High electrical field effects on cell membranes. *Bioelectrochemistry* Vol.70, No.2, pp. 275-282.

Puertolas, E., N. Lopez, S. Condón, I. Alvarez and J. Raso (2010). Potential applications of PEF to improve red wine quality. *Trends in Food Science & Technology* Vol.21, No.5, pp. 247-255.

Rastogi, N. K., M. N. Eshtiagi and D. Knorr (1999). Accelerated Mass Transfer During Osmotic Dehydration of High Intensity Electrical Field Pulse Pretreated Carrots. *Journal of Food Science* Vol.64, No.6, pp. 1020-1023.

Schilling, S., T. Alber, S. Toepfl, S. Neidhart, D. Knorr, A. Schieber and R. Carle (2007). Effects of pulsed electric field treatment of apple mash on juice yield and quality attributes of apple juices. *Innovative Food Science & Emerging Technologies* Vol.8, No.1, pp. 127-134.

Sharma, S. K., S. J. Mulvaney and S. S. H. Rizvi (2000). *Food Process Engineering - Theory and Laboratory Experiments*. Canada, John Wiley & Sons.

Taiwo, K. A., M. N. Eshtiaghi, B. I. O. Ade-Omowaye and D. Knorr (2003). Osmotic dehydration of strawberry halves: influence of osmotic agents and pretreatment methods on mass transfer and product characteristics. *International Journal of Food Science & Technology* Vol.38, No.6, pp. 693-707.

Toepfl, S. (2006). Pulsed Electric Fields (PEF) for Permeabilization of Cell Membranes in Food- and Bioprocessing - Applications, Process and Equipment Design and Cost Analysis. Fakultät III - Prozesswissenschaften der Technischen Universität Berlin. Berlin. PhD.

Toepfl, S. and V. Heinz (2011). Pulsed Electric Field Assisted Extraction - A Case Study. *Nonthermal processing Technologies for Food*. H. Q. Zhang, G. V. Barbosa-Cánovas, V. M. Balasubramaniamet al, Blackwell Publishing Ltd. : pp. 190-200.

Toepfl, S., V. Heinz and D. Knorr (2007). High intensity pulsed electric fields applied for food preservation. *Chemical Engineering and Processing* Vol.46, No.6, pp. 537-546.

Torreggiani, D. (1993). Osmotic dehydration in fruit and vegetable processing. *Food Research International* Vol.26, pp. 59-68.

Tsong, T. Y. (1991). Electroporation of cell membranes. *Biophysical Journal* Vol.60, No.2, pp. 297-306.

Vauck, W. R. A. and H. A. Mueller (2000). *Grundoperationen Chemischer Verfahrenstechnik*. Stuttgart, Deutscher Verlag für Grundstoffindustrie.

Weaver, J. C. (2000). Electroporation of Cells and Tissues. No.28, pp. 24-33.

Wilhelm, C., M. Winterhalter, U. Zimmermann and R. Benz (1993). Kinetics of pore size during irreversible electrical breakdown of lipid bilayer membranes. *Biophysical Journal* Vol.64, No.1, pp. 121-128.

Yin, Y. and G. He (2008). A fast high-intensity pulsed electric fields (PEF)-assisted extraction of dissoluble calcium from bone. *Separation and Purification Technology* Vol.61, No.2, pp. 148-152.

Zhang, Q., G. V. Barbosa-Cánovas and B. G. Swanson (1995). Engineering Aspects of Pulsed Electric Field Pasteurization. *Journal of Food Engineering* Vol.25, pp. 261-281.

Zimmermann, U., G. Pilwat, F. Beckers and F. Riemann (1976). Effects of external electrical fields on cell membranes. *Bioelectrochemistry and Bioenergetics* Vol.3, No.1, pp. 58-83.

# Permissions

The contributors of this book come from diverse backgrounds, making this book a truly international effort. This book will bring forth new frontiers with its revolutionizing research information and detailed analysis of the nascent developments around the world.

We would like to thank Dr. Sina Zereshki, for lending his expertise to make the book truly unique. He has played a crucial role in the development of this book. Without his invaluable contribution this book wouldn't have been possible. He has made vital efforts to compile up to date information on the varied aspects of this subject to make this book a valuable addition to the collection of many professionals and students.

This book was conceptualized with the vision of imparting up-to-date information and advanced data in this field. To ensure the same, a matchless editorial board was set up. Every individual on the board went through rigorous rounds of assessment to prove their worth. After which they invested a large part of their time researching and compiling the most relevant data for our readers. Conferences and sessions were held from time to time between the editorial board and the contributing authors to present the data in the most comprehensible form. The editorial team has worked tirelessly to provide valuable and valid information to help people across the globe.

Every chapter published in this book has been scrutinized by our experts. Their significance has been extensively debated. The topics covered herein carry significant findings which will fuel the growth of the discipline. They may even be implemented as practical applications or may be referred to as a beginning point for another development. Chapters in this book were first published by InTech; hereby published with permission under the Creative Commons Attribution License or equivalent.

The editorial board has been involved in producing this book since its inception. They have spent rigorous hours researching and exploring the diverse topics which have resulted in the successful publishing of this book. They have passed on their knowledge of decades through this book. To expedite this challenging task, the publisher supported the team at every step. A small team of assistant editors was also appointed to further simplify the editing procedure and attain best results for the readers.

Our editorial team has been hand-picked from every corner of the world. Their multi-ethnicity adds dynamic inputs to the discussions which result in innovative outcomes. These outcomes are then further discussed with the researchers and contributors who give their valuable feedback and opinion regarding the same. The feedback is then collaborated with the researches and they are edited in a comprehensive manner to aid the understanding of the subject.

Apart from the editorial board, the designing team has also invested a significant amount of their time in understanding the subject and creating the most relevant covers. They scrutinized every image to scout for the most suitable representation of the subject and create an appropriate cover for the book.

The publishing team has been involved in this book since its early stages. They were actively engaged in every process, be it collecting the data, connecting with the contributors or procuring relevant information. The team has been an ardent support to the editorial, designing and production team. Their endless efforts to recruit the best for this project, has resulted in the accomplishment of this book. They are a veteran in the field of academics and their pool of knowledge is as vast as their experience in printing. Their expertise and guidance has proved useful at every step. Their uncompromising quality standards have made this book an exceptional effort. Their encouragement from time to time has been an inspiration for everyone.

The publisher and the editorial board hope that this book will prove to be a valuable piece of knowledge for researchers, students, practitioners and scholars across the globe.

# List of Contributors

**Vu Trieu Minh and John Pumwa**
Papua New Guinea University of Technology (UNITECH), Lae, Papua New Guinea

**Feng Wang, Ning Zhao, Fukui Xiao and Wei Wei**
State Key Laboratory of Coal Conversion, Institute of Coal Chemistry, Chinese Academy of Sciences, Taiyuan Shanxi

**Yuhan Sun**
State Key Laboratory of Coal Conversion, Institute of Coal Chemistry, Chinese Academy of Sciences, Taiyuan Shanxi
Low Carbon Energy Conversion Center, Shanghai Advanced Research Institute, Chinese Academy of Sciences, Shanghai, P.R. China

**Christopher Enweremadu**
University of South Africa, Florida Campus, South Africa

**José C. Zavala-Loría and Asteria Narváez-García**
Universidad Autónoma del Carmen (UNACAR), DES-DAIT: Facultad de Química, Cd. del Carmen, Cam, México

**Susanne Wagner, Angela Pfleger, Michael Mandl and Herbert Böchzelt**
Joanneum Research Forschungsgesellschaft mbH Graz, Resources, Institute of Water, Energy and Sustainability, Department for Plant Materials Sciences and Utilisation, Austria

**Steven C. Cermak, Roque L. Evangelista and James A. Kenar**
National Center for Agricultural Utilization Research, Agricultural Research Service, United States Department of Agriculture, USA

**Sergio Nicolau Freire Bruno**
Laboratório Nacional Agropecuário, RJ-MG, Ministério da Agricultura, Brazil

**Ana Briones, Juan Ubeda-Iranzo and Luis Hernández-Gómez**
Laboratory of Yeast Biotechnology, University of Castilla La Mancha, Ciudad Real, Spain

**Pelin Onsekizoglu**
Trakya University Department of Food Engineering, Edirne, Turkey

**Poppy Intan Tjahaja and Putu Sukmabuana**
Nuclear Technology Center for Materials and Radiometry, National Nuclear Energy Agency of Indonesia, Bandung, Indonesia

**Toshio Hasegawa**
Saitama University, Japan

**Claudia Siemer, Stefan Toepfl and Volker Heinz**
German Institute of Food Technologies, Germany

Printed in the USA
CPSIA information can be obtained
at www.ICGtesting.com
JSHW011458221024
72173JS00005B/1125